LINEAR AND NONLINEAR SPIN WAVES IN MAGNETIC FILMS AND SUPERLATTICES

LINEAR AND NONLINEAR SPIN WAVES IN MAGNETIC FILMS AND SUPERLATTICES

Editor

Michael G Cottam

Univ. Western Ontario

World Scientific

Singapore • New Jersey • London • Hong Kong

Published by

World Scientific Publishing Co. Pte. Ltd.
P O Box 128, Farrer Road, Singapore 9128
USA office: Suite 1B, 1060 Main Street, River Edge, NJ 07661
UK office: 73 Lynton Mead, Totteridge, London N20 8DH

LINEAR AND NONLINEAR SPIN WAVES IN MAGNETIC FILMS AND SUPERLATTICES

ISBN 981-02-1006-X

Printed in Singapore.

PREFACE

In the past few years, there has been a rapidly growing interest in the properties of spin waves (or magnons) in ordered magnetic materials. These are the low-lying excitations that characterize the dynamical behavior of the magnetization variables in ferromagnets, ferrimagnets, and antiferromagnets, particularly at low temperatures. Many of the recent developments concerning spin waves have been directed towards understanding their behavior in limited magnetic samples (for example, thin films and multilayer systems such as superlattices) and/or in the nonlinear regime. At the same time, there have been dramatic advances in the experimental techniques, both for preparing high-quality magnetic samples in the form of thin films and superlattices and for the study of the spin-wave excitations themselves. Magnetic thin films have long been of technological, as well as scientific, interest (for example, in microwave magnetic and magneto-optical devices and in magnetic recording), and an understanding of both the linear and nonlinear aspects of their magnetic behavior is important.

The objective of this volume is to present an up-to-date and comprehensive account of spin waves in magnetic thin films and superlattices, covering both linear and nonlinear properties. Theoretical and experimental aspects are described, and the connection between them is emphasized. The material, which has been contributed by leading researchers in this field, falls basically into three parts. First, Chap. 1 consists of an extended introduction in which the fundamentals of both linear and nonlinear spin-wave processes in bulk and finite magnetic samples are described in a comprehensive fashion. Next, Chaps. 2 to 6 deal in greater detail with the *linear* spin-wave properties, and finally, the *nonlinear* spin-wave properties are covered in Chaps. 7 to 9. The material in these subsequent chapters is arranged as follows. The linear spin-wave spectra in magnetic films and multilayers (including superlattices) are discussed in Chaps. 2 and 3 respectively. These contain accounts of the localized spin waves at surfaces and interfaces, as well as the bulk (or volume) spin waves that propagate throughout the sample. Chapter 4 concentrates on the electromagnetic regime in which the spin waves couple with photons to form mixed excitations known as spin-wave polaritons. The use of Green's functions as a major theoretical technique for spin waves in multilayers is described in Chap. 5, while some experimental techniques such as light scattering and microwave resonance are detailed in Chap. 6. In the nonlinear region, the spin waves can exhibit a range of additional phenomena, which include instabilities, auto-oscillations, chaotic behavior, and the formation of envelope solitons. These effects are described in Chap. 7 for spherical samples and then in Chaps. 8 and 9 for thin films. It may be noted that each chapter is relatively self-contained, in the sense that most of the concepts and notation are introduced and defined as required, and can therefore be read independently by someone with a general background in spin waves. However, for an overall account of this field, it is appropriate for the reader to follow the sequence of chapters used here in which new ideas are brought in progressively in a logical order.

I wish to thank all the authors for their contributions and for collaborating in the writing of this book. I also gratefully acknowledge the technical assistance and advice from the editorial staff of the World Scientific offices in Singapore and New York.

Michael G. Cottam
London, Ontario
January 1993

CONTENTS

CHAPTER 1

FUNDAMENTALS OF LINEAR AND NONLINEAR SPIN-WAVE PROCESSES IN BULK AND FINITE MAGNETIC SAMPLES

M.G. COTTAM* and A.N. SLAVIN†

* Department of Physics, University of Western Ontario, London, Ontario N6A 3K7, Canada

† Department of Physics, Oakland University, Rochester, Michigan 48309-4401, USA

Contents

1.1. INTRODUCTION

The concept of spin waves as low-lying excitations that occur in ordered magnetic materials was introduced in the 1930s. Initially the experimental evidence for spin waves came from the measurement of thermodynamic properties, while nowadays there are many sensitive and direct techniques for studying linear and nonlinear processes involving spin-wave excitations. These techniques include, for example, the inelastic scattering of light (photons) or particles (e.g., neutrons) from spin waves, microwave spectroscopy, and various forms of magnetic resonance (e.g., spin-wave resonance).

In this book we shall be mainly concerned with spin waves in restricted geometries, that is, with thin films and multilayer systems (including superlattices) in which it is necessary to take account of effects at surfaces and interfaces. The purpose of this chapter is to introduce some of the fundamentals of both linear and nonlinear spin wave processes in a comprehensive fashion, and in the subsequent chapters more detailed treatments are to be found for specific topics. In this chapter we shall find it useful to explain, in the first instance, some of the properties and ideas of spin waves in *bulk* magnetic materials (i.e., in effectively infinite samples) before we generalize to the cases of *films* and *multilayers*. Also we shall introduce the *linear* spin-wave properties first and then extend the treatment to include *nonlinear* processes.

1.1.1. Magnetic Interactions and Ordering

We shall be concerned with three main types of ordered magnetic materials: *ferromagnets*, *simple antiferromagnets*, and *ferrimagnets*, which are shown schematically in Fig. 1.1. Other more complicated schemes of magnetic ordering, such as exhibited by metamagnets and canted antiferromagnets, will not be considered here. In ferromagnets there is an interaction between neighboring electronic spins (or their corresponding magnetic moments) that favors a parallel alignment at low enough temperature. In simple antiferromagnets and ferrimagnets the overall interaction is of the opposite sign, leading to an antiparallel ordering

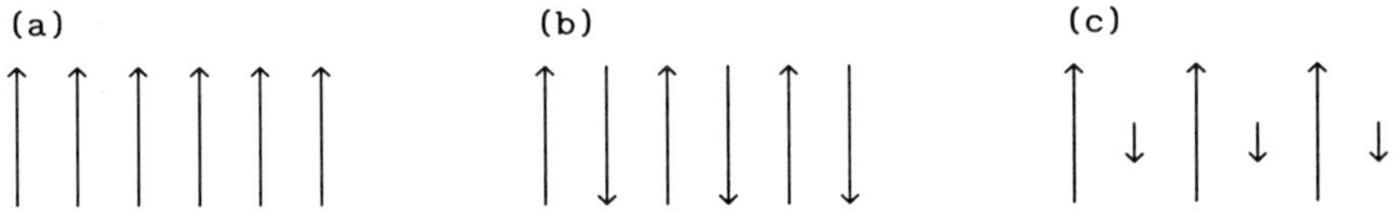

Fig. 1.1. Schematic representation of the spins in three types of ordered magnetic materials: (a) ferromagnets, (b) antiferromagnets, and (c) ferrimagnets.

of the spins. In ferrimagnets the magnetic moments of the "up" and "down" spins are unequal in magnitude, and sometimes there are more than two kinds (or sublattices) of spins. Consequently ferrimagnets exhibit a nonzero spontaneous magnetization at low temperature, and in this respect they behave similarly to ferromagnets. By contrast, in simple antiferromagnets in zero applied magnetic field the symmetry between the sublattices give zero net magnetization. In all three types of magnetic material the long-range magnetic order decreases as the temperature is increased, until eventually there is a phase transition to a paramagnetic state. The critical temperature at which this takes place is called the Curie temperature T_C in ferromagnets and ferrimagnets and the Néel temperature T_N in antiferromagnets.

The nature of the interaction that produces the magnetic ordering was first explained by Heisenberg [1], who showed that it is electrostatic in origin and is due to the quantum-mechanical exchange. General accounts of exchange interactions can be found in most textbooks on magnetism (e.g., see [2-4]). For simplicity, suppose we consider two neighboring magnetic ions, each with one electron. The electrons are spin-$\frac{1}{2}$ particles that obey Fermi-Dirac statistics, and so their total wavefuction, which may be written as the product of a spatial part ψ and a spin part χ_S, must be antisymmetric with respect to interchange of the electrons. The two possible spin states are the symmetric $\chi_{S=1}$ and the antisymmetric $\chi_{S=0}$ (with S denoting the total spin quantum number), and these must be combined with the antisymmetric and symmetric spatial wavefunctions, respectively. These latter quantities, which can be written down in the Heitler-London theory, are different provided there is appreciable overlap of the one-electron wavefunctions. This leads to a difference between the energies of the $S = 1$ and $S = 0$ states. The energy can be expressed in operator form as $-J_{12}\mathbf{S}_1 \cdot \mathbf{S}_2$ (with a suitable choice for the zero of energy) where $\mathbf{S}_1$ and $\mathbf{S}_2$ are spin operators for the two electrons. The quantity J_{12}, which depends on the overlap of the electronic wavefunctions, is called the *exchange interaction*. It is clear that the cases of $J_{12} > 0$ and $J_{12} < 0$ correspond to the spins being parallel and antiparallel, respectively, in their lowest-energy state, as in ferromagnetism and antiferromagnetism.

For a whole system of spins in a magnetic material the above argument generalizes to give the *Heisenberg Hamiltonian*:

$$\mathcal{H}_e = -\sum_{\langle i,j \rangle} J_{ij} \mathbf{S}_i \cdot \mathbf{S}_j , \qquad (1.1)$$

where the summation is over all distinct pairs i and j of magnetic sites. Typically the exchange interaction is short range, and in many cases it is sufficient to consider only nearest-neighbor sites. A proper description of exchange is much more complicated than indicated above, and the reader is referred to reviews by Anderson [5] and Herring [6] for magnetic insulators and metals respectively.

As well as the exchange interactions there are also *magnetic dipole-dipole interactions*. There is a magnetic moment $g\mu_B \mathbf{S}_i$ corresponding to each spin $\mathbf{S}_i$, and the resulting contribution to the magnetic Hamiltonian is

$$\mathcal{H}_d = (g\mu_B)^2 \sum_{\langle i,j \rangle} \left(\frac{\mathbf{S}_i \cdot \mathbf{S}_j}{r_{ij}^3} - \frac{3(\mathbf{S}_i \cdot \mathbf{r}_{ij})(\mathbf{S}_j \cdot \mathbf{r}_{ij})}{r_{ij}^5} \right) \qquad (1.2)$$

where $\mathbf{r}_{ij}$ is the vector connecting sites i and j, g is the Landé g-factor, and μ_B is the Bohr magneton. An individual dipole-dipole interaction is typically very much smaller than the exchange interaction between neighboring sites (often by 2 or 3 orders of magnitude), and so it is the exchange that predominantly determines the magnetic ordering and the *static* properties of magnetic materials. However, we shall see later that the influence of the dipole-dipole interactions becomes important for the *dynamic* properties (i.e., the spin waves) at long enough wavelengths. Essentially this is because the exchange is short range and the dipole-dipole interactions are long range.

In general there are other contributions, apart from the exchange and dipole-dipole terms, that must be included in the Hamiltonian of an ordered magnetic material. In particular, there are effects of magnetic anisotropy, which can arise because of interactions between the magnetic moments of the atoms and the electric field of the crystal lattice. A full description is very complex, but a simplified account can be found, for example, in the book by Akhiezer *et al* [4] where the connection with the spin-orbit interaction is emphasized. Often a phenomenological approach of single-ion anisotropy is followed in which the contribution to the magnetic Hamiltonian is represented by an expansion in powers of components of the spin operator at each site. Only the leading order terms in the expansion that are consistent with the overall symmetry of the crystal lattice need to be retained. For example, in a *uniaxial* (noncubic) ferromagnet the result is of the form

$$\mathcal{H}_a = -\sum_i K(S_i^z)^2 , \tag{1.3}$$

where K is an anisotropy constant. When $K > 0$, the minimum anisotropy energy is achieved when the magnetization is parallel to the uniaxis (the z axis). On the other hand, when $K < 0$, the minimum energy corresponds to the magnetization lying in the xy plane. Different combinations of spin operators are required to represent the magnetic anisotropy in crystal with other symmetries [4]. Also we note that the form of the single-ion anisotropy at a magnetic surface or interface may be different from in the bulk, because the symmetry is modified.

We have framed the preceding discussion in terms of the spins $\mathbf{S}_i$ at the individual sites. However, sometimes an alternative description is employed by dealing instead with the magnetization as a field vector $\mathbf{M}(\mathbf{r})$, which depends on the position $\mathbf{r}$. This *macroscopic approach* may be more convenient in formulating the dynamic properties for wavelengths that are large compared with the microscopic lattice parameter (e.g., as will be the case in ferromagnetic resonance) and/or for including dipole-dipole interactions via Maxwell's equations. The basic connection between the macroscopic and microscopic descriptions is provided by

$$\mathbf{M}(\mathbf{r}) = g\mu_B \sum_i \mathbf{S}_i \, \delta(\mathbf{r}-\mathbf{r}_i) \ . \tag{1.4}$$

We shall make use of both microscopic and macroscopic approaches in this chapter, as appropriate.

1.1.2. Concept of Spin Waves

The idea of spin waves as the elementary excitations in ordered magnetic materials was due to Bloch [7]. He envisaged the spins as deviating slightly from their ground-state orientations, with these disturbances traveling with a wavelike behavior through the crystal. The situation is somewhat analogous to the occurrence of vibrational modes in a solid, where the small-amplitude oscillations of the atoms about the equilibrium positions are the normal modes of the solid. Similarly the spin waves are the normal modes associated with the interacting spins of a magnetic crystal. When the quantum-mechanical nature of the spins is taken into account, the spin waves are quantized with the basic quantum being referred to as the *magnon*, by analogy with the phonon as a quantized lattice vibration.

The mathematical framework of spin waves or magnons will be presented later in this chapter. In fact, various different alternative approaches to spin-wave theory can be followed. The semiclassical theory, such as that due to Heller and Kramers [8], treats the spins as precessing vectors. This viewpoint is particularly helpful in gaining a physical interpretation of spin waves, and it is illustrated schematically for the case of a ferromagnet in Fig. 1.2. Analogous semiclassical representations with precessing spins can be drawn for antiferromagnets and ferrimagnets. Other approaches to spin-wave theory are specifically quantum mechanical, such as the operator method developed by Holstein and Primakoff [9] based on second quantization. In quantum-mechanical terms, a spin wave can be thought of as one quantum of spin reversal spread coherently over the whole crystal.

At low temperatures the spin waves behave, to a good approximation,

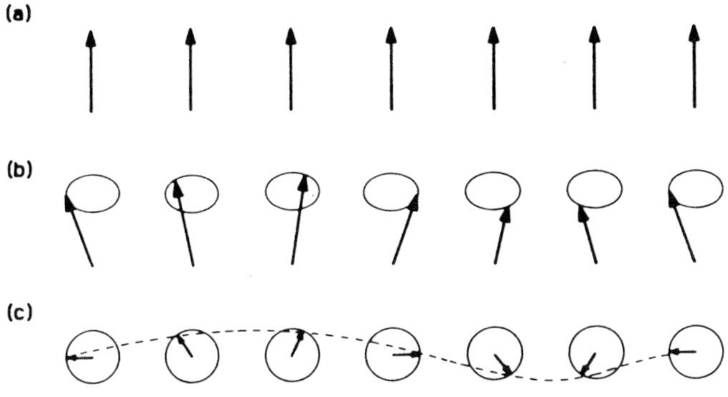

Fig. 1.2. Semiclassical representation of a spin wave in a ferromagnet: (a) the ground state with spin vectors parallel, (b) perspective view of a spin wave of precessing spin vectors, and (c) the same spin wave viewed as from above.

like noninteracting elementary excitations with boson-like characteristics. However, it will be important to realize that the spin waves are not *exact* normal modes of the magnetic system, and it will be shown that this leads to interactions between the spin waves and to other specifically nonlinear effects.

In his original paper, Bloch [7] used spin-wave theory to predict that the magnetization of a Heisenberg ferromagnet at low temperature (compared with T_C) should deviate from its zero-temperature value with a $T^{3/2}$ dependence in zero applied magnetic field. This was confirmed in subsequent measurements of thermodynamic properties. The experimental evidence for spin waves in ferromagnets, antiferromagnets, and ferrimagnets is now very extensive, coming mainly from light scattering, neutron scattering, microwave spectroscopy, and various forms of magnetic resonance. Accounts of the experimental evidence for bulk spin waves are to be found, for example, in Refs. [4] and [10-14].

1.1.3. Effects at Surfaces and Interfaces

We first recall some standard results (e.g., see Kittel [15]) for excitations in a crystalline solid that is effectively infinite. It is well known in this case that any excitation, such as a spin wave, must satisfy symmetry requirements in accordance with *Bloch's Theorem*. This states that the variable $A(\mathbf{r})$ describing the excitation amplitude must have the general form

$$A(\mathbf{r}) = \exp(i\mathbf{q}.\mathbf{r})\, U_{\mathbf{q}}(\mathbf{r}) \;. \tag{1.5}$$

Here $\mathbf{q}$ is a wave vector in the Brillouin zone corresponding to the reciprocal lattice of the crystal, and $U_{\mathbf{q}}(\mathbf{r})$ is a periodic function satisfying the periodicity properties of the crystal lattice. The overall phase factor $\exp(i\mathbf{q}.\mathbf{r})$ gives a plane-wave variation to $A(\mathbf{r})$. For a spin wave $A(\mathbf{r})$ will be an appropriate component of the spin or magnetization. The spin-wave energy, denoted by $\hbar\omega(\mathbf{q})$ where $\omega(\mathbf{q})$ is the excitation frequency, will depend on the three-dimensional wave vector.

Now in the case of a system with one or more surfaces or interfaces, the symmetry arguments that led to the above results must be modified. For simplicity, suppose we consider a semi-infinite magnetic medium with a surface corresponding to the plane $x = 0$ of a Cartesian coordinate system. The thickness of the crystal and its dimensions in the y and z directions are taken to be sufficiently large (effectively infinite) that the medium may be regarded as filling the half-space $x \leq 0$. The system still has translational symmetry in the yz plane, corresponding to the the two-dimensional (2D) symmetry group of the atomic layers parallel to the surface, but there are no translational symmetry operations in the x direction. As a consequence, Eq. (1.5) becomes replaced by (e.g., see [16])

$$A(x,\mathbf{R}) = \exp(i\mathbf{Q}.\mathbf{R})\, U_{\mathbf{Q}}(x,\mathbf{R}) \;, \tag{1.6}$$

where $\mathbf{R} = (y,z)$ is a 2D vector parallel to the surface, and $\mathbf{Q} = (q_y,q_z)$ is the in-plane wave vector. There is the factor $\exp(i\mathbf{Q}.\mathrm{R})$ giving a plane-wave variation to $A(x,\mathbf{R})$, but only for propagation parallel to the surface. It is not possible from symmetry arguments alone to make a definite statement about the dependence of $A(x,\mathbf{R})$ on the coordinate x perpendicular to the surface. Instead, it is necessary to obtain the x dependence explicitly by solving an equation of motion for $A(x,\mathbf{R})$. This gives rise to the possibility of solutions that are localized to a particular region of x, rather than spread through the whole crystal as for the *bulk spin waves* of an infinite medium. If the solution is localized near the surface, we have a *surface spin wave*. In some simple cases, as we shall see later, the x dependence of a surface spin wave may take an exponential form $\exp(-\kappa|x|)$, where κ is a positive, real decay constant. Bulk spin waves (i.e, spin waves having a wavelike behavior in 3D) may also exist in the vicinity of a surface, but they will need to satisfy boundary conditions at the surface.

When the above arguments are generalized to a magnetic film or multilayer, where there are two or more surfaces or interfaces, we shall again find the possibility of localized spin waves (surface or interface spin waves) as well as bulk spin waves. For multilayers, a case of special interest is provided by *superlattices*. These are structures in which the composition and thicknesses of the layers are such as to give an overall quasi-periodicity. For example, in Fig. 1.3 we illustrate a two-component superlattice consisting of a large number of alternate layers of media that are labeled 1 and 2 with thicknesses L_1 and L_2 respectively. For spin-wave studies we shall be interested in cases where one or both of the media are magnetic. The periodicity length (or "unit-cell" length) of the superlattice in Fig. 1.3 is equal to $L_1 + L_2$, and translations through this distance in the x direction constitute a symmetry operation of the structure. This gives rise to distinctive properties of superlattices, as we shall discuss later.

1.2. SPIN WAVES IN BULK MATERIALS

It will be helpful, before discussing the effects of surfaces and interfaces, to review some calculations for spin waves in *bulk* magnetic materials (i.e., effectively infinite crystals). For the most part we shall derive the results for ferromagnets, since these materials are

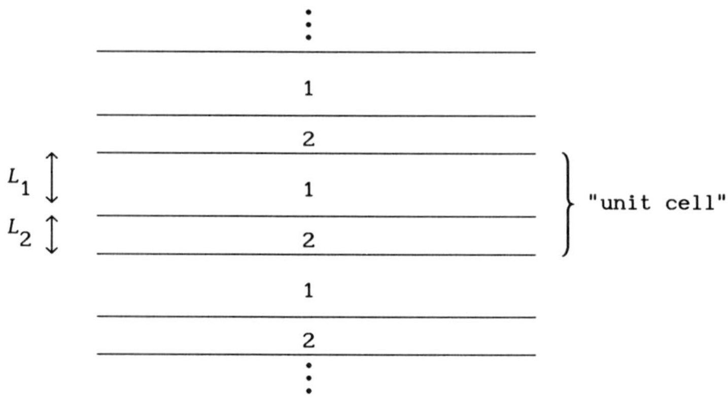

Fig. 1.3. Schematic representation of an infinite two-component super lattice with alternating layers (labeled by 1 and 2) of thicknesses L_1 and L_2.

more straightforward and also the results apply in some cases to ferri magnets. However, analogous calculations can be carried out specifically for antiferromagnets and ferrimagnets. We shall also illustrate both the microscopic and macroscopic approaches to spin-wave theory.

1.2.1. Regimes of Wave Vector

In the preceding section we identified exchange interactions and dipole-dipole interactions as both being important for spin-wave properties. The relative importance is governed largely by the magnitude of the wave vector $\mathbf{q}$ of the spin wave. Typically exchange interactions are dominant for spin waves with $|\mathbf{q}| > 10^6$ cm^{-1}; this is the *exchange region* which includes most of the Brillouin zone (a zone boundary wave vector has a magnitude of about 10^8 cm^{-1}). For smaller values of $|\mathbf{q}|$ (i.e., for longer wavelengths) the dipole-dipole interactions become significant because they extend over a longer range than the exchange interactions. Eventually, for small enough $|\mathbf{q}|$ (typically $|\mathbf{q}| < 10^5$ cm^{-1}), the dipole-dipole interactions dominate. There is an intermediate regime, known as the *dipole-exchange region* and typically corresponding to 10^5 cm^{-1} < $|\mathbf{q}|$ < 10^6 cm^{-1}, in which the exchange and dipolar terms may be comparable. The dipole-dipole region can be treated microscopically, but it is usually more convenient to follow a macroscopic approach in terms of Maxwell's equations of electromagnetism. A simplification occurs if the inequality

$$|\mathbf{q}| \gg \omega/c \tag{1.7}$$

holds, where ω is the frequency and c is the velocity of light, because the effects of retardation can then be neglected. This defines the so-called *magnetostatic region*, which typically corresponds to the wave-vector range $30\ \text{cm}^{-1} < |\mathbf{q}| < 10^5\ \text{cm}^{-1}$. At smaller $|\mathbf{q}|$ the full form of Maxwell's equations, including retardation, must be used; this is the *electromagnetic region*.

These different regions of behavior are summarized in Table 1.1. However, it is important to emphasize that the numbers may vary for different magnetic materials. Also, when we come to surface geometries, we shall see that the interplay between exchange and dipolar effects may sometimes be more subtle, although the same regions as shown in Table 1.1 can usually be identified.

Table 1.1. The different regions of spin-wave excitations in terms of the magnitude $|\mathbf{q}|$ of their wave vector. The numbers are approximate for ferromagnetic materials. For comparison, a Brillouin zone boundary wave vector is approximately of magnitude $10^8\ \text{cm}^{-1}$.

Region	Wave-vector range
Exchange region	$|\mathbf{q}| > 10^6\ \text{cm}^{-1}$
Dipole-exchange region	$10^6\ \text{cm}^{-1} > |\mathbf{q}| > 10^5\ \text{cm}^{-1}$
Magnetostatic region	$10^5\ \text{cm}^{-1} > |\mathbf{q}| > 30\ \text{cm}^{-1}$
Electromagnetic region	$|\mathbf{q}| < 30\ \text{cm}^{-1}$

1.2.2. Microscopic Theory of Spin Waves

We first consider a Heisenberg ferromagnet with an applied magnetic field H_o along the z direction (the direction of net magnetization). The spin Hamiltonian is

$$\mathcal{H} = -\frac{1}{2}\sum_{i,j} J_{ij}\mathbf{S}_i\cdot\mathbf{S}_j - g\mu_B H_o \sum_i S_i^z , \tag{1.8}$$

where $J_{ij} > 0$ and the notation is the same as in the preceding section. The summations are over all magnetic sites, and the factor $\frac{1}{2}$ in the first term ensures that each distinct pair of sites is counted once.

There are several alternative ways of proceeding with the calculation. Here we follow a method based on the *Holstein-Primakoff transformation* [9] from spin operators to boson operators. The Hamiltonian (1.8) involves the three components S_j^x, S_j^y, and S_j^z of each spin $\mathbf{S}_j$, and the required transformation can be expressed as

$$S_j^+ \equiv S_j^x + iS_j^y = (2S)^{1/2}(1 - a_j^\dagger a_j/2S)^{1/2} a_j ,$$

$$S_j^- \equiv S_j^x - iS_j^y = (2S)^{1/2} a_j^\dagger (1 - a_j^\dagger a_j/2S)^{1/2} ,$$

$$S_j^z = S - a_j^\dagger a_j . \tag{1.9}$$

Here S is the spin quantum number, and $a_j^\dagger$ and a_j are boson creation and annihilation operators satisfying the usual commutation relations

$$[a_i, a_j^\dagger] = \delta_{ij} . \tag{1.10}$$

It is easily verified that Eqs. (1.9) and (1.10) reproduce the correct commutation properties of the spin operators.

At low temperatures ($T \ll T_C$) the fractional spin reversal will be small, so that $S_j^z \simeq S$ and we expand the square roots in Eq. (1.9) by treating $a_j^\dagger a_j/2S$ as a small quantity. This yields, for example,

$$S_j^+ = (2S)^{1/2}[a_j - (a_j^\dagger a_j a_j/4S) + \dots] , \tag{1.11}$$

and the Hamiltonian becomes in terms of the boson operators

$$\mathcal{H} = - NS(Jz_oS/2 + g\mu_B H_o) + \mathcal{H}_2 + \mathcal{H}_{int} \, . \tag{1.12}$$

where J denotes the exchange interaction to the z_o nearest neighbors, and N is the total number of magnetic sites. The first term on the right of Eq. (1.12) represents the ground-state energy of the system, $\mathcal{H}_2$ is a term bilinear in the boson operators, and $\mathcal{H}_{int}$ represents the higher-order terms. For the present we neglect $\mathcal{H}_{int}$, but we shall come back to it later in considering nonlinear effects. If we define Fourier transformations of the boson operators from a position to a wave vector representation by

$$a_j = N^{-1/2} \sum_{\mathbf{q}} e^{-i\mathbf{q}.\mathbf{r}_j} a_{\mathbf{q}} \, , \qquad a_j^{\dagger} = N^{-1/2} \sum_{\mathbf{q}} e^{i\mathbf{q}.\mathbf{r}_j} a_{\mathbf{q}}^{\dagger} \, , \tag{1.13}$$

it is straightforward to show that the term $\mathcal{H}_2$ takes the simple form

$$\mathcal{H}_2 = \sum_{\mathbf{q}} \omega(\mathbf{q}) \, a_{\mathbf{q}}^{\dagger} a_{\mathbf{q}} \, . \tag{1.14}$$

In the language of second quantization (e.g., see [17]), $a_{\mathbf{q}}^{\dagger}$ and $a_{\mathbf{q}}$ are the operators that create and annihilate, respectively, a spin wave of wave vector $\mathbf{q}$ and frequency $\omega(\mathbf{q})$. The spin-wave dispersion relation is found to be

$$\omega(\mathbf{q}) = g\mu_B H_o + Jz_oS(1 - \gamma_{\mathbf{q}}) \, , \tag{1.15}$$

where $\gamma_{\mathbf{q}}$ is a structure factor defined by

$$\gamma_{\mathbf{q}} = z_o^{-1} \sum_{\mathbf{d}} \exp(i\mathbf{q}.\mathbf{d}) \tag{1.16}$$

with $\mathbf{d}$ denoting a vector connecting any spin site to its nearest neighbors. The spin-wave frequency is equal to $g\mu_B H_o$ at $\mathbf{q} = 0$, and it generally increases with increasing $|\mathbf{q}|$. For example, for a simple cubic lattice, $z_o = 6$ and $\gamma_{\mathbf{q}} = [\cos(q_x a) + \cos(q_y a) + \cos(q_z a)]/3$ where a is the lattice parameter.

A slightly different way of obtaining the above results is to form the equation of motion for $a_{\mathbf{q}}$ (or $a_{\mathbf{q}}^{\dagger}$) by using the standard equation for any quantum-mechanical operator X (e.g., see Schiff [18]):

$$i\frac{dX}{dt} = [X, \mathcal{H}] \qquad (\hbar = 1) \ , \tag{1.17}$$

Taking $X = a_{\mathbf{q}}$ and evaluating the commutator within the approximation of neglecting the higher-order $\mathcal{H}_{\text{int}}$ terms in the Hamiltonian, we find

$$i\frac{d}{dt} a_{\mathbf{q}} = \omega(\mathbf{q})\, a_{\mathbf{q}} \tag{1.18}$$

where $\omega(\mathbf{q})$ is the same as in Eq. (1.15). This describes a simple harmonic oscillator with $\omega(\mathbf{q})$ as the angular frequency of excitation.

The above results for a Heisenberg ferromagnet may readily be generalized to include the dipole-dipole interactions as well as the exchange interactions [10]. This requires the addition of a term like Eq. (1.2) to the Hamiltonian. The expression for the bilinear term $\mathcal{H}_2$ becomes

$$\mathcal{H}_2 = \sum_{\mathbf{q}} \left(A_{\mathbf{q}}\, a^{\dagger}_{\mathbf{q}} a_{\mathbf{q}} + \tfrac{1}{2}B_{\mathbf{q}}\, a_{\mathbf{q}} a_{-\mathbf{q}} + \tfrac{1}{2}B^{*}_{\mathbf{q}}\, a^{\dagger}_{\mathbf{q}} a^{\dagger}_{-\mathbf{q}} \right) \tag{1.19}$$

with

$$A_{\mathbf{q}} = g\mu_B H_0 + Jz_0 S(1 - \gamma_{\mathbf{q}}) + S(D^{zz}_{\mathbf{0}} + \tfrac{1}{2}D^{zz}_{\mathbf{q}}) \ , \tag{1.20}$$

$$B_{\mathbf{q}} = \tfrac{1}{2}S(D^{xx}_{\mathbf{q}} - D^{yy}_{\mathbf{q}} - 2iD^{xy}_{\mathbf{q}}) \ . \tag{1.21}$$

The quantities $D^{\mu\nu}_{\mathbf{q}}$ are dipole-dipole sums defined by

$$D^{\mu\nu}_{\mathbf{q}} = (g\mu_B)^2 \sum_{\mathbf{r}} \frac{(3r_\mu r_\nu - |\mathbf{r}|\delta_{\mu\nu})}{|\mathbf{r}|^5} \exp(-i\mathbf{q}.\mathbf{r}) \ . \tag{1.22}$$

The above summation is over all vectors connecting a spin site to other spins in the lattice. The Hamiltonian (1.19) can be solved by a "diagonalization" procedure (e.g., see [10]) to show that the elementary excitations, the spin waves, have frequencies given by

$$\omega(\mathbf{q}) = [A^2_{\mathbf{q}} - |B_{\mathbf{q}}|^2]^{1/2} \ . \tag{1.23}$$

This generalizes the previous result in Eq. (1.15) for the case of a Heisenberg ferromagnet. Also, in this case, the operator which creates a spin wave of wave vector $\mathbf{q}$ is no longer simply $a^{\dagger}_{\mathbf{q}}$, but instead is a linear combination of the $a^{\dagger}_{\mathbf{q}}$ and $a_{-\mathbf{q}}$ operators [10,17].

The summations in Eq. (1.22) are complicated to evaluate because of the long-range nature of the dipole-dipole interactions [10]. For a simple cubic lattice and small wave vector such that $|\mathbf{q}|a \ll 1$ where a is the lattice parameter (but $|\mathbf{q}|L \gg 1$ where L is of the order of the sample dimensions) the spin-wave dispersion relation approximates to

$$\omega(\mathbf{q}) = [(g\mu_B H_i + Dq^2)(g\mu_B H_i + Dq^2 + 4\pi g\mu_B M_o \sin^2\theta_{\mathbf{q}})]^{1/2} . \qquad (1.24)$$

Here $D = SJa^2$ is an exchange factor, M_o is the static magnetization, and $H_i = H_o - N^z M_o$ is just the internal field in the sample, where N^z is the classical demagnetizing factor. The spin-wave frequency depends not only on the magnitude of $\mathbf{q}$ but also on its direction through the angle $\theta_{\mathbf{q}}$, which denotes the angle between $\mathbf{q}$ and the z axis. The result in the magnetostatic region is obtained when Dq^2 becomes negligibly small in Eq. (1.24).

1.2.3. Macroscopic Theory of Spin Waves

We now describe spin-wave theory from the point of view of a macroscopic, continuum approach, which may be useful at relatively long wavelengths (such that $|\mathbf{q}|a \ll 1$). Instead of spin operators at discrete lattice sites, we work in terms of a magnetization field vector $\mathbf{M}(\mathbf{r})$ defined as in Eq. (1.4). The exchange and dipole-dipole contributions to the total energy must now be re-expressed in terms of $\mathbf{M}(\mathbf{r})$.

For a ferromagnet the macroscopic exchange energy, corresponding to the Hamiltonian of Eq. (1.1), is given by

$$W_e = -\frac{1}{2(g\mu_B)^2}\int d^3\mathbf{r}\int d^3\mathbf{r}'\; J(\mathbf{r}-\mathbf{r}')\mathbf{M}(\mathbf{r})\cdot\mathbf{M}(\mathbf{r}') . \qquad (1.25)$$

Because the exchange is short range, we can choose to expand $\mathbf{M}(\mathbf{r}')$ about the point $\mathbf{r}$ in a Taylor series in powers of $\mathbf{r}'-\mathbf{r}$ and retain only terms up to the second order:

$$\mathbf{M}(\mathbf{r}') = \mathbf{M}(\mathbf{r}) + (r'_\mu - r_\mu)\frac{\partial\mathbf{M}(\mathbf{r})}{\partial r_\mu} + \frac{1}{2}(r'_\mu - r_\mu)(r'_\nu - r_\nu)\frac{\partial^2\mathbf{M}(\mathbf{r})}{\partial r_\mu \partial r_\nu} + \ldots \qquad (1.26)$$

where repeated Greek indices are understood to be summed over components x, y, and z. The total exchange energy can now be written as

$$W_e = \int d^3\mathbf{r}\left[-\tfrac{1}{2}\lambda\mathbf{M}^2(\mathbf{r}) + w_e(\mathbf{r})\right] \tag{1.27}$$

where the integral is over the volume of the ferromagnet. The first term would represent the energy if the magnetic moments were aligned parallel, and $w_e(\mathbf{r})$ is the exchange energy density for deviations in the alignment; it is given in leading order by

$$w_e(\mathbf{r}) = -\tfrac{1}{2}\,\beta_{\mu\nu}\mathbf{M}(\mathbf{r})\cdot\frac{\partial^2\mathbf{M}(\mathbf{r})}{\partial r_\mu\partial r_\nu} \quad \text{or} \quad \tfrac{1}{2}\,\beta_{\mu\nu}\frac{\partial\mathbf{M}(\mathbf{r})}{\partial r_\mu}\cdot\frac{\partial\mathbf{M}(\mathbf{r})}{\partial r_\nu}\ . \tag{1.28}$$

The second form of $w_e(\mathbf{r})$ above can be obtained from the first by using integration by parts in Eq. (1.27). The above quantities λ and $\beta_{\mu\nu}$ are related to the exchange by

$$\lambda = \frac{1}{(g\mu_B)^2}\int d^3\mathbf{r}'\ J(\mathbf{r}')\ , \qquad \beta_{\mu\nu} = \frac{1}{2(g\mu_B)^2}\int d^3\mathbf{r}'\ J(\mathbf{r}')r'_\mu r'_\nu\ . \tag{1.29}$$

In a cubic crystal, the off-diagonal elements of $\beta_{\mu\nu}$ will be zero and the diagonal elements will all be equal. We then have $\beta_{\mu\nu} \equiv \beta\ \delta_{\mu\nu}$, and Eq. (1.28) simplifies to

$$w_e(\mathbf{r}) = -\tfrac{1}{2}\beta\mathbf{M}(\mathbf{r})\cdot\nabla^2\mathbf{M}(\mathbf{r}) \quad \text{or} \quad \tfrac{1}{2}\beta\left[\left(\frac{\partial\mathbf{M}(\mathbf{r})}{\partial x}\right)^2 + \left(\frac{\partial\mathbf{M}(\mathbf{r})}{\partial y}\right)^2 + \left(\frac{\partial\mathbf{M}(\mathbf{r})}{\partial z}\right)^2\right]. \tag{1.30}$$

There is another energy density $w_d(\mathbf{r})$ corresponding to the dipole-dipole interactions, and this can similarly be derived [4] starting from the Hamiltonian $\mathcal{H}_d$ of Eq. (1.2). This turns out to be rather more complicated than the exchange case, and it is more convenient to use Maxwell's equations instead. We first describe the procedure in the absence of retardation effects. The total magnetization is written as the sum of a static part and a fluctuating part:

$$\mathbf{M}(\mathbf{r},t) = M_0\hat{\mathbf{z}} + \mathbf{m}(\mathbf{r})e^{-i\omega t}\ , \tag{1.31}$$

where $\hat{\mathbf{z}}$ is a unit vector along the z axis, the direction of static magnetization, and we assume a Fourier component of angular frequency ω for the fluctuating part, the spin wave. The magnetic field $\mathbf{H}_d$ due to the dipole-dipole interactions can likewise be expressed as

$$\mathbf{H}_d(\mathbf{r},t) = -N^z M_0\hat{\mathbf{z}} + \mathbf{h}_d(\mathbf{r})e^{-i\omega t}\ , \tag{1.32}$$

where the first part represents the demagnetizing field and the second part is the fluctuating term. These fields must satisfy the magnetostatic form of Maxwell's equations:

$$\mathrm{div}\,[\mathbf{h}_d(\mathbf{r}) + 4\pi\mathbf{m}(\mathbf{r})] = 0\ , \qquad \mathrm{curl}\ \mathbf{h}_d(\mathbf{r}) = 0\ . \tag{1.33}$$

We introduce wave-vector Fourier transformations by

$$\mathbf{m}(\mathbf{r}) = \int d^3\mathbf{q}\ \mathbf{m}_{\mathbf{q}}\ e^{i\mathbf{q}\cdot\mathbf{r}}\ , \qquad \mathbf{h}_d(\mathbf{r}) = \int d^3\mathbf{q}\ \mathrm{h}_{d\mathbf{q}}\ e^{i\mathbf{q}\cdot\mathbf{r}}\ , \tag{1.34}$$

and by substituting these into Eq. (1.33) we obtain the following relationship between the Fourier components:

$$\mathbf{h}_{d\mathbf{q}} = -\ 4\pi\mathbf{q}(\mathbf{q}\cdot\mathbf{m}_{\mathbf{q}})/q^2\ . \tag{1.35}$$

To solve for the spin-wave frequencies, we use the torque equation of motion for the magnetization:

$$\frac{\partial\mathbf{M}}{\partial t} = -\ \gamma(\mathbf{M}\times\mathbf{H}_{\mathrm{eff}})\ , \tag{1.36}$$

where $-\gamma$ is the gyromagnetic ratio (with $\gamma > 0$). This is the semi-classical analog of the quantum-mechanical operator equation (1.17). $\mathbf{H}_{\mathrm{eff}}$ is the total effective field:

$$\mathbf{H}_{\mathrm{eff}}(\mathbf{r},t) = H_o\hat{\mathbf{z}} + \mathbf{H}_e(\mathbf{r},t) + \mathbf{H}_d(\mathbf{r},t)\ , \tag{1.37}$$

composed of an applied field term H_o in the z direction, an exchange field term $\mathbf{H}_e$, and a dipole field term $\mathbf{H}_d$. We have shown above that $\mathbf{H}_d$ has a static part, representing the demagnetizing field, and a fluctuating part. The exchange field can be found from the exchange energy W_e using the functional derivative relationship $\mathbf{H}_e = -\ \delta W_e/\delta\mathbf{M}$ (see [4]), together with Eqs. (1.27), (1.30), and (1.31):

$$\mathbf{H}_e(\mathbf{r},t) = \lambda\mathbf{M}(\mathbf{r},\mathrm{t}) + \beta\nabla^2\mathbf{m}(\mathbf{r})e^{-i\omega t}\ . \tag{1.38}$$

For simplicity, we have assumed a cubic crystal. Also we have omitted the effects of magnetic anisotropy, which could straightforwardly be included as an extra term in Eq. (1.37). The next step is to substitute for $\mathbf{M}$ and $\mathbf{H}_{\mathrm{eff}}$ in Eq. (1.36) and to linearize the equation of motion by neglecting terms that are second order in the fluctuating terms:

$$-i\omega\mathbf{m}(\mathbf{r}) = \gamma[H_i\mathbf{m}(\mathbf{r}) - M_o\beta\nabla^2\mathbf{m}(\mathbf{r}) - M_o\mathbf{h}_d(\mathbf{r})] \times \hat{\mathbf{z}} \ . \qquad (1.39)$$

As before, $H_i = H_o - N^zM_o$ is the internal field in the ferromagnet.

In the absence of exchange (i.e., when $\beta = 0$), Eq. (1.39) can be re-expressed as a susceptibility relation:

$$\mathbf{m}(\mathbf{r}) = \bar{\bar{\chi}}(\omega)\ \mathbf{h}_d(\mathbf{r}) \qquad (1.40)$$

where

$$\bar{\bar{\chi}}(\omega) = \frac{1}{4\pi}\begin{pmatrix} \chi_1 & i\chi_2 & 0 \\ -i\chi_2 & \chi_1 & 0 \\ 0 & 0 & 0 \end{pmatrix}, \qquad (1.41)$$

$$\chi_1 = \omega_M\omega_H/(\omega_H^2 - \omega^2)\ , \qquad \chi_2 = \omega_M\omega/(\omega_H^2 - \omega^2)\ . \qquad (1.42)$$

Here we have introduced the characteristic frequencies ω_M and ω_H defined by

$$\omega_M = 4\pi\gamma M_o\ , \qquad \omega_H = \gamma H_i \qquad (1.43)$$

It is clear from Eqs. (1.40) and (1.41) that $m_{\mathbf{q}}^z = 0$ in the linear spin-wave approximation. We now have two expressions connecting $\mathbf{m}$ and $\mathbf{h}_d$, namely Eqs. (1.35) and (1.40). This allows us to solve for the spin-wave frequency ω, giving $\omega = \omega(\mathbf{q})$ where

$$\omega(\mathbf{q}) = [\omega_H(\omega_H + \omega_M\sin^2\theta_{\mathbf{q}})]^{1/2}\ . \qquad (1.44)$$

In this limit (corresponding to the magnetostatic region) it can be seen that $\omega(\mathbf{q})$ is independent of $|\mathbf{q}|$ but depends on the orientation of $\mathbf{q}$ (through the angle $\theta_{\mathbf{q}}$ defined earlier). The band of spin-wave frequencies has a range from ω_H (when $\theta_{\mathbf{q}} = 0$) to $[\omega_H(\omega_H + \omega_M)]^{1/2}$ (when $\theta_{\mathbf{q}} = \pi/2$).

When exchange effects are included, we need to take account of the $\nabla^2\mathbf{m}(\mathbf{r})$ term in Eq. (1.39). This can be achieved by using Eq. (1.34) to rewrite Eq. (1.39) as a relationship between the Fourier components $\mathbf{m}_{\mathbf{q}}$ and $\mathbf{h}_{d\mathbf{q}}$. Using Eq. (1.35) and denoting $\alpha = \beta/4\pi$, we then find

$$\omega(\mathbf{q}) = [(\omega_H + \omega_M\alpha q^2)(\omega_H + \omega_M\alpha q^2 + \omega_M\sin^2\theta_{\mathbf{q}})]^{1/2}\ . \qquad (1.45)$$

Hence we obtain the same result as in Eq. (1.24) from the microscopic theory (noting that $\omega_M \alpha = D$).

1.2.4. Effects of Retardation

The previous parts of Sec. 1.2 were concerned with spin-wave modes in which the coupling to the electromagnetic field can be described within the magnetostatic approximation. We now turn to the case where retardation effects are included by using the full form of Maxwell's equations. As can be seen from Table 1.1, this becomes important only at very small wave vectors. In this electromagnetic region the influence of exchange can be neglected. The resulting modes, known as *magnetic polaritons*, arise due to a mixing of the crystal excitation (a spin wave) with a photon. The dispersion relations of bulk magnetic polaritons were first studied by Auld [19].

Maxwell's equations for the fluctuating part $\mathbf{h}_d$ of the total field $\mathbf{H}_{eff}$ yield the wave equation [20,21]

$$\nabla^2 \mathbf{h}_d - \nabla(\nabla \cdot \mathbf{h}_d) - (\varepsilon/c^2)\,\bar{\bar{\mu}}\,\partial^2 \mathbf{h}_d/\partial t^2 = 0 \tag{1.46}$$

and

$$\nabla \cdot (\bar{\bar{\mu}}\,\mathbf{h}_d) = 0 \tag{1.47}$$

where ε is the dielectric function (assumed to be a scalar constant) and $\bar{\bar{\mu}}$ is the permeability tensor at the mode frequency ω. From the susceptibility relationship obtained in Eqs. (1.40) and (1.41) we have a gyromagnetic permeability tensor of the form

$$\bar{\bar{\mu}} = \begin{pmatrix} 1+\chi_1 & i\chi_2 & 0 \\ -i\chi_2 & 1+\chi_1 & 0 \\ 0 & 0 & 0 \end{pmatrix}, \tag{1.48}$$

where χ_1 and χ_2 are defined in Eq. (1.42).

The bulk magnetic polaritons can be studied by seeking plane-wave $\exp[i(\mathbf{q}.\mathbf{r}-\omega t)]$ solutions of Eqs. (1.46) and (1.47). This gives

$$q^2 \mathbf{h}_d - \mathbf{q}(\mathbf{q}\cdot\mathbf{h}_d) - (\varepsilon\omega^2/c^2)\bar{\bar{\mu}}\,\mathbf{h}_d = 0\,, \qquad \mathbf{q}\cdot(\bar{\bar{\mu}}\,\mathbf{h}_d) = 0\,. \tag{1.49}$$

Since the applied magnetic field is along the z direction, there is no loss of generality in choosing $q_y = 0$, i.e., we take $\mathbf{q} = (q_x, 0, q_z)$ with $q_x^2 + q_z^2 = q^2$. By substituting for the permeability tensor in Eq. (1.49) we eventually obtain the implicit polariton dispersion relation as

$$(\omega_H - \xi_1\omega_M)^2 - \omega^2 = (\xi_2\omega_M)^2 \tag{1.50}$$

where ξ_1 and ξ_2 depend on the frequency and the wave-vector components:

$$\xi_1 = \tfrac{1}{2}(2\varepsilon\omega^2/c^2 - q_x^2)/(q^2 - \varepsilon\omega^2/c^2) \ , \tag{1.51}$$

$$\xi_2 = \tfrac{1}{2}q_x^2/(q^2 - \varepsilon\omega^2/c^2) \ . \tag{1.52}$$

It can easily be checked that, in the unretarded region (effectively the limiting case of $c/\omega \longrightarrow \infty$), the solution of Eq. (1.50) reduces to the magnetostatic result of Eq. (1.44) as expected. In general, in the retarded region there are two solutions of Eq. (1.50) for a given value of q, one branch joining up with the magnetostatic mode at larger q and the other branch tending to the photon line. This is illustrated in Fig. 1.4 for values of $\theta_{\mathbf{q}}$ corresponding to two different propagation directions. The change-over from the electromagnetic region to the magnetostatic region to the exchange region is also indicated.

1.2.5. Antiferromagnets and Ferrimagnets

Analogous types of calculations can be carried out for bulk spin waves in antiferromagnets and ferrimagnets for the different regions of magnetic behavior. The main difference is that, because an antiparallel alignment of spins is favored at low temperatures, a description in terms of two (or sometimes more) sublattices is required. This makes the calculations more complicated and also there will be more than one branch to the spin-wave spectrum. We very briefly outline some of the results here.

For *antiferromagnets* in the exchange-dominated region we can write the Heisenberg Hamiltonian as

$$\mathcal{H} = \sum_{i,j} J_{ij}\mathbf{S}_i\cdot\mathbf{S}_j - g\mu_B H_A \sum_i S_i^z + g\mu_B H_A \sum_j S_j^z \ , \tag{1.53}$$

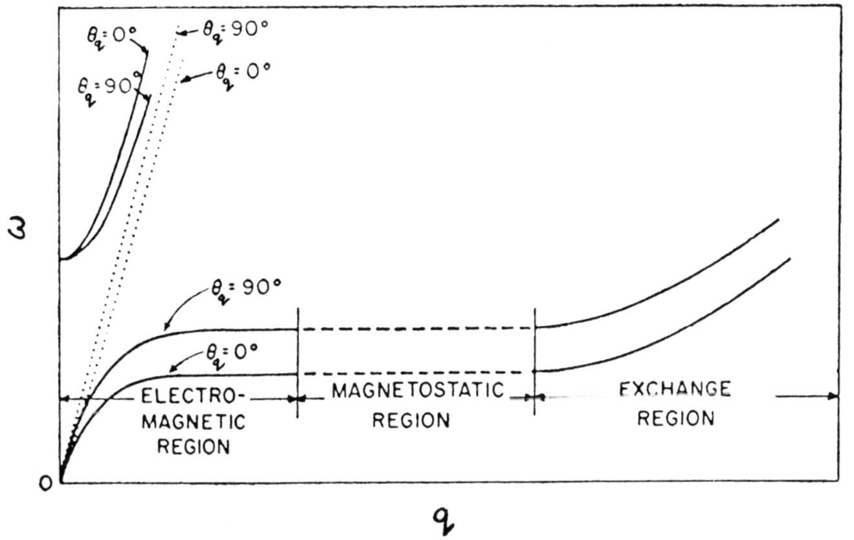

Fig. 1.4. Form of the dispersion curves for spin-wave modes propagating parallel ($\theta_{\mathbf{q}} = 0°$) and perpendicular ($\theta_{\mathbf{q}} = 90°$) to the static magnetization direction in a bulk ferromagnet. The axis for q is schematic and the different regions of magnetic behavior for different wave vectors are as indicated in Table 1.1.

where i and j refer to sites on sublattice 1 (spin-up) and 2 (spin-down) respectively. The last two terms in $\mathcal{H}$ describe a uniaxial magnetic anisotropy, which is represented here in terms of an effective magnetic field H_A which is equal in magnitude but opposite in direction on the two sublattices. In general, the anisotropy might have a more complicated single-ion form as mentioned in Sec. 1.1. Also, for simplicity, we have taken the applied magnetic field to be zero and the exchange J_{ij} to couple only the nearest-neighbor sites on opposite sublattices.

A microscopic theory for the spin-wave dispersion relation can be carried out by generalizing the calculation described in Sec. 1.2.2. The reader is referred to Keffer [10] or Kittel [17] for details.

Briefly, one may introduce boson operators by defining a Holstein-Primakoff transformation separately for each sublattice. By diagonalizing the terms in the Hamiltonian that are bilinear in the operators, one then obtains the spin-wave operators and the spin-wave frequency $\omega(\mathbf{q})$. The result for the frequency is

$$\omega(\mathbf{q}) = [(g\mu_B H_A + Jz_o S)^2 - (Jz_o S\gamma_{\mathbf{q}})^2]^{1/2} , \tag{1.54}$$

where the notation is the same as in Sec. 1.2.2. At wave vector $\mathbf{q} = 0$ this becomes

$$\omega(0) = [\omega_A(\omega_A + 2\omega_E)]^{1/2} , \tag{1.55}$$

where we introduce the notation $\omega_A = g\mu_B H_A$ and $\omega_E = Jz_o S$ for the anisotropy field and exchange frequencies respectively. Equation (1.55) gives the well-known antiferromagnetic resonance (AFMR) frequency. There are, in fact, two degenerate spin-wave modes for each $\mathbf{q}$. This degeneracy is removed if there is an applied magnetic field [10].

The effect of dipole-dipole interactions in antiferromagnets can most conveniently be included by using a macroscopic theory analogous to that of Sec. 1.2.3. It is necessary to write down a torque equation of motion of the form of Eq. (1.36) for each sublattice. We denote by $\mathbf{m}_1(\mathbf{r})$ and $\mathbf{m}_2(\mathbf{r})$ the fluctuating parts of the magnetization for each sublattice, and $\mathbf{h}_{d1}(\mathbf{r})$ and $\mathbf{h}_{d2}(\mathbf{r})$ are the fluctuating dipolar fields. The appropriate susceptibility relationship, analogous to Eq. (1.40), is the response of the overall fluctuating magnetization $\mathbf{m}(\mathbf{r}) = \mathbf{m}_1(\mathbf{r}) + \mathbf{m}_2(\mathbf{r})$ to a dipolar field such that $\mathbf{h}_{d1}(\mathbf{r}) = \mathbf{h}_{d2}(\mathbf{r}) \equiv \mathbf{h}_d(\mathbf{r})$. It can be shown (e.g., see [22,23]) that the susceptibility tensor has the same gyromagnetic form as in Eq. (1.41), but the definitions of χ_1 and χ_2 are changed to

$$\chi_1 = \frac{2\omega_A\omega_M}{\omega_A(\omega_A + 2\omega_E) - \omega^2} , \qquad \chi_2 = 0 \tag{1.56}$$

in the case of zero applied magnetic field. When the spin-wave frequencies are derived, it is found that the two branches are non-degenerate (due to the dipole-dipole interactions) even in zero applied field (e.g., see [10]).

The spin-wave calculations for two-sublattice *ferrimagnets* are similar to those for antiferromagnets with the difference that the spin quantum numbers are unequal for the two sublattices of a ferrimagnet. Details can be found in Ref. [10]. There are two branches to the spin-wave spectrum, which are well separated in frequency for small $|\mathbf{q}|$ and for $T \ll T_C$, and the behavior of the lower branch is similar to that of a spin wave in a ferromagnet. For other ferrimagnets a multi-sublattice must be employed, and there are additional branches to the spin-wave spectrum.

1.3. SPIN WAVES IN FILMS

We now describe the properties of linear spin waves propagating in magnetic films of finite thickness. One has to take into account the boundary conditions on the film surfaces, and that leads to qualitative changes in the spin-wave spectrum. First, the spin-wave spectrum in films becomes discrete, consisting of separate dispersion branches corresponding to spin waves with different distributions of variable magnetization across the film thickness. A second effect of the boundaries is in terms of the dependence of the spin-wave eigenfrequencies in films on the *magnitude* of the wave vector. This dependence occurs for films even in the non-exchange case, when in unlimited magnetic media the spin-wave frequency depends only on the *direction* of propagation defined by the angle $\theta_{\mathbf{q}}$ (e.g., as in Eq. (1.44)). The existence of surface spin waves (i.e., spin-wave modes localized near the film surfaces) is another important qualitative effect caused by the presence of the boundaries, as mentioned in Sec. 1.1.3. We shall discuss all these effects below mostly in the framework of macroscopic spin-wave theory for long-wavelength excitations, but we also outline the microscopic approach for the case of purely exchange spin waves in films.

1.3.1. The Boundary Problem

We consider here a magnetic film of finite thickness L. In order to describe the case of *general orientation* of the static magnetization,

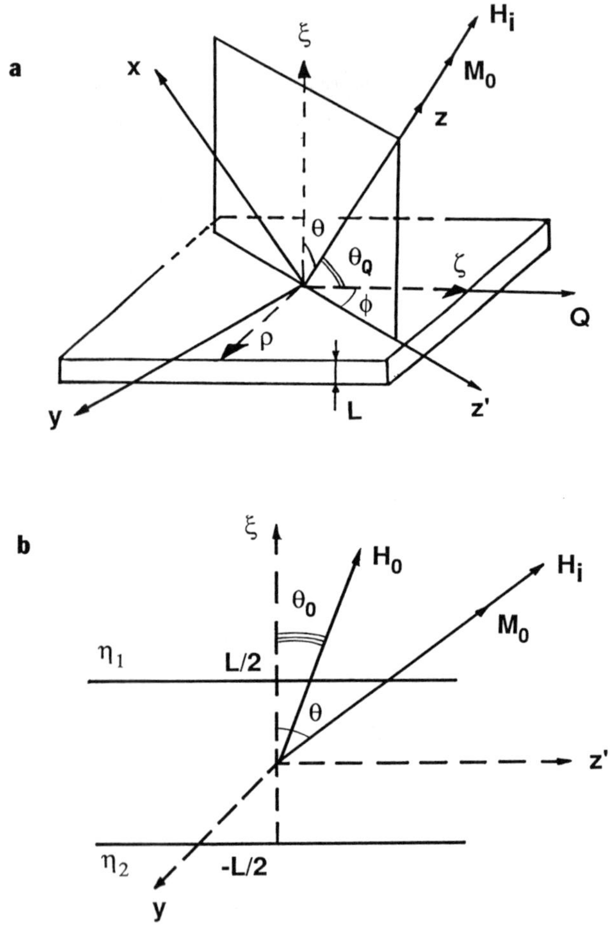

Fig. 1.5. The assumed thin-film geometry: (a) the orientations of the $\xi\rho\zeta$ and xyz coordinate axes, and (b) the directions of the saturation magnetization and magnetic field vectors.

we introduce a $\xi\rho\zeta$ Cartesian coordinate system with the surfaces corresponding to the planes $\xi = L/2$ and $\xi = -L/2$ (see Fig. 1.5). The ρ and ζ dimensions of the film are sufficiently large that they may be regarded as infinite. The film is magnetized to saturation by a static external magnetic field $\mathbf{H}_o$ in an arbitrary direction. The ρ axis is taken to be perpendicular to the direction of propagation of the spin wave. For convenience, we shall also employ another coordinate system xyz, in which the axis z is parallel to both the direction of the saturation magnetization $\mathbf{M}_o$ and the internal static magnetic field $\mathbf{H}_i$.

The variable magnetization and the variable dipole field in the film, written as $\mathbf{m}(\xi,\mathbf{R},t)$ and $\mathbf{h}_d(\xi,\mathbf{R},t)$ respectively, can be Fourier-transformed in two dimensions as

$$\mathbf{m}(\xi,\mathbf{R},t) = \int d^2\mathbf{Q}\ \mathbf{m}_{\mathbf{Q}}(\xi)\ \exp[i(\mathbf{Q}\cdot\mathbf{R} - \omega t)] \ , \tag{1.57}$$

$$\mathbf{h}_d(\xi,\mathbf{R},t) = \int d^2\mathbf{Q}\ \mathbf{h}_{d\mathbf{Q}}(\xi)\ \exp[i(\mathbf{Q}\cdot\mathbf{R} - \omega t)] \ , \tag{1.58}$$

where $\mathbf{Q}$ is a two-dimensional (2D) wave vector for the spin wave and $\mathbf{R}$ is a 2D position vector, both vectors lying in the film plane. To obtain a dispersion relation for the spin waves in a film using the macroscopic theory, one has to solve the boundary problem in the film geometry. Specifically, one has to integrate the system of equations consisting of: (i) the equation of motion for the magnetization (the Landau-Lifshitz equation) as in Eq. (1.36), and (ii) Maxwell's equations in the magnetostatic form of Eq. (1.33), along with the appropriate boundary conditions on the film surfaces.

In the zero-exchange limit ($\alpha = 0$) the only boundary conditions are the usual electrodynamic boundary conditions requiring continuity of the tangential components of the magnetic field, H_t, and the normal components of the magnetic induction, $B_n = H_n + 4\pi M_n$ (e.g., see [21]):

$$H_t^e = H_t^i \ , \qquad H_n^e = H_n^i + 4\pi M_n \ , \tag{1.59}$$

where the superscripts e and i denote the fields that are external and internal to the film, respectively, and M_n is the normal component of the film magnetization.

Equation (1.59) enables us to determine the magnitude H_i and orientation θ of the internal static magnetic field $\mathbf{H}_i$ when the magnitude H_o and orientation θ_o of the external field $\mathbf{H}_o$ and the magnitude M_o of the saturation magnetization $\mathbf{M}_o$ of the film are given (see Fig. 1.5b):

$$H_o \sin\theta_o = H_i \sin\theta , \qquad H_o \cos\theta_o = (H_i + 4\pi M_o) \cos\theta . \qquad (1.60)$$

It is clear from Eq. (1.60) that in the case of parallel magnetization ($\theta = \pi/2$) the internal and external static magnetic fields are equal. By contrast, in the case of perpendicular magnetization ($\theta = 0$) the internal magnetic field is smaller than the external field because of the demagnetization effect.

When the exchange is nonzero, the exchange field $\mathbf{h}_e = 4\pi\alpha \nabla^2\mathbf{m}$ must be included in the equation of motion, Eq. (1.36). The order of the differential equations, obtained in the course of simultaneous solution of Eqs. (1.33) and (1.36), is then increased. Also the electrodynamic boundary conditions given in Eq. (1.59) are no longer sufficient for the solution of the boundary problem for the film, and *additional boundary conditions* determining the values of magnetization on the film surfaces are necessary. These boundary conditions are obtained by integration of the equation of motion for the magnetization in the surface layer near the boundary with the non-magnetic medium (see [24,25]). This is carried out in the same manner as the electrodynamic boundary conditions are obtained by integration of Maxwell's equations in the boundary layer [21]. For our geometry (shown in Fig. 1.5) these additional "exchange" boundary conditions at each surface have the form [26]:

$$\frac{\partial m^x}{\partial n} + \eta \cos 2\theta \, m^x = 0 ,$$
$$\frac{\partial m^y}{\partial n} + \eta \cos^2\theta \, m^y = 0 . \qquad (1.61)$$

Here η, which is called the *pinning parameter*, is proportional to the surface anisotropy constant K_s [10], and the derivatives are taken in the direction of the normal to the surface. The pinning parameter may have a different value at each film surface. We note that α and η are related to the parameters A and K_s used by other researchers (e.g., see [26]) by $\alpha = A/2\pi M_o^2$ and $\eta = K_s/A$.

There are two major approaches available for the solution of the system of Eqs. (1.33) and (1.36) with the boundary conditions (1.59) and (1.61) in our chosen film geometry: we refer to these as the *method of magnetostatic potential* and the *method of tensorial Green functions*. Both methods are described in detail in Chapters 2 and 3 of the present book. We show below some simple examples of their application to the calculations of spin-wave spectra in magnetic films.

1.3.2. Magnetostatic Spin Waves in Ferromagnets

Here we shall consider spin waves with values of the in-plane wave number Q that satisfy the inequality $\alpha^{-1/2} \gg Q \gg \omega/c$. Following the discussion in Sec. 1.2.1, this condition means that the effects of electromagnetic retardation and inhomogeneous exchange interactions are negligible for these waves. Spin waves satisfying the above criterion are usually called dipolar spin waves or *magnetostatic* waves.

Solutions for magnetostatic waves in a film geometry were found in [4,27-29] using a magnetostatic potential method. The magnetostatic potential ψ is introduced by the equation

$$\mathbf{h}_d = \nabla\psi \ . \tag{1.62}$$

Substituting Eq. (1.62) into Eq. (1.33), we find

$$\nabla[(1 + 4\pi\overline{\overline{\chi}})\nabla\psi] = 0 \ . \tag{1.63}$$

Inside the magnetic film, which is magnetized along the z direction, Eq. (1.63) will reduce to the Walker equation [30]:

$$(1 + \chi_1)\left(\frac{\partial^2\psi^i}{\partial x^2} + \frac{\partial^2\psi^i}{\partial y^2} \right) + \frac{\partial^2\psi^i}{\partial z^2} = 0 \ . \tag{1.64}$$

Outside the magnetic medium Eq. (1.63) yields the Laplace equation:

$$\nabla^2\psi^e = 0 \ . \tag{1.65}$$

The electrodynamic boundary conditions relating ψ^i and ψ^e at the film surfaces can be obtained using Eqs. (1.40), (1.59), and (1.62). For a film with perpendicular magnetization (the case of $\theta = 0$ in Fig. 1.5a) we get

$$\psi^i = \psi^e \ , \qquad \frac{\partial\psi^i}{\partial z} = \frac{\partial\psi^e}{\partial z} \ . \tag{1.66}$$

In the case of a film with parallel magnetization ($\theta = \pi/2$) we have

$$\psi^{i} = \psi^{e} , \qquad (1 + \chi_1) \frac{\partial\psi^{i}}{\partial x} + i\chi_2 \frac{\partial\psi^{i}}{\partial y} = \frac{\partial\psi^{e}}{\partial x} . \tag{1.67}$$

We now discuss these two cases separately.

a. Case of static magnetization perpendicular to the film surfaces.

We can write the solutions of Eqs. (1.64) and (1.65) for the magnetostatic potential in the general form

$$\psi^{i} = e^{i(q_x x + q_y y)}[A \cos(q_z z) + B \sin(q_z z)] , \tag{1.68}$$

$$\psi^{e} = e^{i(q_x x + q_y y)} \begin{cases} C\, e^{-Qz} , & z > L/2 , \\ D\, e^{Qz} , & z < -L/2 . \end{cases} \tag{1.69}$$

In this case the z axis is perpendicular to the film surface, and the in-plane wave vector is $\mathbf{Q} = (q_x, q_y)$. Relationships between the unknown constants A, B, C, and D are obtained from the two boundary conditions of Eq. (1.66) applied at $z = L/2$ and $z = -L/2$. From the resulting 4×4 determinantal condition we get the dispersion relation for magnetostatic waves in a perpendicularly magnetized film in the form [4,28,29]:

$$\omega^2 = \omega_H(\omega_H + \omega_M \sin^2\theta_{\mathbf{q}}) , \tag{1.70}$$

where ω_H and $\theta_{\mathbf{q}}$ are as defined before. In the present case we can write

$$\sin^2\theta_{\mathbf{q}} = Q^2/(Q^2 + q_z^2) , \qquad Q^2 = q_x^2 + q_y^2 . \tag{1.71}$$

The expressions determining the wave-vector component q_z are different for the modes with symmetric [$\psi^{i} \sim \cos(q_z z)$] and antisymmetric [$\psi^{i} \sim \sin(q_z z)$] magnetostatic potential across the film thickness [4,28,29]:

$$\text{symmetric modes:} \qquad q_z \tan(q_z L/2) = Q , \tag{1.72}$$

$$\text{antisymmetric modes:} \qquad q_z \cot(q_z L/2) = -Q . \tag{1.73}$$

Since $Q > 0$ the quantity q_z is always real and Eqs. (1.70)-(1-73) specify the spectrum of bulk magnetostatic waves. It is clear that the wave frequency ω increases with increasing in-plane wave number Q, so the group velocity $V_g = \partial\omega/\partial Q$ is positive. The bulk magnetostatic waves propagating in perpendicularly magnetized films are sometimes called *magnetostatic forward volume waves* (MSFVW). From the above equations it is easy to prove that the frequencies of these waves lie within the limits of the magnetostatic wave spectrum in an infinite ferromagnet [see Eq. (1.44)]. Therefore we have

$$\omega_H < \omega_{MSFVW} < [\omega_H(\omega_H + \omega_M)]^{1/2} . \tag{1.74}$$

b. Case of the static magnetization parallel to the film surfaces.

We now have $\theta = \pi/2$ in Fig. 1.5a, so the x axis is perpendicular to the film surfaces and $\mathbf{Q} = (q_y, q_z)$. Following an analogous procedure (see [4,27]), the dispersion relation can be expressed in the same form as Eq. (1.70) where now

$$\sin^2\theta_{\mathbf{q}} = (q_x^2 + q_y^2)/(q_x^2 + Q^2) , \qquad Q^2 = q_y^2 + q_z^2 , \tag{1.75}$$

and the expression determining the values of the perpendicular wave-vector component q_x has the form [4]:

$$2q_x\cot(2q_xL) = \frac{1}{Q}\left[q_x^2 - Q^2 + (\omega_H/\omega_M)(q_x^2 + Q^2)(q_y/q_z)^2\right] . \tag{1.76}$$

The magnetostatic waves propagating in an arbitrary direction in a parallel magnetized film are neither symmetric nor antisymmetric. However, for the waves propagating along the direction of static magnetic field ($q_y = 0$) it is possible to split Eq. (1.76) into two separate equations for symmetric and antisymmetric modes [4]:

$$\text{symmetric modes:} \qquad q_x\cot(q_xL/2) = -|q_z| , \tag{1.77}$$

$$\text{antisymmetric modes:} \qquad q_x\tan(q_xL/2) = |q_z| . \tag{1.78}$$

Equations (1.77) and (1.78) are quite similar to Eqs. (1.72) and (1.73). They allow only real values of q_x and describe the spectrum of bulk magnetostatic waves with frequencies in the range from ω_H to $[\omega_H(\omega_H + \omega_M)]^{1/2}$, c.f., Eq. (1.74). It is easy to verify that the frequency of these waves decreases with increasing $|q_z|$ and the group velocity of these waves is negative. These waves are sometimes called *magnetostatic backward volume waves* (MSBVW).

However, the interesting case for a parallel magnetized film occurs when q_y is nonzero and satisfies $q_y^2 > (\omega_H/\omega_M)q_z^2$. Apart from the infinite number of real solutions for q_x corresponding to bulk waves, Eq. (1.76) has one imaginary solution for q_x corresponding to a surface wave. In the special case of propagation perpendicular to the direction of static magnetic field (the Voigt geometry, corresponding to $q_z = 0$) it is possible to write the dispersion equation for this *magnetostatic surface wave* (MSSW) or *Damon-Eshbach wave* [27] as

$$\omega_S^2 = \omega_H\left\{\omega_H + \omega_M\left[1 + \omega_M\frac{(1 - e^{-2|q_y|L})}{4\omega_H}\right]\right\} . \tag{1.79}$$

It can be shown from Eq. (1.79) that the frequency of surface magnetostatic waves starts (for $|q_y|L = 0$) at the upper boundary of the bulk spin-wave band and increases with increasing $|q_y|L$, so that

$$[\omega_H(\omega_H + \omega_M)]^{1/2} < \omega_s < \omega_H + \omega_M/2 \tag{1.80}$$

in the Voigt geometry. The group velocity of these waves is positive. Depending on the direction of propagation (i.e., on the sign of q_y), the magnetostatic surface wave in Eq. (1.79) is localized either near the top or the bottom surface of the film. Such behavior is termed *"field-displacement" nonreciprocity* (see [31]) and will be discussed further in Chapters 2 and 3.

1.3.3. Exchange-Dominated Spin Waves in Ferromagnets

Next we briefly consider the opposite case when the inhomogeneous exchange interactions play the dominant role in the spin-wave propagation in films and the dipole-dipole interactions are negligible, allowing us to neglect the variable dipole magnetic field $\mathbf{h}_d(\mathbf{r})$ in Eq. (1.39). Typically this assumption is relevant for propagating spin waves with short wavelengths (such that the in-plane wave number satisfies $Q \gg \alpha^{-1/2}$), but we shall see in the next subsection that the exchange may also be important for the case of standing spin waves (when

Q = 0).

For simplicity we take a perpendicularly magnetized film ($\theta = 0$) where the direction of static magnetization (the z axis) coincides with the film normal (see Fig. 1.5a). In this case the exchange boundary conditions, as in Eq. (1.61), are identical for the two in-plane components, m^x and m^y, of the variable magnetization. Defining $m^+ = m^x + im^y$ and $m^- = m^x - im^y$ as usual, we get the equation for the circularly polarized amplitudes of variable magnetization in the form

$$\left(\frac{d^2}{dz^2} + \lambda_\pm^2\right) m^\pm(\mathbf{r}) = 0 \ , \tag{1.81}$$

where

$$\lambda_\pm^2 = - (\omega_H \pm \omega)/\alpha\omega_M \ . \tag{1.82}$$

The amplitudes $m^+(\mathbf{r})$ and $m^-(\mathbf{r})$ have the same boundary conditions which, from Eq. (1.61), can be expressed as

$$\left(\frac{\partial}{\partial z} + \eta\right) m^\pm(\mathbf{r}) = 0 \ , \qquad \text{when } z = L/2 \ , \tag{1.83}$$

$$\left(-\frac{\partial}{\partial z} + \eta'\right) m^\pm(\mathbf{r}) = 0 \ , \qquad \text{when } z = -L/2 \ , \tag{1.84}$$

where η and η' are the pinning parameters at the upper and lower surfaces of the film. We assume the coordinate dependence of the variable magnetization in the form of inhomogeneous spin waves

$$m^\pm(\mathbf{r}) = e^{i\mathbf{Q}\cdot\mathbf{R}} \, \Phi(z) \ . \tag{1.85}$$

with $\mathbf{Q} = (q_x, q_y)$ and $\mathbf{R} = (x, y)$. The solution of Eqs. (1.81), (1.83), and (1.84) then yields the spin-wave dispersion equation

$$\omega^2 = \Omega_{nQ}^2 \equiv [\omega_H + \alpha\,\omega_M(Q^2 + q_{zn})]^2 \tag{1.86}$$

and the expression for the eigenfunctions $\Phi_n(z)$ in the form

$$\Phi_n(z) = A_n \left\{\cos[q_{zn}(z+L/2)] + \frac{\eta'}{q_{zn}} \sin[q_{zn}(z+L/2)]\right\} , \tag{1.87}$$

where the integer $n = 1,2,3...$ is a mode number. The allowed discrete values of the wave number q_{zn} are determined by the equation [32,33]:

$$(q_{zn}^2 - \eta\,\eta')\,\tan(q_{zn}L) = q_{zn}(\eta + \eta')\ . \tag{1.88}$$

In the $Q = 0$ case of standing spin waves, Eq. (1.86) yields the eigenfrequencies of the *spin-wave resonance* (SWR) modes [32-34]:

$$\omega = |\Omega_n| \equiv |\omega_H + \alpha\,\omega_M q_{zn}|\ , \tag{1.89}$$

and Eq. (1.89) describes the variable magnetization distributions for these modes. In Eq. (1.89) the branches $\omega = \Omega_n > 0$ and $\omega = -\Omega_n < 0$ correspond respectively to the SWR modes with right-handed ($m^- \neq 0$) and left-handed ($m^+ \neq 0$) circular polarizations [34]. It is clear from Eqs. (1.39)-(1.41) that the frequencies in Eq. (1.89) are the poles of the susceptibility tensor $\bar{\bar{\chi}}$ in the case when the inhomogeneous exchange interaction is taken into account.

Equation (1.88) has an infinite number of real solutions for q_{zn} corresponding to the bulk modes of SWR. In addition, there may be one or two imaginary solutions for q_z corresponding to the *surface modes* of SWR [35]. The surface modes of SWR can exist only when η and/or η' are negative. The negative sign of a pinning parameter means that the surface anisotropy in the film is of the "easy plane" type [32,33]. A complete account of the conditions for the existence of the surface solutions of the Eq.(1.88) is given in [36]. By contrast with the case of surface magnetostatic waves, the frequencies of the surface modes of SWR are situated *below* the spectrum of bulk SWR modes. It is worth noting that, for standing spin waves ($Q = 0$), the electrodynamic boundary conditions (1.59) on the film surfaces are satisfied automatically [4], and the above equations give the full solution to the problem of determination of SWR modes in perpendicularly magnetized films. The eigenfunctions (1.87) form a complete set of orthogonal functions, and each of them satisfies the exchange and the electrodynamic boundary conditions.

The phenomenon of *spin-wave resonance* (i.e., the excitation of standing spin waves in ferromagnetic films by a variable microwave magnetic field) was predicted by Kittel [34] in 1958 and was observed experimentally by Seavey and Tannenwald [37]. Since that time SWR has attracted a lot of attention from experimentalists as it provides direct evidence of the existence of spin waves and a simple experimental method for determining the exchange constant α (see [32,33] and references therein; see also Chapter 5).

An alternative formulation of exchange spin waves is given later in Sec. 1.3.5 in terms of a microscopic approach.

1.3.4. Dipole-Exchange Spin Waves in Ferromagnets

The solution of the boundary problem for spin waves in a ferromagnetic film in the case when *both* dipole-dipole and exchange interactions are taken into account is much more complicated than the preceding magnetostatic (non-exchange) and pure exchange (non-dipolar) limits. The problem of calculating the dipole-exchange spin-wave spectrum in films was addressed by many authors in the framework of the magnetostatic potential method [38-44]. The alternative approach based on the tensorial Green function method was developed in [45-47, 36] (see also Chapter 2). The exact results obtained by both these methods are rather cumbersome and usually are analyzed numerically. In the framework of the second method, however, it is possible to obtain an explicit *approximate* dispersion equation for dipole-exchange spin waves in a film magnetized in an arbitrary direction.

To give a qualitative idea of the structure of the spectrum we quote below this approximate dispersion equation for the case of unpinned surface spins ($\eta = \eta' = 0$) [36]:

$$\omega_{nQ}^2 = [\omega_H + \alpha\,\omega_M(Q^2 + q_{\xi n}^2)]\,[\omega_H + \alpha\,\omega_M(Q^2 + q_{\xi n}^2) + \omega_M F_{nn}]\ . \quad (1.90)$$

Here we are assuming a general geometry as in Fig. 1.5a, with $q_{\xi n} = n\pi/L$ and $n = 0,1,2\ldots$. The quantity F_{nn} has a complicated form given by

$$F_{nn} = (1 - P_{nn}\cos^2\phi)\sin^2\theta + P_{nn}\left[\cos^2\theta + \frac{\omega_M}{\Omega_{nQ}}(1 - P_{nn})\sin^2\theta\,\sin^2\phi\right]\ , \quad (1.91)$$

where $P_{nn}(QL)$ is a matrix element of the dipole-dipole interaction defined in [47]. When QL varies in the range $0 \le QL < \infty$, the matrix element P_{nn} remains in the interval 0 to 1. For the lowest spin-wave mode ($n = 0$) the result is

$$P_{00} = 1 - \frac{[1 - \exp(-QL)]}{QL} \; . \tag{1.92}$$

Equation (1.90) was obtained using standard perturbation theory [48] in first order. To get this result it was assumed that: (i) each spin wave mode $\omega_{n\mathbf{Q}}$ has the distribution of the variable magnetization along the film thickness $m_{n\mathbf{Q}}(\xi)$ described by one of the functions Φ_n (as in Eq. (1.87)) for an arbitrary value of the in-plane wave number Q, and (ii) the spin-wave modes $\omega_{n\mathbf{Q}}$ do not interact with each other. Both these assumptions are violated near points of crossing (occurring when $\omega_{n\mathbf{Q}} = \omega_{n'\mathbf{Q}}$ for $n \ne n'$) of the dispersion branches determined by (1.90), and near these points higher order approximations of the perturbation theory should be taken into account.

We note that the dispersion equation (1.90) has a form similar to that of the spin wave dipole-exchange dispersion equation (1.45) in a bulk ferromagnet. The main distinctions arising due to the film geometry are:

(i) the component $q_{\xi n}$ of the spin-wave wave number assumes different discrete values for different spin wave modes. This gives the exchange splitting of the spectrum;

(ii) the function F_{nn} (which is dipole-dipole in origin and plays an analogous role to $\sin^2\theta_q$) also assumes discrete values and depends upon the direction of magnetization θ, the direction of spin-wave propagation ϕ, values of in-plane Q and transverse $q_{\xi n}$ wave numbers, and the film thickness L. For certain values of θ and ϕ (e.g., $\theta = \pi/2$, $\phi = \pi/2$) F_{nn} exceeds unity, and this corresponds to quasi-surface spin-wave modes.

When $Q = 0$, Eq. (1.90) gives the spectrum of SWR in the film. In particular, for the perpendicularly magnetised film ($\theta = 0$) we recover Eq. (1.89). In the limiting case of large in-plane wave numbers $Q \to \infty$ (or thick films $L \to \infty$) we get $q \to Q$, $P_{nn} \to 1$, and the function F_{nn} reduces to

$$F_{nn} = 1 - \sin^2\theta \cos^2\phi = \sin^2\theta_{\mathbf{Q}} , \qquad (1.93)$$

where $\theta_{\mathbf{Q}}$ is the angle between the direction of wave propagation **Q** and the direction of the internal static magnetic field $\mathbf{H}_i$ (see Fig. 1.5a). It is easy to verify, that in the non-exchange limit ($\alpha = 0$), Eq. (1.90) yields results which agree closely with the results of Eqs. (1.70) and (1.79) for the magnetostatic theory (taking the same magnetization orientations and spin-wave propagation directions) in the spectral regions where the dispersion branches described by Eq. (1.90) are not crossing.

In conclusion, Eq. (1.90) shows that in a ferromagnetic film the relative contributions of the dipole-dipole and exchange interactions to the spin-wave spectrum depend critically on the film thickness L and the angles θ and ϕ. In the region of relatively small Q the slopes of the dispersion curves are determined by the dipole-dipole interaction, but the exchange splitting at $Q = 0$ (as in SWR modes) creates the conditions for mode crossings and significant dipole-exchange modifications in the spectrum, even in the long-wavelength ($QL \ll 1$) part of the spectrum where in bulk ferromagnets the exchange interaction is practically negligible. In the short-wavelength ($Q \geq \alpha^{-1/2}$) part of the spectrum the exchange interaction is dominant.

1.3.5. Exchange Spin Waves in Ferromagnets: Microscopic Theory

We now re-examine the exchange-dominated spin waves in ferromagnetic films from the viewpoint of the microscopic theory. This will enable some of the results in Sec. 1.3.3 (which are restricted to small wave vectors such that $Q^2a^2 << 1$) to be extended to larger **Q**. The main differences occurring in the microscopic theory are: (i) we have to solve a set of finite-difference equations rather than a differential equation, and (ii) the boundary conditions are incorporated within the equations of motion.

The spin Hamiltonian is assumed to be the same as in Eq. (1.8) except that the sites i and j are now restricted to a finite-thickness film. We consider the example of a simple cubic lattice structure and

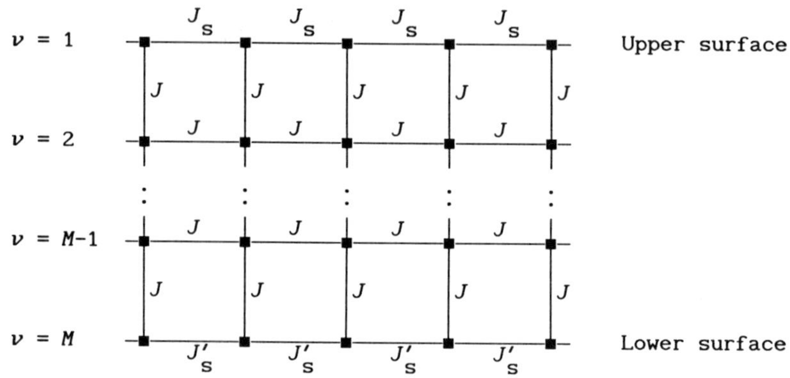

Fig. 1.6. Simplified microscopic model for a ferromagnetic film of simple cubic structure and (001) surfaces. The film has M atomic layers (labeled with the index ν) and the nearest neighbor exchange constants J, J_s, and J'_s are indicated.

(001) film surfaces. It is assumed that there are M atomic layers and the exchange interactions J_{ij} couple nearest neighbors only, having the values J_s and J'_s between spins in the upper surface and the lower surface respectively, and the bulk value J otherwise (see Fig. 1.6). Also we introduce an index ν to label the layers in the xy plane, where $\nu = 1$ and $\nu = M$ label the upper and lower surfaces respectively.

We make a Holstein-Primakoff transformation from spin operators to boson operators, as in Eq. (1.9), and neglect terms in the Hamiltonian that are higher order than quadratic in the boson operators. The boson operators can be 2D Fourier transformed according to

$$a_j = N_2^{-1/2} \sum_{\mathbf{Q}} e^{-i\mathbf{Q}\cdot\mathbf{R}_j} a_{\mathbf{Q},\nu_j}, \qquad a_j^{\dagger} = N_2^{-1/2} \sum_{\mathbf{Q}} e^{i\mathbf{Q}\cdot\mathbf{R}_j} a_{\mathbf{Q},\nu_j}^{\dagger} \tag{1.94}$$

where $\mathbf{Q} = (q_x, q_y)$, $\mathbf{R}_j = (x_j, y_j)$, and ν_j is the layer index for site j. N_2 is the number of magnetic sites in each layer. When the equation of motion for $a_{\mathbf{Q},\nu_j}$ is formed for different values of ν_j using Eq. (1.17) and the quadratic part of the boson Hamiltonian, we obtain the following set of coupled equations (e.g., see [49]):

$$[\omega - g\mu_B(H_o+H_a) - SJ - 4SJ_s(1 - \Gamma_Q)]\, a_{Q,1} + SJ\, a_{Q,2} = 0 , \tag{1.95}$$

$$[\omega - g\mu_B H_o - 2SJ - 4SJ(1 - \Gamma_Q)]\, a_{Q,\nu} + SJ(a_{Q,\nu+1} + a_{Q,\nu-1}) = 0 ,$$
$$(2 \le \nu \le M-1) \tag{1.96}$$

$$[\omega - g\mu_B(H_o-H'_a) - SJ - 4SJ'_s(1 - \Gamma_Q)]\, a_{Q,M} + SJ\, a_{Q,M-1} = 0 . \tag{1.97}$$

The effects of anisotropy fields H_a and H'_a at the upper and lower film surfaces have been included. Also we have assumed a time dependence like $\exp(-i\omega t)$ and have defined

$$\Gamma_Q = [\cos(q_x a) + \cos(q_y a)]/2 . \tag{1.98}$$

Various rigorous mathematical techniques are available to solve the set of equations (1.95)-(1.97) for the frequencies ω of the spin waves [49,50]. A simple approach is to note that the bulk spin waves correspond to solutions for $a_{Q,\nu}$ made up of two waves (one in the z direction and the other in the -z direction):

$$a_{Q,\nu} = U_Q \exp(-iq_z\nu a) + V_Q \exp(iq_z\nu a) . \tag{1.99}$$

where q_z is a real wave-vector component. When this is substituted into Eq. (1.96) it is easily verified that the dispersion relation of bulk spin waves is the same as in Eq. (1.15), and the amplitudes U_Q and V_Q are undetermined. However, the substitution of Eq. (1.99) into the boundary equations (1.95) and (1.97) gives, in each case, a restriction on the ratio U_Q/V_Q, leading to the consistency condition:

$$\tan(q_z L) = \frac{i[b(Q,q_z) - b(Q,-q_z)]}{[b(Q,q_z) + b(Q,-q_z)]} , \tag{1.100}$$

where

$$b(Q,q_z) = [\Delta_s - \exp(iq_z a)][\Delta'_s - \exp(iq_z a)] . \tag{1.101}$$

The quantities Δ_s and Δ'_s depend on the surface exchange constants and on the in-plane wave vector **Q**. For example,

$$\Delta_s = [1 - g\mu_B H_a/SJ + 4(1 - J_s/J)(1 - \Gamma_Q)]^{-1} , \tag{1.102}$$

and there is a similar expression for Δ'_s with J_s and H_a replaced by J'_s and H'_a. The bulk spin-wave dispersion relation is formally the same as Eq. (1.15) for an infinite ferromagnet, except that the component q_z is now "quantized". The set of discrete values of q_z is determined by Eq. (1.100), and this result is a generalization of Eq. (1.88) to larger wave numbers.

The surface spin waves of the film can similarly be analysed. In this case we seek solutions of the form

$$a_{\mathbf{Q},\nu} = W_{\mathbf{Q}} \exp(-\kappa\nu a) + W'_{\mathbf{Q}} \exp(\kappa\nu a) \ , \tag{1.103}$$

where $\mathrm{Re}(\kappa) > 0$. On substituting into Eqs. (1.95)-(1.97) we eventually obtain $\omega = \omega_s(\mathbf{Q})$ where

$$\omega_s(\mathbf{Q}) = g\mu_B H_0 + 4SJ(1 - \Gamma_{\mathbf{Q}}) + SJ[2 - \exp(\kappa a) - \exp(-\kappa a)] \tag{1.104}$$

and κ satisfies

$$\tanh(\kappa L) = \frac{[b(\mathbf{Q},-i\kappa) - b(\mathbf{Q},i\kappa)]}{[b(\mathbf{Q},-i\kappa) + b(\mathbf{Q},i\kappa)]} \ , \tag{1.105}$$

In the limit of the film thickness $L \longrightarrow \infty$ we have $\tanh(\kappa L) \to 1$, and Eq. (1.105) has two solutions corresponding to $e^{-\kappa a} = \Delta_s$ and $e^{-\kappa a} = \Delta'_s$, *provided* $|\Delta_s| < 1$ and $|\Delta'_s| < 1$. These correspond to decoupled surface spin waves, one localized at the upper surface and one at the lower surface. For general values of L, it can be shown [50,51] that there may be zero, one, or two surface spin waves, depending on the parameters Δ_s, Δ'_s, and L.

It is staightforward to demonstrate that, at long wavelengths (such that $|\mathbf{Q}a| \ll 1$, $|q_z a| \ll 1$, and $|\kappa a| \ll 1$), the above results of the microscopic theory are consistent [16] with those obtained from a macroscopic continuum approach as in Sec. 1.3.3. The connection between the surface pinning parameter and the parameters of the microscopic theory at any surface is provided by

$$\eta = g\mu_B H_a / aSJ - (1 - J_s/J)aQ^2 \ . \tag{1.106}$$

1.4. SPIN WAVES IN MULTILAYER SYSTEMS

Next we consider spin-wave excitations in multilayer magnetic systems. Calculations relevant to the different regimes of magnetic behavior (exchange-dominated, dipole-exchange, magnetostatic, polariton, etc.) can be carried out in an analogous fashion to those already described in Sec. 1.3 with the essential difference that in multilayers there are additional surfaces or interfaces at which the boundary conditions must be applied. The appropriate boundary conditions at an interface between two magnetic media are a direct generalization of those discussed previously for a single magnetic film In general, the occurrence of additional interfaces in a multilayer system may lead to the spin-wave characteristics being more complicated than in single films.

We concentrate here on two cases using the magnetostatic approximation: a double-layer magnetic system and a magnetic superlattice. These provide relatively straightforward examples, and yet they are also relevant to experimental studies. In this section we comment only briefly on exchange effects. Other cases and approximations are described in much more detail in some of the later chapters in this book; in particular, see Chapters 3 and 4. Some topics are also covered in recent review articles by Mills [52] and Grünberg [53].

1.4.1. Double-Layer Magnetic Systems

We consider a double-layer magnetic system as represented in Fig. 1.7, where there are two ferromagnetic layers with thicknesses L_1 and L_2 separated by a nonmagnetic spacer layer of thickness d. Because of this separation the magnetic coupling between the two ferromagnets is expected to come about largely through the long-range dipole-dipole interactions, rather than through exchange interactions, and we shall assume this to be the case here (but see the later discussion for important exceptions). For simplicity, it is assumed that both magnetic layers are composed of the same material and are magnetized in the same direction (the z direction) parallel to the surfaces. In this calculation we

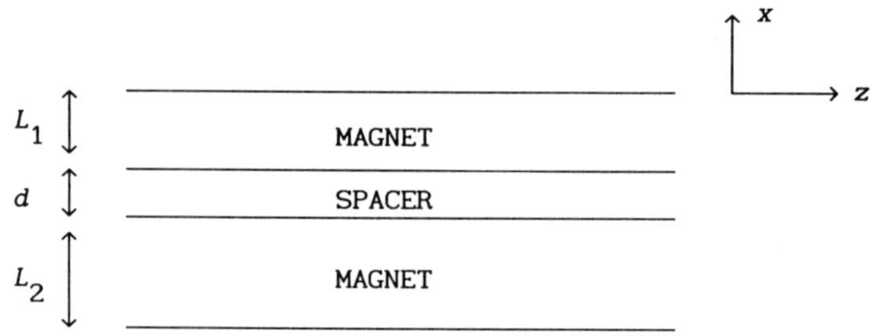

Fig. 1.7. Assumed geometry for a double-layer magnetic system consisting of two ferromagnetic films with thicknesses L_1 and L_2 separated by a nonmagnetic spacer of thickness d.

make the magnetostatic approximation of neglecting exchange interactions within each ferromagnet and neglecting retardation.

Equation (1.64) holds for the magnetostatic potential ψ inside each magnetic layer, and Eq. (1.65) applies in all other regions. By analogy with Eqs. (1.68) and (1.69) we have as the general form of the solution for ψ in the present case:

$$\psi = e^{i(q_y y + q_z z)}\, C_1 e^{-Qx}, \qquad x > 0,$$

$$\psi = e^{i(q_y y + q_z z)}[A_1\cos(q_x x) + B_1\sin(q_x x)], \qquad -L_1 < x < 0,$$

$$\psi = e^{i(q_y y + q_z z)}[C_2 e^{-Qx} + D_2 e^{Qx}], \qquad -(L_1+d) < x < L_1,$$

$$\psi = e^{i(q_y y + q_z z)}[A_2\cos(q_x x) + B_2\sin(q_x x)], \qquad -(L_1+L_2+d) < x < -(L_1+D),$$

$$\psi = e^{i(q_y y + q_z z)}\, D_1 e^{Qx}, \qquad x < -(L_1+L_2+d). \tag{1.107}$$

where q_x is a complex wave-vector component in the x-direction (perpendicular to the layers) and satisfies Eq. (1.76). The application of the boundary conditions given in Eq. (1.67) at each of the the interfaces $x = 0$, $x = -L_1$, $x = -(L_1+d)$, and $x = -(L_1+L_2+d)$, leads to linear relationships between the coefficients A_i, B_i, C_i, and D_i (i = 1, 2). From the resulting 8×8 determinantal condition we obtain expressions for the dispersion relations of the coupled spin-wave modes in the multilayer system. The results are summarized below: the reader may find details of the calculations in [54].

The bulk magnetostatic modes are characterized by q_x real in Eq. (1.107), corresponding to a wave-like variation with distance x across the thickness of each magnetic film. The bulk-mode frequencies are identical to those in a single film with the same transverse magnetization M_0, i.e., they are given by Eqs. (1.70) and (1.75).

By contrast, the coupled surface magnetostatic modes correspond to q_x being imaginary. In the special case when the in-plane propagation is perpendicular to the magnetization direction (i.e., when $q_z = 0$), the frequencies can be expressed as

$$\omega = [(\omega_H + \omega_M/2)^2 - \omega_M^2 \beta/4]^{1/2} \tag{1.108}$$

where ω_H and ω_M are characteristic frequencies related to the applied magnetic field and the static magnetization, as defined in Sec. 1.3, and the dimensionless parameter β satisfies

$$[\beta - e^{-2|q_y|L_1}]\ [\beta - e^{-2|q_y|L_2}]$$
$$- \beta\, e^{-2|q_y|d}\ [1 - e^{-2|q_y|L_1}]\ [1 - e^{-2|q_y|L_2}] = 0\ . \tag{1.109}$$

Equation (1.109) is a quadratic equation in β, giving two solutions (as expected) for the coupled surface modes in the two ferromagnetic layers. In the limit of large separation ($d \longrightarrow \infty$) it can be seen that the solutions for β become $e^{-2|q_y|L_1}$ and $e^{-2|q_y|L_2}$. The corresponding frequencies, obtained from Eq. (1.108), are then just the surface magnetostatic mode frequencies of decoupled films with thicknesses L_1 and L_2. As d is reduced, the magnetic layers become coupled and the frequencies are modified. For comparison, the bulk-mode frequency corresponds to $\beta = 1$ in Eq. (1.108). The behavior of the bulk and surface modes is illustrated in Fig. 1.8.

Note that the surface-mode frequencies both lie *above* the bulk-mode frequency in the magnetostatic calculation just described. If exchange interactions are included (either within the ferromagnetic layers or

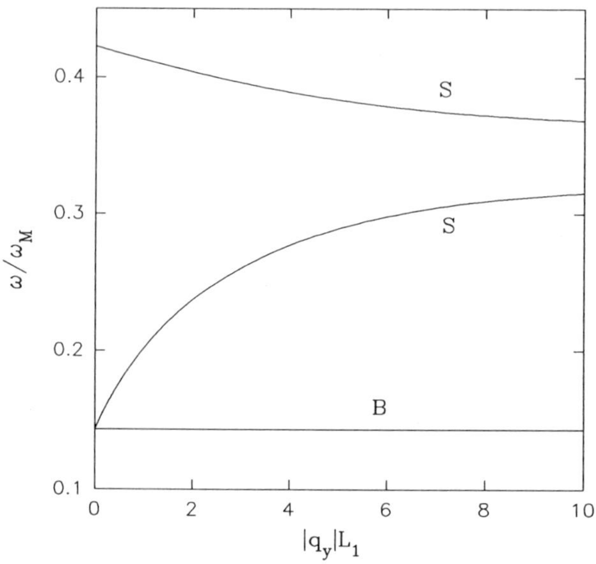

Fig. 1.8. The frequencies of the bulk (B) and surface (S) modes of a double-layer magnetic system as a function of spacer thickness d, using Eqs. (1.108) and (1.109) and reduced units. The assumed parameters correspond to $L_2/L_1 = 1.5$, $\omega_H/\omega_M = 0.02$, and $|q_y|L_1 = 0.2$.

across the spacer layer, or both), then the above conclusions are modified. This is currently an active field of research, both experimentally and theoretically (see Chapter 3). For example, there have been extensive studies of the exchange coupling between two Fe films separated by several atomic layers of Cr, and it was established that the net coupling may oscillate in value as a function of the Cr thickness d and may be overall antiferromagnetic or ferromagnetic in character depending on the value of d [55,56].

1.4.2. Magnetic Superlattices

We have already introduced the concept of superlattices in Sec. 1.1 as multilayer systems in which there are (usually) two components of different compositions and thicknesses. These are deposited alternately to form a periodic (or quasi-periodic) structure, as shown, for example, in Fig. 1.3. It is nowadays feasible to fabricate high-quality superlattices in which one or both of the component materials are ferromagnetic or antiferromagnetic (e.g., see [53]), and this has stimulated many experimental and theoretical studies of their magnetic properties.

For simplicity, we next briefly describe some magnetostatic calculations for superlattices in which a magnetic layer (a ferromagnet or an antiferromagnet) alternates with a nonmagnetic layer. The reader is referred to [57-59] for mathematical details; further discussion and other cases are also dealt with in Chapters 3 and 4 of this book.

First we consider the case where the magnetic layers are identical ferromagnets of thickness L_1 and with static magnetization in a direction parallel to the surfaces. In an infinite superlattice (i.e., one that effectively extends indefinitely in both directions along the x axis perpendicular to the layers), the periodicity length is $L = L_1 + L_2$ where L_2 is the spacer-layer thickness. Because the system still has translational symmetry parallel to the surfaces (in the yz plane) the magnetostatic potential can be expressed as

$$\psi = e^{i(q_y y + q_z z)} \psi(x) . \tag{1.110}$$

However, because of the additional periodicity in the x direction we also have

$$\psi(x + L) = e^{iq_x L} \psi(x) \tag{1.111}$$

in accordance with Bloch's theorem (e.g., see [15,16]). Here q_x is a real component of wave vector for the superlattice excitation. It satisfies $|q_x| \le \pi/L$, which is just the condition for q_x to lie in the first Brillouin zone associated with the periodicity length L. The $e^{iq_x L}$ factor in Eq. (1.111) describes the modulation of $\psi(x)$ from one period to another. The calculations for this case lead to the prediction of a band of collective spin-wave excitations of the superlattice [57]. When the in-plane wave vector is perpendicular to the magnetization ($q_y \ne 0$, $q_z = 0$) the frequencies can be written in the same form as Eq. (1.108) but with

$$\beta = 1 - \left(\frac{2\sinh(q_y L_1)\sinh(q_y L_2)}{\cosh(q_y L) - \cos(q_x L)} \right) \tag{1.112}$$

in the present case. We note that β, and therefore the mode frequency, are *even* functions of the in-plane wave number q_y. Hence the modes may propagate in either the +y or -y directions with the same frequency. In one of these cases (depending on the magnetic field direction) they can be shown to be localized at the upper surfaces of each of the ferromagnetic layers and in the other case they are localized at the lower surfaces [57].

For a semi-infinite superlattice (i.e., one that is terminated at one end as in Fig. 1.9) the possibility arises of there being one (or more) surface modes of the superlattice, in addition to the band of bulk modes discussed above. Bloch's theorem, as quoted in Eq. (1.111), is not applicable in this case because a translation through distance L in the x direction is no longer a symmetry operation. In consequence, surface modes of the superlattice can occur. For these, the $e^{iq_x L}$ modulation factor in Eq. (1.111) becomes replaced by an attenuation factor $e^{-\kappa L}$ where Re(κ) > 0. The theory is described in [57], where it is shown (quoting here the case of propagation such that q_z = 0) that there is a single surface mode and it exists only provided $L_1 > L_2$, i.e., the thickness of the ferromagnet must exceed that of the non-

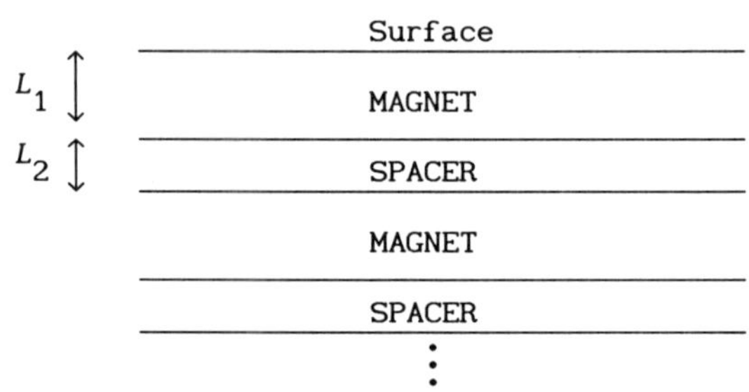

Fig. 1.9. Schematic representation of a semi-infinite magnetic superlattice with alternating layers of a ferromagnet of thickness L_1 and a nonmagnetic spacer of thickness L_2.

magnetic spacer. The attenuation factor κ in this case is then given by

$$\kappa = |q_y| \left(\frac{L_1 - L_2}{L_1 + L_2} \right) . \tag{1.113}$$

Remarkably, the frequency of the above surface mode is independent of both the propagation wave number q_y and the thicknesses L_1 and L_2. It is equal to $\omega_H + \omega_M/2$, which is just the same expression as for the surface-mode frequency in a semi-infinite ferromagnet, e.g., see Eq. (1.79). A generalization of the above theory to include the case of a finite superlattice is carried out in [58].

Magnetostatic calculations have also been performed for infinite and semi-infinite superlattices with alternating ferromagnetic/nonmagnetic layers in the case of the static magnetization *perpendicular* to the surface [59]. The spectrum of bulk and surface modes turns out to be drastically different from the preceeding case of magnetization parallel to the surface. The analogous theory has also been developed for superlattices incorporating antiferromagnetic layers (in this case with the sublattice magnetizations perpendicular to the surface) alternating with nonmagnetic spacer layers [59]. For this orientation of the static magnetization (or sublattice magnetizations) there can be many surface modes of the semi-infinite superlattice.

Recently there has been much interest in the role of exchange interactions in magnetic superlattices, just as described in Sec. 1.4.1 for the double-layer magnetic systems. This has included exchange interactions within each magnetic layer and also exchange (or effective exchange) between different magnetic layers. For example, if both superlattice materials are magnetic (as in FeF_2/MnF_2 superlattices [60]), there will be exchange coupling across the interfaces. On the other hand, in Y/Gd superlattices [61] there is evidence for an oscillatory coupling (having the character of Rudermann-Kittel interactions) between the ferromagnetic Gd layers via the paramagnetic Y layers. The behavior of magnetic superlattices in the exchange-dominated regime is discussed in Chapter 3 (see also [16]).

Finally, we note that in the very long-wavelength regime of

magnetic polaritons, where retardation effects become important, an *effective-medium approach* to magnetic superlattices has proved to be a useful simplification (e.g. see [62]). This method depends on showing that the superlattice behaves, in some respects, like a single medium but with an effective (or averaged) permeability tensor; it is described in detail in Chapter 4.

1.5. FUNDAMENTALS OF MAGNETIC NONLINEARITIES

Up to this point we have restricted the discussion to *linear* spin-wave processes in magnetically-ordered substances, assuming that the level of spin-wave excitation in the magnetic media is sufficiently low (with $|\mathbf{m}| \ll M_0$) that the spin waves do not interact appreciably with each other. This involved linearizing the nonlinear equation of motion for the magnetization (1.36), effectively using $\varepsilon \equiv |\mathbf{m}|/M_0 \ll 1$ as a parameter of smallness. In the absence of this linearization, Eq. (1.36) describes all kinds of *interactions* between the excited spin waves. These interactions are responsible for the very rich and interesting nonlinear dynamics of spin waves as well as for the relaxation processes in the non-equilibrium system of excited spin waves.

Nonlinear spin-wave processes in ferromagnets first attracted the attention of physicists when Bloembergen and Damon [63] and Bloembergen and Wang [64] published the results of ferromagnetic resonance experiments at high power levels. The "saturation" of the main ferromagnetic resonance, as well as the appearance of "subsidiary" (compared with the case of small amplitudes) absorption of the energy of uniform precession of the magnetization, was observed in these experiments. These nonlinear effects were seen to have clearly pronounced power thresholds.

The results of the above experiments were explained by Suhl in his classical paper [65] as a *parametric instability* of uniform precession (of frequency ω_0 and zero wave vector) with respect to decay into pairs of spin waves with frequencies ω_1, ω_2 and wave vectors $\mathbf{q}_1$, $\mathbf{q}_2$ satisfying the conservation conditions

$$n\,\omega_0 = \omega_1 + \omega_2\,, \qquad 0 = \mathbf{q}_1 + \mathbf{q}_2\,, \tag{1.114}$$

where n is a positive integer. This was the first formulation of the conditions of parametric resonance in continuous media, and n is known here as the *order* of the parametric instability.

Soon afterwards, the effect of *parallel pumping* (or the parametric excitation of spin-wave pairs by the variable magnetic field polarized along the direction of static magnetization) was predicted and observed by Schlömann et al [66]. This effect was also independently predicted by Morgenthaler [67]. A review account of the pioneering work in this field is given in [68]. It took almost another decade to get a better understanding of the nonlinear stage of the process of parallel pumping. By developing an idea by Schlömann [69], Zakharov et al. [70] went on to show that the four-wave interaction

$$\omega_{\mathbf{q}} + \omega_{-\mathbf{q}} = \omega_{\mathbf{q}'} + \omega_{-\mathbf{q}'} \tag{1.115}$$

between the spin wave pairs parametrically excited by the parallel pumping field has the leading role in the limitation of the field amplitude. The so-called *S-theory* of nonlinear interaction of parametrically excited spin waves was subsequently elaborated by the same authors in the framework of a classical Hamiltonian formalism [71]. This theory provided the explanation of several new nonlinear effects as well as the effect of *auto-oscillations* of magnetization discovered earlier by Hartwick et al [72]. Further generalizations and developments of the S-theory are presented in [73,74].

More recently, interest in nonlinear spin-wave processes in ferromagnets has been revived by progress in the theory of nonlinear dynamics of *dissipative* systems (e.g., see [75]) and by new results obtained in the theory of *magnetic solitons* (see [74,76,77] and references therein).

We next present a brief outline of the macroscopic nonlinear theory of spin waves in ferromagnets [71,73]. Subsequently, for completeness, we also discuss the alternative microscopic approach to the problem of spin-wave interactions. Further applications relating to nonlinear spin-wave processes are given in Sec. 1.6 and in Chapters 7, 8, and 9 of this book.

1.5.1. Origins of the Nonlinear Effects

The equation of motion for the magnetization, Eq. (1.36), preserves the magnitude of the vector of the total magnetization **M**:

$$|\mathbf{M}|^2 = m_x^2 + m_y^2 + M_z{}^2 = M_0{}^2 \ . \tag{1.116}$$

It follows that only two components of the magnetization vector are independent, and so the third component M_z can be expressed in terms of m_x and m_y:

$$M_z = M_0\left[1 - \frac{(m_x^2 + m_y^2)}{M_0^2} \right]^{\frac{1}{2}} \simeq M_0\left[1 - \frac{|\mathbf{m}|^2}{2M_0^2} \right] = M_0(1 - |\psi|^2) \tag{1.117}$$

where $\mathbf{m} = (m_x, m_y)$ is the transverse magnetization and we denote

$$|\psi|^2 = |\mathbf{m}|^2/2M_0{}^2. \tag{1.118}$$

Equation (1.117) shows that, in the case of $|\mathbf{m}| << M_0$, the variable part m_z of the z component of magnetization, M_z, is of the second order of smallness in terms of the small parameter $\varepsilon = |\mathbf{m}|/M_0$. That is why the m_z was previously excluded from the linearized equations of motion for the magnetization [see Eq. (1.39)]. It is also clear from Eq. (1.117) that, if the transverse variable magnetization **m** varies in time with frequency ω, then the component m_z of the variable magnetization varies in time with a frequency 2ω.

The above dependence of M_z on the amplitude of the linearly excited transverse variable magnetization **m** implies a dependence of the spin-wave eigenfrequences in Eqs. (1.45), (1.70), and (1.75) on the dimensionless spin-wave amplitude ψ. By substituting $M_z = M_0(1 - |\psi|^2)$ for M_0 in the expressions for ω_M and ω_H in Eq. (1.43), we get from each of the dispersion equations the explicit "nonlinear" dispersion equation $\omega_{\mathbf{q}} \equiv f(\mathbf{q}, |\psi|^2)$ for the spin waves propagating in ferromagnetic media.

As we shall discuss later, this simple and approximate argument is valid only when the three-wave parametric decay processes, which correspond to $n = 1$ in Eq. (1.114), are prohibited by conservation laws. In other words, it applies when the frequency $\omega_{\mathbf{q}}$ of the linearly excited spin wave is less than $2\omega_H$.

1.5.2. Classical Hamiltonian Formalism for Spin Waves

To get a more accurate and detailed picture of the nonlinear spin-wave processes we now use an approach based on the classical Hamiltonian formalism for ferromagnets as developed in [71,73]. The quantum-mechanical occupation numbers for the long-wavelength spin waves excited in ferromagnets are usually macroscopically large, and consequently the classical approach will be adequate in this case for a theoretical description of linear and nonlinear spin wave phenomena. As shown in [78], the method using a classical Hamiltonian formalism is natural and convenient for describing spin-wave interactions in different media. This method allows one to eliminate non-essential specific properties of the medium from the equations of motion for the wave amplitudes. In effect, the problems of wave interactions are similar for different media, and the specific "medium" information is contained in the coefficients of the Hamiltonian function.

The total energy W in an *unlimited* ferromagnetic medium, when account is taken of the Zeeman interaction of the magnetization with the external magnetic field, the dipole-dipole interaction, and the inhomogeneous exchange interaction, can be written in the form:

$$W = \int d^3\mathbf{r}\ (w_H + w_d + w_e)\ . \tag{1.119}$$

Here $w_H = -\mathbf{M}\cdot\mathbf{H}$, $w_d = -\mathbf{M}\cdot\mathbf{H}_d/2$, and w_e and is given by Eq. (1.30). For simplicity we ignore here the elastic and magnetoelastic energy contributions and the energy of the magnetic crystallographic anisotropy. The total external magnetic field $\mathbf{H}$ consists of the static bias magnetic field $\mathbf{H}_o$ and the pumping variable magnetic field $\mathbf{h}_p(\mathbf{r},t)$:

$$\mathbf{H} = \mathbf{H}_o + \mathbf{h}_p(\mathbf{r})e^{-i\omega_p t}\ . \tag{1.120}$$

The total magnetization $\mathbf{M}(\mathbf{r},t)$, which consists of the static part $M_o\hat{\mathbf{z}}$ and the variable part $\mathbf{m}\exp(-i\omega t)$, is determined by Eq. (1.31).

Using the expression for the effective magnetic field $\mathbf{H}_{eff}$ [4] as

$$\mathbf{H}_{\mathrm{eff}} = - \frac{\delta W}{\delta \mathbf{M}(\mathbf{r},t)} , \tag{1.121}$$

the torque equation of motion (1.36) for the magnetization $\mathbf{M}$ can be rewritten in a form containing the variational derivative of the total energy W:

$$\frac{\partial \mathbf{M}}{\partial t} = \gamma \left[\mathbf{M} \times \frac{\delta W}{\delta \mathbf{M}} \right] . \tag{1.122}$$

Equation (1.122) can now be re-expressed as a system of canonical Hamiltonian equations for the spin-wave amplitudes by making a sequence of canonical transformations (e.g., see [70,71,73]). First, we introduce the quantities $a_{\mathbf{q}}$ and $a^*_{\mathbf{q}}$ as the classical analogs of the operators $a_{\mathbf{q}}$ and $a^\dagger_{\mathbf{q}}$ of the well-known Holstein-Primakoff transformation when written in a wave-vector representation [see Eqs. (1.9) and (1.13)]. Then, in general, a canonical transformation is made from $a_{\mathbf{q}}$ and $a^*_{\mathbf{q}}$ to new variables $b_{\mathbf{q}}$ and $b^*_{\mathbf{q}}$, which describe the *normal modes* in the linear spin-wave approximation. These canonical transformations were first used in the classical theory of spin waves in ferromagnets by Schlömann [69]. The energy W of a ferromagnet becomes a Hamiltonian function in terms of the variables $b_{\mathbf{q}}$ and $b^*_{\mathbf{q}}$, and the equation of motion for the individual spin-wave amplitudes $b_{\mathbf{q}}$ has the form

$$\frac{\partial b_{\mathbf{q}}}{\partial t} = -i \, \frac{\delta \mathcal{H}(b_{\mathbf{q}}, b^*_{\mathbf{q}})}{\delta b^*_{\mathbf{q}}} . \tag{1.123}$$

For a general description of spin-wave interactions, it may sometimes be necessary to consider, in addition to the interactions between one spin wave and another (see later), the possibility of interactions between spin waves and nonmagnetic excitations such as phonons. These latter interactions, which contribute to the damping of the spin waves, can in most cases be taken into account phenomenologically by introducing a dissipative term into the canonical equation of motion (1.123) to give

$$\frac{\partial b_{\mathbf{q}}}{\partial t} + \Gamma_{\mathbf{q}} b_{\mathbf{q}} = -i \, \frac{\delta \mathcal{H}(b_{\mathbf{q}}, b_{\mathbf{q}}^{*})}{\delta b_{\mathbf{q}}^{*}} , \tag{1.124}$$

where $\Gamma_{\mathbf{q}} = \gamma \Delta H_{\mathbf{q}}$ with $2\Delta H_{\mathbf{q}}$ denoting a spin-wave full linewidth. This procedure has been justified, for example, by means of diagrammatic techniques [74]. The procedure does not apply, however, when the spin-wave damping is due to scattering by inhomogeneities [79].

Usually the energy of the spin-wave excitations in ferromagnets is sufficiently small that $\varepsilon = |\mathbf{m}|/M_0 \ll 1$, and so the Hamiltonian function can be expanded in a Taylor series in powers of $b_{\mathbf{q}}$ and $b_{\mathbf{q}}^{*}$:

$$\mathcal{H}(b_{\mathbf{q}}, b_{\mathbf{q}}^{*}) = \sum_{\mathbf{q}} \omega(\mathbf{q}) b_{\mathbf{q}} b_{\mathbf{q}}^{*} + \mathcal{H}_{\mathrm{p}} + \mathcal{H}_{\mathrm{int}} , \tag{1.125}$$

where

$$\mathcal{H}_{\mathrm{p}} = \mathcal{H}_{\mathrm{p1}} + \mathcal{H}_{\mathrm{p2}} + \dots , \qquad \mathcal{H}_{\mathrm{int}} = \mathcal{H}_{3} + \mathcal{H}_{4} + \dots . \tag{1.126}$$

The significance of the terms $\mathcal{H}_{\mathrm{p}}$ and $\mathcal{H}_{\mathrm{int}}$, and their expansions, will be explained below.

First, the quadratic part of the spin-wave Hamiltonian function $\mathcal{H}$ in Eq. (1.125) corresponds to the "kinetic energy" of non-interacting spin waves and has a diagonal form in terms of $b_{\mathbf{q}}$ and $b_{\mathbf{q}}^{*}$. The bulk spin-wave eigenfrequency $\omega(\mathbf{q})$ is determined by Eq. (1.23), where $A_{\mathbf{q}}$ and $B_{\mathbf{q}}$ are now given by the expressions

$$A_{\mathbf{q}} = \omega_{\mathrm{H}} + \alpha \omega_{\mathrm{M}} q^{2} + |B_{\mathbf{q}}| , \qquad B_{\mathbf{q}} = \tfrac{1}{2}\omega_{\mathrm{M}} \sin^{2}\theta_{\mathbf{q}} \exp(2i\phi_{\mathbf{q}}) . \tag{1.127}$$

Here $\theta_{\mathbf{q}}$ and $\phi_{\mathbf{q}}$ are the polar and azimuthal angles of the spin-wave wave vector $\mathbf{q}$ in spherical polar coordinates, taking the z axis parallel to the direction of saturation magnetization $\mathbf{M}_0$. We note that $\omega(\mathbf{q})$, as determined by Eqs. (1.23) and (1.127), coincides with the expression (1.45) obtained previously in the linear theory. The quantities $b_{\mathbf{q}}$ and $b_{\mathbf{q}}^{*}$, which are the normal variables of the linear theory, are also important in the development of nonlinear spin-wave theory. All the "linear" information about the system under investigation is contained in the dispersion relation $\omega(\mathbf{q})$, and all the additional information about the interaction of the spin waves with the external pumping field

$\mathbf{h}_p(\mathbf{r}, t)$ and between themselves is contained in the remaining coefficients of the expansion in Eq. (1.125).

Explicit expressions for the terms in $\mathcal{H}_p$ and $\mathcal{H}_{int}$ for the case of a ferromagnetic medium are presented in [71,73]. However, the physical meaning of each term in Eq. (1.126) can be easily understood by recalling that the canonical variables (or spin wave amplitudes) $b_{\mathbf{q}}$ and $b^*_{\mathbf{q}}$ are classical analogs of the quantum boson operators of creation and annihilation (see [9] and also Sec. 1.2.2 in this chapter).

a. Terms contributing to $\mathcal{H}_p$.

For example, the terms $\mathcal{H}_{p1}$ and $\mathcal{H}_{p2}$, which originate from the Zeeman part w_H of the total energy W in Eq. (1.119), describe the excitation of spin waves by the external pumping magnetic field $\mathbf{h}_p$, which is assumed to have a spatial variation like $\exp(i\mathbf{q}_p \cdot \mathbf{r})$ where $\mathbf{q}_p$ is the pumping wave vector. The first term $\mathcal{H}_{p1}$, which contains one spin-wave amplitude $b_{\mathbf{q}}$, describes the process of linear excitation of spin waves by the *transverse* (perpendicular to $\mathbf{M}_o$) component $h_p^{\perp}$ of the pumping field. For spatially homogeneous pumping (the special case of $\mathbf{q}_p = 0$) the term $\mathcal{H}_{p1}$ gives the well-known phenomenon of ferromagnetic resonance, i.e., the excitation of the homogeneous precession of magnetization corresponding to $\mathbf{q} = 0$. The amplitude b_o of the homogeneous precession is (see [65,70]):

$$|b_o| = h_p^{\perp} U / [(\omega_o - \omega_p)^2 + \Gamma_o^2]^{1/2} , \tag{1.128}$$

where ω_o is the eigenfrequency of homogeneous precession, $\Gamma_o = \gamma \Delta H_o$ with $2\Delta H_o$ denoting the ferromagnetic resonance linewidth, and U is the strength coefficient for the Hamiltonian $\mathcal{H}_{p1}$.

The second term $\mathcal{H}_{p2}$ contains two spin-wave amplitudes $b_{\mathbf{q}_1}$ and $b_{\mathbf{q}_2}$. It describes the phenomenon of *parallel pumping*, i.e., the parametric excitation of a pair of spin waves by the longitudinal (parallel to $\mathbf{M}_o$) component h_p^z of the pumping field. This process is the parametric instability of the first order ($n = 1$) and it satisfies the following conservation laws:

$$\omega_p = \omega_1 + \omega_2 , \qquad \mathbf{q}_p = \mathbf{q}_1 + \mathbf{q}_2 . \tag{1.129}$$

where ω_1 is a shorthand for $\omega(\mathbf{q}_1)$, etc. In the case of homogeneous pumping ($\mathbf{q}_p = 0$) the two excited spin waves propagate in opposite directions ($\mathbf{q}_1 = -\mathbf{q}_2 \equiv \mathbf{q}$) and have the same frequency ($\omega_1 = \omega_2 = \omega_p/2$). The threshold value of h_p^z is then easily calculated [66, 67] to be

$$(h_p^z)_{th} = \frac{\Gamma_{\mathbf{q}}}{|V_{\mathbf{q}}|} = \frac{\omega_p \Gamma_{\mathbf{q}}}{\gamma |B_{\mathbf{q}}|} = \frac{2\omega_p \Delta H_{\mathbf{q}}}{\omega_M \sin^2\theta_{\mathbf{q}}} , \qquad (1.130)$$

where $V_{\mathbf{q}}$ is the coefficient of the Hamiltonian $\mathcal{H}_{p2}$. It is clear from Eq. (1.130) that the lowest threshold corresponds to $\theta_{\mathbf{q}} = \pi/2$, i.e., to the excitation of spin waves propagating perpendicular to $\mathbf{M}_0$.

In general, the subsequent higher-order terms in the expansion of $\mathcal{H}_p$ correspond to explosive instabilities of the spin-wave amplitudes, and we do not discuss them here.

b. Terms contributing to $\mathcal{H}_{int}$.

The first term $\mathcal{H}_3$ in Eq. (1.126) for the expansion of $\mathcal{H}_{int}$ contains three spin-wave amplitudes ($b_{\mathbf{q}}$ or $b_{\mathbf{q}}^*$) and describes interaction of three spin waves, each of which satisfies the dispersion relation $\omega(\mathbf{q})$ of Eq. (1.45). The conservation laws for this three-wave interaction process can be written in the form

$$\omega_1 = \omega_2 + \omega_3 , \qquad \mathbf{q}_1 = \mathbf{q}_2 + \mathbf{q}_3 . \qquad (1.131)$$

The simplest special case of the above process is the *first-order Suhl instability* of homogeneous procession [65]. This occurs when $\mathbf{q}_1 = 0$ and $\mathbf{q}_2 = -\mathbf{q}_3 \equiv \mathbf{q}$, and corresponds to the parametric excitation of "half-frequency" spin waves with $\omega_2 = \omega_3 = \omega(\mathbf{q}) = \omega_p/2$, together with the "subsidiary" absorption of power in high-amplitude ferromagnetic resonance experiments [63,64]. The instability occurs when the amplitude of homogeneous precession b_0 exceeds the threshold level given by [65,70]

$$|b_0|_{th} = \frac{[(\omega(\mathbf{q}) - \omega_p/2)^2 + \Gamma_{\mathbf{q}}^2]^{1/2}}{|V_{0,\mathbf{q},-\mathbf{q}}|} , \qquad (1.132)$$

where $V_{0,\mathbf{q},-\mathbf{q}}$ is the coefficient of the Hamiltonian term $\mathcal{H}_3$. The threshold level of the transverse component $h_p^\perp$ of pumping field for this process can be calculated using Eqs. (1.128) and (1.132), and is given by [65]

$$(h_p^{\perp})_{th} = \frac{2\omega_p\{[(\omega_0 - \omega_p)^2 + \Gamma_0^2][(\omega(\mathbf{q}) - \omega_p/2)^2 + \Gamma_{\mathbf{q}}^2]\}^{1/2}}{\gamma\omega_M[\omega/2 + \omega_H + \alpha\omega_M q^2]\sin(2\theta_{\mathbf{q}})} . \tag{1.133}$$

It follows from Eq. (1.133) that the threshold level for this process is especially low in the case when the "subsidiary resonance" ($\omega_p/2 = \omega(\mathbf{q})$) coincides with the "main" ferromagnetic resonance ($\omega_p = \omega_0$). The minimization of the expression in Eq. (1.133) shows that the parametrically-excited spin waves with $\mathbf{q} \simeq 0$ are largely responsible for the subsidiary absorption.

The next term $\mathcal{H}_4$ contains four spin-wave amplitudes and describes four-wave interaction process with the following conservation laws:

$$\omega_1 + \omega_2 = \omega_3 + \omega_4 , \qquad \mathbf{q}_1 + \mathbf{q}_2 = \mathbf{q}_3 + \mathbf{q}_4 . \tag{1.134}$$

The simplest process of this kind is the *second-order Suhl instability* of homogeneous precession ($\mathbf{q}_1 = \mathbf{q}_2 = 0$; $\mathbf{q}_3 = -\mathbf{q}_4 \equiv \mathbf{q}$), which is responsible for the saturation of the main resonance (when $\omega_1 = \omega_2 = \omega_3 = \omega_4 = \omega_p \cong \omega_0$) and for the foldover effects observed in ferromagnetic resonance experiments at high input power levels [63,64,68]. The threshold value of the amplitude of homogeneous precession for this process is given by [65,70]

$$|b_0|^2_{th} = \frac{[(\omega(\mathbf{q}) - \omega_p)^2 + \Gamma_{\mathbf{q}}^2]^{1/2}}{|S_{0,0,\mathbf{q},-\mathbf{q}}|} , \tag{1.135}$$

where $S_{0,0,\mathbf{q},-\mathbf{q}}$ is the coefficient of the Hamiltonian $\mathcal{H}_4$. The minimum threshold value of the pumping field h_p for this process corresponds to the excitation of z-directed spin waves (having $\theta_{\mathbf{q}} = 0$ or π) with $\omega(\mathbf{q}) = \omega_0 = \omega_p$, and is given by [65,68]

$$(h_p^{\perp})_{th} = \frac{\Gamma_0}{\gamma}\left(\frac{\Gamma_{\mathbf{q}}}{\omega_M}\right)^{1/2} . \tag{1.136}$$

It is reasonable to ask up to which order of $b_{\mathbf{q}}$ should we expand the spin-wave interaction Hamiltonian $\mathcal{H}_{int}$. It is shown in [73] that $\mathcal{H}_5$

and the higher order terms need not, in general, be considered. The argument is that, since $\varepsilon = |\mathbf{m}|/M_0 \ll 1$, each successive term in the expansion in Eq. (1.126) is smaller than the preceding one and the dynamics of the system will be determined, in the first instance, by the lowest term $\mathcal{H}_3$. The three-wave processes described by $\mathcal{H}_3$ may, however, be prohibited in some cases by the conservation laws. For example, in the spectrum $\omega(\mathbf{q})$ there is a "gap" equal to ω_H, and for small values of $\mathbf{q}$ the three-wave decay processes in Eq. (1.131) are prohibited because the wave having the lowest energy cannot experience further decay. In this "non-decay" part of the spin-wave spectrum the next term $\mathcal{H}_4$ in the expansion (1.126) should then be taken into account.

The fact that three-wave processes are prohibited by the conservation laws in certain wave-vector regions does *not* mean that the three-wave Hamiltonian $\mathcal{H}_3$ can be ignored totally. The matrix elements of $\mathcal{H}_3$ are large compared with the matrix elements of $\mathcal{H}_4$, and therefore their contribution to the amplitude of four-wave processes must be taken into account *in the second order* of perturbation theory [71,73]. This leads to a renormalization of the coefficients of the four-wave Hamiltonian: their renormalized value $T_{\mathbf{q}_1,\mathbf{q}_2,\mathbf{q}_3,\mathbf{q}_4}$ is equal to the unrenormalized quantity $S_{\mathbf{q}_1,\mathbf{q}_2,\mathbf{q}_3,\mathbf{q}_4}$ plus correction terms of order $|V_{\mathbf{q}_1,\mathbf{q}_2,\mathbf{q}_3}|^2/\omega_p$. For cubic ferromagnets the contribution of the three-wave processes can be of the same order of magnitude as the amplitudes $S_{\mathbf{q}_1,\mathbf{q}_2,\mathbf{q}_3,\mathbf{q}_4}$ of the four-wave processes themselves [80].

An important feature of the four-wave processes in Eq. (1.134), and described by the Hamiltonian $\mathcal{H}_4$, is that these processes are allowed for any general form of the spin-wave dispersion law $\omega(\mathbf{q})$. Consequently, the term $\mathcal{H}_4$ will determine, to leading order, the spin-wave dynamics for all "non-decay" spin-wave spectra. The higher order terms $\mathcal{H}_5$, $\mathcal{H}_6$, ... will describe smaller and, as a rule, negligible corrections.

We emphasize that, for a general description of the spin-wave dynamics in ferromagnets, the four-wave term $\mathcal{H}_4$ should be taken into account even in the decay part of the spectrum where the three-wave processes in Eq. (1.131) are allowed. This is because the Hamiltonian $\mathcal{H}_3$ originates from the long-range magnetic dipole-dipole interaction whereas $\mathcal{H}_4$ comes mainly from the exchange interaction. Also it should be noted that when the magnetic anisotropy energy is taken into account in Eq. (1.119) the four-wave interaction of spin waves turns out to be strongly anisotropic, e.g. see calculations in [71,73].

The above classical Hamiltonian formalism for spin waves *in unlimited ferromagnetic media* is one of the main theoretical methods currently used for spin-wave nonlinear dynamics (see, e.g., [81-83]). In many cases it gives adequate results as confirmed by a number of experiments [84-86]. However, this formalism, which does not take into account the boundary conditions on the surfaces of a ferromagnetic sample, cannot explain several dimension-dependent nonlinear effects observed in YIG spheres and films [87-90]. Presently attempts are being made by several authors to incorporate the boundary conditions (both electrodynamic and exchange) into this formalism to achieve a fuller understanding of spin-wave nonlinear dynamics in limited ferromagnetic samples and films [91-96] (see also Sec. 1.6 of this Chapter).

1.5.3. Microscopic Approach to Spin-Wave Interactions

We next reconsider the effects of spin-wave interactions treated in a *microscopic* approach by using a spin Hamiltonian. This involves generalizing the linear theory described in Sec. 1.2.2 to include the interaction term $\mathcal{H}_{int}$. We shall, for simplicity, omit the nonlinear effects due to a pumping magnetic field $\mathbf{h}_p$ since this has been dealt with in the preceeding subsection. Instead we focus on the respective roles of the exchange and dipole-dipole interactions in contributing to the processes described by $\mathcal{H}_{int}$.

We begin by considering an *unlimited*, or bulk, ferromagnet represented by a spin Hamiltonian with an exchange part, as in Eq. (1.8), and a dipole-dipole part, as in Eq. (1.2). Using the Holstein-Primakoff transformation at low temperatures $T \ll T_C$, we showed earlier that the spin-wave dispersion relation is given by Eq. (1.23). This holds in the approximation of using the bilinear Hamiltonian $\mathcal{H}_2$ of Eq. (1.22) and ignoring higher-order products of the operators. Before proceeding further, it is convenient to 'diagonalize' $\mathcal{H}_2$, which mixes operators with labels $\mathbf{q}$ and $-\mathbf{q}$, by applying the canonical transformation [9,97]:

$$a^\dagger_{\mathbf{q}} = u^*_{\mathbf{q}}\, b^\dagger_{\mathbf{q}} + v_{\mathbf{q}}\, b_{-\mathbf{q}}\,, \qquad a_{\mathbf{q}} = u_{\mathbf{q}}\, b_{\mathbf{q}} + v^*_{\mathbf{q}}\, b^\dagger_{-\mathbf{q}}\,. \tag{1.137}$$

If the factors $u_{\mathbf{q}}$ and $v_{\mathbf{q}}$ are chosen such that

$$u_{\mathbf{q}} = \left(\frac{A_{\mathbf{q}} + \omega(\mathbf{q})}{2\omega(\mathbf{q})}\right)^{1/2}, \qquad v_{\mathbf{q}} = \frac{B_{\mathbf{q}}}{|B_{\mathbf{q}}|}\left(\frac{A_{\mathbf{q}} - \omega(\mathbf{q})}{2\omega(\mathbf{q})}\right)^{1/2}, \tag{1.138}$$

then it is easy to show (e.g., see [4,10]) that

$$\mathcal{H}_2 = \sum_{\mathbf{q}} \omega(\mathbf{q})\, b^\dagger_{\mathbf{q}} b_{\mathbf{q}}\,. \tag{1.139}$$

Hence $b^\dagger_{\mathbf{q}}$ and $b_{\mathbf{q}}$ are the operators that create and annihilate a spin wave of wave vector $\mathbf{q}$ and frequency $\omega(\mathbf{q})$. Notice that, in the limit of the dipole-dipole terms becoming negligibly small compared with the exchange effects ($|B_{\mathbf{q}}| \ll A_{\mathbf{q}}$), we have $u_{\mathbf{q}} \longrightarrow 1$ and $v_{\mathbf{q}} \longrightarrow 0$ and so the a and b operators become identical.

Next the terms in the interaction Hamiltonian $\mathcal{H}_{\text{int}}$ can be obtained by expanding the spin operators in terms of the Holstein-Primakoff operators as in Eq. (1.11), transforming to the $b^\dagger$ and b operators, and identifying the contributions to the full Hamiltonian that are cubic, quartic, etc., in these spin-wave operators. The lowest order contributions to $\mathcal{H}_{\text{int}}$ are found (e.g. see [4,10] and references therein) to be of the form

$$\mathcal{H}_3 = \sum_{\mathbf{q}_1,\mathbf{q}_2} \left[F_3(\mathbf{q}_1,\mathbf{q}_2)\, b^\dagger_{\mathbf{q}_1} b^\dagger_{\mathbf{q}_2} b_{\mathbf{q}_1+\mathbf{q}_2} + \text{h.c.}\right], \tag{1.140}$$

$$\begin{aligned}\mathcal{H}_4 = \sum_{\mathbf{q}_1,\mathbf{q}_2,\mathbf{q}_3} \Big[&F_4(\mathbf{q}_1,\mathbf{q}_2,\mathbf{q}_3)\, b^\dagger_{\mathbf{q}_1} b^\dagger_{\mathbf{q}_2} b_{\mathbf{q}_3} b_{\mathbf{q}_1+\mathbf{q}_2-\mathbf{q}_3} \\ &+ \left(F'_4(\mathbf{q}_1,\mathbf{q}_2,\mathbf{q}_3)\, b^\dagger_{\mathbf{q}_1} b^\dagger_{\mathbf{q}_2} b^\dagger_{\mathbf{q}_3} b_{\mathbf{q}_1+\mathbf{q}_2+\mathbf{q}_3} + \text{h.c.}\right)\Big]\,,\end{aligned} \tag{1.141}$$

where h.c. denotes the Hermitian conjugate. The amplitudes F_3, F_4, and F'_4 depend on the wave-vector labels and on the exchange and dipole-dipole interactions; detailed expressions are to be found in the literature cited above. However, the main characteristics of these terms are summarized below.

The contribution $\mathcal{H}_3$ describes *three-magnon processes* in which there is a *splitting* (i.e., two magnons are created and one is annihilated) or a *confluence* (the conjugate process in which two magnons are annilated and one is created). It arises purely from the dipole-dipole interactions (e.g., from the expansion of terms like S^+S^z or S^-S^z in the spin Hamiltonian), and so it is absent in the limiting case of the Heisenberg model.

The contribution $\mathcal{H}_4$ describes *four-magnon processes*: the term involving F_4 corresponds to *scattering* (in which two magnons are annihilated and two are created) while the term involving F_4' (and its conjugate) correspond to four-magnon *splitting* and *confluence*. The splitting and confluence processes again arise only from the dipole-dipole interactions (e.g., in this case from the expansion of terms like S^+S^+ or S^-S^- in the spin Hamiltonian), whereas both exchange and dipole-dipole interactions contribute to the scattering process (e.g. from the expansion of terms like S^+S^- or S^zS^z). In the limiting case of the Heisenberg model, where the dipolar interactions are neglected, we have (e.g., see [4]):

$$F_4(\mathbf{q}_1,\mathbf{q}_2,\mathbf{q}_3) = \frac{Jz_o}{4N}\,[2\gamma_{\mathbf{q}_1-\mathbf{q}_3} + 2\gamma_{\mathbf{q}_2-\mathbf{q}_3} - \gamma_{\mathbf{q}_1} - \gamma_{\mathbf{q}_2} - \gamma_{\mathbf{q}_3} - \gamma_{\mathbf{q}_1+\mathbf{q}_2-\mathbf{q}_3}] \tag{1.142}$$

where the notation is explained in Sec. 1.2.2. This result, which has been obtained by retaining the leading term in a $1/S$ expansion, is consistent with the rigorous theory of spin-wave interactions in Heisenberg ferromagnets developed by Dyson [98]. Three-magnon and four-magnon processes in $\mathcal{H}_{int}$ are represented schematically in Fig. 1.10.

Fig. 1.10. Schematic representation of (a) the three-magnon and (b) the four-magnon interaction vertices.

An important effect of the interaction terms of $\mathcal{H}_{\text{int}}$ is to renormalize the spin-wave excitations leading, in general, to a frequency shift and a damping (or reciprocal lifetime). These properties have been thoroughly studied in unlimited ferromagnets with exchange or dipole-dipole interactions (or both), particularly at low temperatures (e.g., see [4,10,17,98] and references therein). Extensions to higher temperatures have been carried out by diagrammatic perturbation theory (see [99-101]). Some of the above references also describe the analogous calculations for unlimited *antiferromagnetic* samples.

The microscopic theory of spin-wave interactions in *limited* magnetic media is currently a topic where further work is required, just as in the case of macroscopic theories. The main difficulty in a microscopic theory comes from the long-range nature of the dipole-dipole terms. Dipole-dipole sums can be defined as in Eq. (1.22), but the sums no longer extend over an infinite lattice and this leads to a more complicated behavior in most cases. On the other hand, several calculations have been reported for the spin-wave interactions in *semi-infinite Heisenberg ferromagnets*. The interaction term is a four-magnon scattering process, analogous to the F_4 term in Eq. (1.141), except that each of the four magnons may be a bulk or surface mode. The renormalization of the bulk and surface modes can be investigated for this model and geometry, and results have been obtained for the frequency shifts [102,103] and the damping [103,104] at $T \ll T_C$.

1.6. NONLINEAR DYNAMICS OF SPIN WAVES

In this section we develop further some specific aspects of the nonlinear dynamics of spin waves, particularly in the cases of limited geometries, using the basic concepts already introduced in Sec. 1.5. Here we follow the classical Hamiltonian formalism in order to incorporate more readily both the dipolar and exchange effects.

1.6.1. Spin-Wave Parametric Processes in Limited Ferromagnetic Samples

The classical Hamiltonian formalism described in Sec. 1.5 is valid for plane spin waves propagating in unlimited ferromagnetic media. However, the experimental investigations on parametric excitation of spin waves are necessarily performed in limited ferromagnetic samples (spheres, thin films, etc.). If the reciprocal dimension $1/L$ of the ferromagnetic sample is comparable to the wave number q of the spin waves involved in a parametric process, the plane spin waves cannot be considered the eigenfunctions of magnetization, and the results based on the plane-wave expansion (1.34) are, strictly speaking, inaccurate.

To make a quantitative comparison between between theory and experiment in this case it is necessary to take into account the electrodynamic and exchange boundary conditions, Eqs. (1.59) and (1.61), at the sample surfaces. The solution of this boundary problem yields a system of complete orthogonal functions: these are eigenmodes of magnetization which satisfy the boundary conditions and involve three indices (depending on the sample shape). In effect, the sample acts as a resonator and the magnetization of the sample may be expanded in these "resonator" eigenmodes. Such an expansion can be used, where appropriate, instead of the plane spin-wave representation (1.34) in the framework of the above described classical Hamiltonian formalism (e.g., see [94,95]).

For a ferromagnetic ellipsoid in the non-exchange limit, the eigenmodes of magnetization are the magnetostatic, or Walker, modes [30] (see Sec. 1.3). The theory of spin-wave parametric processes based on the above type of expansion for the Walker modes of an ellipsoid was developed in [105,106]. It gives the thresholds of parametric instability and the selection rules for Walker-mode pairs parametrically excited by the uniform precession of magnetization (or by the uniform pumping field in the case of parallel pumping). The results are quite similar to Eqs. (1.130), (1.131), (1.133), but since the spectrum of Walker modes is discrete, the dependence of the threshold amplitude h_{th} of the pumping field on the value of the external magnetic field H_0 is not smooth: it consists of a series of sharp minima, each corresponding to the excitation of a certain pair of Walker modes. This work received little attention at first because Suhl's theory [65], based on the plane wave expansion (1.34), gave results which were in good agreement with all the experiments performed by that time (even for modes with $q \simeq 0$). In fact, Suhl [65] had correctly pointed out that good results obtained from the plane spin-wave model are due to the fact that the frequencies of the Walker modes (which are in reality excited at the threshold) all fall between the upper and the lower limits of the spin wave spectrum $\omega(\mathbf{q})$ [see Eq. (1.45)] for $q \simeq 0$. Later experiments, performed using higher resolution microwave spectrometers, have indeed shown a series of sharp dips on a threshold curve in the case of "subsidiary" absorption in a ferromagnetic sphere [87,107,108]. This fine structure of the absorption is associated with the parametric excitation of Walker-mode pairs [109].

Recently the theory of parametric excitation of Walker modes has attracted the attention of researchers [94,96,110] in connection with the problem of chaotic dynamics of auto-oscillations of magnetization observed in spherical ferromagnetic samples above the first order Suhl instability threshold [111].

When at least one of the dimensions of a finite ferromagnetic sample is comparable to the exchange length $\ell_{ex} = \alpha^{1/2}$, the exchange interaction and the exchange boundary conditions (1.61) should be taken into account. In particular, the system of dipole-exchange magnetization eigenmodes for a thin ferromagnetic disk was obtained in [90,95]. The theory of auto-oscillations of magnetization in disks, developed in [95] on the basis of an expansion in the dipole-exchange eigenmodes, gives a very good quantitative agreement between the predictions of the model and the measured auto-oscillation frequencies and threshold fields (see [95] and Chapter 8 of this book).

In the nonlinear theory of spin wave processes in thin ferromagnetic films the exchange interaction has to be included, in general, for the same reasons as in the linear theory (see Sec. 1.3). The corresponding theory of spin-wave parametric instability in films was developed in [46,91-93,112-116]. The case of an infinite film is analytically much simpler than the case of a sphere or disk as only one dimension in the medium is finite. A ferromagnetic film acts as a waveguide for spin waves, and the eigenmodes of magnetization in a film are the inhomogeneous plane waves with magnetization distribution $\mathbf{m}_{\mathbf{Q}}(\xi)$ across the film thickness [see Eq. (1.57)]. When both electrodynamic and exchange boundary conditions on the film surfaces are taken into account and the spin-wave eigenfunctions $\Phi_n(\xi)$ of Eq. (1.87) are found, the magnetization $\mathbf{m}(\mathbf{r})$ of the film can be written as [91,92]:

$$\mathbf{m}(\mathbf{r},t) = \mathbf{m}(\xi,\mathbf{R},t) = \int d^2\mathbf{Q} \sum_n \mathbf{m}_{n\mathbf{Q}}(t)\, \Phi_n(\xi)\, \exp(i\mathbf{Q}\cdot\mathbf{R}) \ . \tag{1.143}$$

The Hamiltonian formalism based on the above expansion is developed in [92,116]. The Hamiltonian function, Eq. (1.125), is written in terms of the amplitudes $b_{n\mathbf{Q}}$ of the spin waves in the film, which are analogous to the spin wave-amplitudes $b_{\mathbf{q}}$ in the case of the unlimited medium. The coefficients of the Hamiltonian function in this case depend on the in-plane wave vector $\mathbf{Q}$, the index n (which determines the value of the transverse wave number $q_{\xi n}$), and the magnetization distribution of mode n across the film thickness.

By contrast with the case of an unlimited ferromagnetic medium [70,71,73], the coefficients of the Hamiltonian function in the case of a ferromagnetic film turn out to be analytical functions of Q even in the region $QL \ll 1$, where the long-range dipole-dipole interaction is dominant [92]. These analytical properties of the Hamiltonian coefficients in the film are connected with the fact that the diagonal matrix element P_{nn} of the dipole-dipole interaction in the film [see Eq. (1.92) and also Chapter 2] vanishes when $Q \longrightarrow 0$, while the analogous quantity $\sin^2\theta_{\mathbf{q}}$ for an unlimited ferromagnetic medium (see Eq. (1.24)) does not have a definite limit when $q \longrightarrow 0$. The analytical properties of the Hamiltonian coefficients make possible the description of four-wave parametric processes in films in terms of envelopes of spin-wave packets and enable us to calculate the parameters of the spin-wave envelope solitons (e.g., see Chapter 9).

The approximate "diagonal" expression for the discrete spin wave eigenfrequencies in a film $\omega_{n\mathbf{Q}}$ is analogous to Eq. (1.23):

$$\omega_{n\mathbf{Q}}^2 = A_{n\mathbf{Q}}^2 - |B_{n\mathbf{Q}}|^2 = \Omega_{n\mathbf{Q}}(\Omega_{n\mathbf{Q}} + \omega_M F_{nn}) \tag{1.144}$$

where $A_{n\mathbf{Q}}$, $B_{n\mathbf{Q}}$, and $\Omega_{n\mathbf{Q}}$ are determined in [92]:

$$2A_{n\mathbf{Q}} = 2\Omega_{n\mathbf{Q}} + \omega_M \sin^2\theta + \omega_M P_{nn}[1 - \sin^2\theta\ (1 + \cos^2\phi)] \ , \tag{1.145}$$

$$B_{nQ} = A_{nQ} - \Omega_{nQ} - \omega_M P_{nn} \sin^2\phi + i(\omega_M/2)P_{nn} \cos\theta \sin 2\phi \ , \qquad (1.146)$$

$$\Omega_{nQ} = \omega_H + \alpha\, \omega_M (Q^2 + q_{\xi n}^2) \ . \qquad (1.147)$$

As expected, Eq. (1.144) is equivalent to Eq. (1.90) obtained previously in the linear theory [36,47].

It is worth noting that $|B_{nQ}|$ provides a measure of the ellipticity of the spin-wave modes. The value of this coefficient is very significant in the theory of parametric instability as the spin-wave modes having circular polarization ($|B_{nQ}| = 0$) cannot be excited parametrically. For example, in the case of parallel pumping it is easy to determine, using Eqs. (1.130) and (1.146), which spin-wave modes b_{nQ} are excited at the threshold and hence the form of the threshold h_{th} as a function of H_0. When the film is magnetized perpendicularly to its surface ($\theta = 0$), we have [114]:

$$|B_{nQ}| = \omega_M P_{nn}/2 \ . \qquad (1.148)$$

The dipole matrix element P_{nn} has its maximum value (unity) when $Q \to \infty$, and this corresponds to the minimum value of the threshold field in Eq. (1.130). So, the spin wave modes with the large values of Q are excited at the threshold. These modes are quite similar to the plane spin waves and therefore the threshold curve in this case is smooth and similar to the usual "butterfly" threshold curve obtained in unlimited ferromagnetic media [66]. The standing spin waves ($Q = 0$) are circularly polarized ($|B_{nQ}(0)| = 0$) and therefore cannot be excited parametrically.

On the other hand, when the film is magnetized parallel to its surface ($\theta = \pi/2$), we have [46]:

$$|B_{nQ}| = \omega_M |1 - P_{nn}(1 + \sin^2\phi)|/2 \ . \qquad (1.149)$$

In this case the standing spin waves ($Q = 0$, $P_{nn} = 0$) have the largest ellipticity and are excited at the threshold. The threshold curve consists of a series of sharp minima corresponding to excitation of the standing modes with different n. So, in this case, parallel pumping can be used to observe a fine structure of microwave absorption due to the discreteness of the spin-wave spectrum in the film. This is confirmed experimentally [117-119].

It is also necessary to take into account the possibility of inhomogeneities of the microwave pumping field in the film plane. In many experiments performed on magnetic films, the spin waves were excited by a strip-line transducer of width W, so the microwave pumping magnetic field of this transducer was localized to a width of order W. In particular, the influence of the pumping field localization on the threshold of spin-wave parametric instability in films was studied theoretically in [91,116,120-122] and experimentally in [123-129]. When the pumping field is localized in the width W, the wave-vector conservation laws in Eqs. (1.129), (1.131), and (1.134) are fulfilled, for the in-plane components, to an accuracy of order [116,121]:

$$\Delta \mathbf{Q}_{\mathrm{p}} = (2\pi\delta/W)\, \mathbf{e}_{W} , \qquad (1.150)$$

Here $\delta \leq 1$ is a dimensionless parameter depending on the form of the localized field distribution in the film plane, and $\mathbf{e}_W$ is an in-plane unit vector along the width of the localization zone.

The above result means that a localized pumping field can directly excite spin waves having *nonzero* in-plane wave number Q in the range from 0 to approximately $|\Delta \mathbf{Q}_{\mathrm{p}}|$ and propagating in the direction of $\mathbf{e}_W$. This linear excitation of propagating spin waves is described by the term $\mathcal{H}_{\mathrm{p1}}$ in the Hamiltonian in Eq. (1.126). Naturally, the spin wave satisfying the "resonant" condition $\omega(\mathbf{Q}) = \omega_{\mathrm{p}}$ has the biggest amplitude (see, e.g., Eq. (1.128)) and this can be observed experimentally.

In the case of parametric excitation of the spin waves by localized pumping, we should again take account of Eq. (1.150). For example, for localized parallel pumping (described by the term $\mathcal{H}_{\mathrm{p2}}$ in Eq. (1.126)) we obtain $|\mathbf{Q}_{\mathrm{p}} - \mathbf{Q}_1 - \mathbf{Q}_2| \leq |\Delta \mathbf{Q}_{\mathrm{p}}|$ from Eqs. (1.129) and (1.150). As a consequence, the frequency conservation law in Eq. (1.129) becomes approximately

$$\omega_{\mathrm{p}} - \omega_1 - \omega_2 \simeq (\mathbf{v}_{g1} + \mathbf{v}_{g2}) \cdot \Delta \mathbf{Q}_{\mathrm{p}} . \qquad (1.151)$$

Here $\mathbf{v}_{gi} = \partial\omega_i/\partial\mathbf{Q}_i$ ($i = 1,2$) is the spin-wave group velocity calculated with $\Delta\mathbf{Q}_p$ set equal to zero.

By substituting Eq. (1.151) and similar formulae into Eqs. (1.130) and (1.133), it can be shown that the parametrically excited spin waves may carry energy out of the zone of pumping localization, leading to the considerable increase in the threshold amplitude of the pumping field. For example, in the simplest case of *degenerate* parametric excitation of spin waves under localized parallel pumping (i.e., $\mathbf{Q}_1 = -\mathbf{Q}_2 \equiv \mathbf{Q}$, $v_{g1} = v_{g2} \equiv v_{gn\mathbf{Q}}$, $\Gamma_1 = \Gamma_2 \equiv \Gamma_{n\mathbf{Q}}$) we obtain

$$(h_p^z)_{th} = (\omega_p/\gamma|B_{n\mathbf{Q}}|) \left[\left(v_{gn\mathbf{Q}}\frac{2\pi\delta}{W}\cos\varphi\right)^2 + \Gamma_{n\mathbf{Q}}^2\right]^{1/2}, \qquad (1.152)$$

where φ is the angle between the direction of the spin wave group velocity $\mathbf{v}_{gn\mathbf{Q}}$ and the unit vector $\mathbf{e}_W$. An analogous expression for the case of perpendicular pumping was obtained in [116]. It is clear from Eq. (1.152) that in the case of localized pumping the threshold depends not only on the spin-wave relaxation parameter $\Gamma_{n\mathbf{Q}}$ and the ellipticity characterized by $|B_{n\mathbf{Q}}|$, but also on the magnitude and direction of the group velocity $\mathbf{v}_{gn\mathbf{Q}}$. All these factors should be taken into account in the minimization procedure for the threshold amplitudes. In general, the propagating spin waves with the smallest group velocity will have the lowest threshold of excitation under localized pumping. This conclusion was confirmed experimentally in [127-129].

In summary, the main differences between parametric spin-wave processes in limited and unlimited ferromagnetic media are:

(i) The discreteness of the spin-wave spectrum in limited samples creates a fine structure of microwave absorption corresponding to parametric excitation of individual spin-wave modes. It also leads to additional selection rules for the parametric processes and thus increases the stability of the spin system under the external pumping.

(ii) Localization of the pumping field in films leads to the possibility of direct linear excitation of spin waves having non-zero in-plane wave vector $\mathbf{Q}$. The localization also leads to an increase in threshold values for spin-wave parametric processes, because parametrically excited travelling spin waves carry energy out of the excitation zone.

(iii) Coefficients of the spin wave Hamiltonian function, Eq. (1.125), in films are analytical functions of the in-plane wave vector **Q**.

1.6.2. Dynamics of Spin-Wave Auto-Oscillations: Transition to Chaos

It is often the case that above the threshold of parametric excitation of spin waves (for parallel or perpendicular pumping) the new stationary state is not established, and the magnetization in a ferromagnetic sample performs complicated *auto-oscillations* [72,107,130]. We summarize some of the results here; a fuller account is given in Chapter 7 of this book.

The main experimental properties of auto-oscillations of magnetization in YIG samples are (see [71,107]):

1. Auto-oscillations appear at pumping levels which are typically 0.1 to 1 dB higher than the threshold of parametric excitation of spin waves. For spin waves with small wave numbers the threshold of auto-oscillations is considerably higher. The threshold also increases when inhomogeneities are introduced into the crystal.
2. Auto-oscillations appear having frequencies in the 10 to 100 kHz range depending on the bias magnetic field. For a small excess above the threshold the spectrum of auto-oscillations consists of a single line. With an increase in the pumping power level the frequency of auto-oscillations increases and new lines appear in the spectrum of auto-oscillations. For larger excess above the threshold a noise-like chaotic spectrum of auto-oscillations is observed.
3. The auto-oscillations demonstrate giant crystallographic anisotropy which exceeds the crystallographic anisotropy of the spin-wave spectrum by at least one order of magnitude. In the case of parallel pumping the intensity of auto-oscillations for bias magnetic field orientation along the <111> axis exceeds the intensity for bias field orientation along the <100> axis by factor of roughly 100. Under perpendicular pumping the maximum and minimum intensities of auto-oscillations correspond to the bias magnetic field oriented along the <100> and <111> axes, respectively.

The theoretical interpretation of these auto-oscillations has proved to be very difficult as their spectrum and intensity are sensitive to virtually all the experimental parameters, such as amplitude of pumping field, magnitude of bias magnetic field, size, shape, orientation of sample, etc. Over the years several theoretical models were proposed to explain the nature of auto-oscillations [71,107,131-132]. These theories explain certain of the experimental observations, but fail to give a full account of this complicated physical phenomenon.

The simplest linear model [131] assumes that several discrete frequencies, corresponding to the eigenmodes of magnetization of a sample are present in the spectrum of parametrically excited spin waves. Beats between these frequencies lead to the appearance of auto-oscillations. This model explains the dependence of the auto-oscillation frequency on the sample size and on crystallographic anisotropy, but ignores the frequency dependence on the pump power. Another model [107,132] was based on the idea that parametrically excited spin waves can influence the mean magnetization of the sample, and if the magnetization amplitude follows the amplitude of parametric spin waves with a certain delay, the auto-oscillations of magnetization can develop in the sample.

The nonlinear nature of the auto-oscillations under parallel pumping was first recognized by Zakharov et al. [71]. In the framework of their *S*-theory, they interpreted auto-oscillations of magnetization as a result of the internal instability of a stationary distribution of parametrically excited spin waves against the excitation of collective modes of oscillations in the system of these waves. The eigenfrequencies Ω_m of these collective oscillations were calculated in [71] as

$$\Omega_m = 2N_{\mathbf{q}} \sqrt{S_m(2T_m + S_m)} \ , \qquad (1.153)$$

in an unlimited sample. Here $N_{\mathbf{q}}$ is the integral amplitude of parametrically excited spin waves (proportional to the excess of the pumping field amplitude h_p^z above its threshold value given in Eq. (1.130)), S_m and T_m are the Fourier amplitudes of the coefficients of the Hamiltonian term $\mathcal{H}_4$ in Eq. (1.126) describing four-wave interactions, and $m = 0, \pm1, \pm2, \ldots$ is the mode number. For cubic ferromagnets magnetized in the directions <111> or <110>, $\Omega_m \neq 0$ only for the modes having numbers $m = 0, \pm2$ [71]. The criterion $S_m(2\,T_m + S_m) < 0$ gives the condition of instability of collective oscillations in the system of parametric spin waves and therefore the condition of existence of auto-oscillations of magnetization under parallel pumping [71]. This condition was verified experimentally in [133].

Calculations [71,73,74] show that the coefficients T_m and S_m are strongly dependent on the crystallographic anisotropy even in the case of materials where the spectrum of spin waves has a very weak dependence on anisotropy (such as in YIG). The theory of auto-oscillations based on Eq. (1.153) successfully explains the giant crystallographic anisotropy of auto-oscillations and the dependence of the auto-oscillation frequency on the pumping field amplitude. At the same time this theory predicts that the power threshold for the appearance of auto-oscillations coincides with the threshold of parametric excitation of spin waves [see Eq. (1.130)] and that the frequency of auto-oscillations is zero at the threshold ($\Omega_m = 0$ when $N_q = 0$). Neither of these conclusions are confirmed by experimental data. The theory [71] also ignores the observed sample-size dependence of the auto-oscillation frequency [88,89].

Nevertheless, the main idea of the *S*-theory [71] was to identify the four-wave nonlinear interaction, Eq. (1.115), as the origin of auto-oscillations. This has inspired nearly all later theoretical approaches to the nonlinear dynamics of spin wave auto-oscillations. The authors of [71] realized that auto-oscillations of magnetization in their model could be both periodic and stochastic, and in the latter case these auto-oscillations can be interpreted as secondary turbulence in a system of parametric spin waves.

The early numerical simulations [81,134] of the nonlinear dynamics of unstable collective oscillations, which were performed on a simplified model of a stationary state consisting of two "beams" (or two groups of pairs of parametrically excited spin waves) have shown that at low supercriticalities the regime of periodic auto-oscillations is established. This regime corresponds to the formation of a limit cycle in the phase space of the system. With the increase of supercriticality the spectrum of auto-oscillations widens and chaotic auto-oscillations take place. Both of these effects were previously observed in experiments [107,130].

Further developments of the theory of auto-oscillations are connected with progress in nonlinear mechanics and the introduction of the notion of a "strange" (or stochastic) attractor. After the pioneering work by Lorenz [135], it became clear that purely dynamical systems with a small number of degrees of freedom can demonstrate chaotic and unpredictable behavior. It is well known that, if the phase space of a nonlinear dissipative system is two-dimensional, there are only two types of attracting trajectories (or attractors) in this phase space: these are a *stable focus* and a *stable limit cycle*, corresponding to an equilibrium state and periodic motion, respectively. Lorenz showed that, if the phase space of a dynamical dissipative system is at least three-dimensional, then another type of attracting trajectory, corresponding to a very complicated and unpredictable motion of the system, is possible. This new type of of attracting trajectory was called a *strange attractor* [135]. For the formation of a strange attractor (and therefore for the chaotic behavior of a dissipative dynamical system) it is necessary that [136]: (i) all (or nearly all) neighboring trajectories in a certain region of a phase space are unstable and diverge from one other; and (ii) all these trajectories remain in the limited phase volume.

For example, in a three-dimensional phase space it is easy to imagine such a behavior. Trajectories may be unstable and may diverge in a certain two-dimensional plane and may return to the origin when they go out of the plane. The trajectory in this case has the form of an uncoiling spiral with a tail which returns to the origin and starts to uncoil again (see Fig. 1.11). The trajectory lies in the limited phase volume, never crosses itself, and behaves in a very complicated and unpredictable way.

Soon it was discovered that, apart from the rigid excitation of chaos described by the Lorenz model [135], there are several typical scenarios of strange attractor formation (or transition to chaotic behavior) in dynamical systems with dissipation [75,136]:

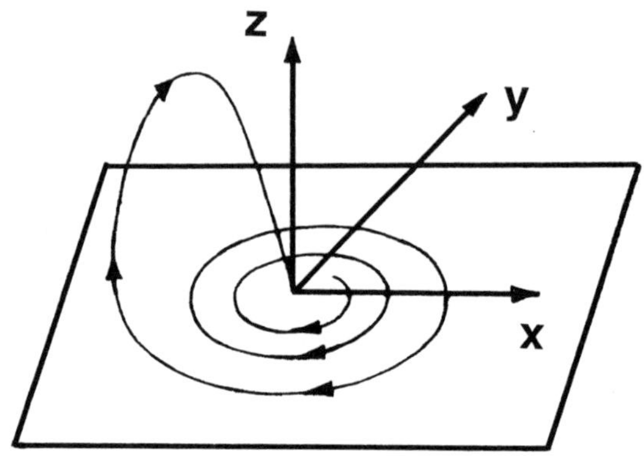

Fig. 1.11. Example of an unstable returning trajectory, corresponding to chaotic motion in a dynamical system [136].

(i) A series of Hopf bifurcations from the equilibrium state to single-frequency periodic, then to two- or three-frequency quasi-periodic, and finally to chaotic behavior (the Ruelle-Takens scenario [137]);

(ii) A series of period-doubling bifurcations leading from single-frequency periodic to finally chaotic behavior (the Feigenbaum scenario [138]);

(iii) Intermittency, characterized by a complex cascade of bifurcations with an alternation of periodic and chaotic motions [139].

The above ideas were extremely stimulating for the theoretical and experimental investigations of the nonlinear dynamics of auto-oscillations. Nakamura et al. [82], using the two-beam numerical model developed for the case of parallel pumping in [81,134], have shown that for certain values of the parameters this model demonstrates the Feigenbaum scenario [138].

The experimental discovery of period-doubling bifurcations in spin-wave auto-oscillations under parallel [140,141] and perpendicular [110] pumping created a boom of interest in the subject and several groups of researchers started to investigate the transition to chaos in auto-oscillations (see Chapter 7 and references therein).

A major effort to create a more consistent theory of periodic and chaotic spin-wave auto-oscillations was undertaken by Zhang and Suhl [142]. The ideology of this work is similar to that of Zakharov et al [71], i.e., the physical origin of auto-oscillations is considered to be the instability of collective oscillations in the system of parametrically excited spin waves. However, the scope of this work is wider and covers not only the case of parallel pumping but also high-power ferromagnetic resonance (or perpendicular pumping). Using the center manifold theorem [143], the authors of [142] have calculated the threshold of the appearance of periodic spin-wave auto-oscillations (or the threshold of Hopf bifurcation [143] of creation of a limit cycle in the phase space of a system). This threshold turned out to be about 10 dB higher than the threshold of parametric excitation of spin waves. The expression for the auto-oscillation frequency was also obtained and the frequency was estimated to be of the order of the spin-wave relaxation $\Gamma_{\mathbf{q}}$. The theory in [143] shows that only a few collective spin-wave eigenmodes are active in nonlinear spin-wave dynamics. Therefore it provides support for the "two-beam" models [81-84,134] which were used in numerical simulations and have explained many experimentally-observed features of periodic and chaotic auto-oscillations. However, later experiments [88,89,144,145] have shown that the quantitative predictions of the theory [143] are not confirmed. As recently pointed out by Rezende [88], spin-wave auto-oscillations are still in need of a good model which will explain all the experimentally observed features of this interesting nonlinear phenomenon.

1.6.3. Dynamics of Propagating Spin Waves in Films: Envelope Solitons

The preceding discussion of auto-oscillations referred to situations where the spin waves were excited parametrically by a *homogeneous* external pumping (with most experiments being performed on ferromagnetic spheres). We now consider the case of a ferromagnetic film where the pumping field can be localized and can therefore directly excite propagating spin waves having nonzero in-plane wave vector $\mathbf{Q}$.

When the amplitudes of these linearly excited spin waves are large enough, the three- and four-wave interaction processes [see Eqs. (1.131) and (1.134)] can take place. The film geometry makes possible the direct observation of nonlinear spin-wave processes at microwave frequencies by placing the receiving antenna on the film surface at a certain distance from the zone of wave excitation [123,124]. Also the film geometry makes the study of nonlinear spin-wave processes accessible by Brillouin light scattering [126]. Hence the nonlinear dynamics of spin waves may be investigated by two independent experimental techniques.

In the following we restrict our attention to the case when the three-wave decay processes are prohibited by conservation laws and so only the four-wave processes contribute. The theory of four-wave instability of propagating spin waves was developed in [70,73] and later extended to the case of a ferromagnetic film [92]. The equations of motion for the amplitudes $b_{n\mathbf{Q}}$ of spin wave modes taking part in the four-wave process (1.134) in a magnetic film can be written in the form of Eq. (1.123) as

$$\frac{\partial b_{n\mathbf{Q}}}{\partial t} + i\omega_{n\mathbf{Q}} b_{n\mathbf{Q}} + i \sum_{n_1, n_2, n_3} \int T_{n\mathbf{Q}, n_1\mathbf{Q}_1, n_2\mathbf{Q}_2, n_3\mathbf{Q}_3}\, b^*_{n_1\mathbf{Q}_1} b_{n_2\mathbf{Q}_2} b_{n_3\mathbf{Q}_3}$$
$$\times\, \delta(\mathbf{Q} + \mathbf{Q}_1 - \mathbf{Q}_2 - \mathbf{Q}_3)\, d^2\mathbf{Q}_1 d^2\mathbf{Q}_2 d^2\mathbf{Q}_3 \,, \qquad (1.154)$$

where $T_{n\mathbf{Q}, n_1\mathbf{Q}_1, n_2\mathbf{Q}_2, n_3\mathbf{Q}_3}$ is the coefficient of the four-wave part $\mathcal{H}_4$ of the Hamiltonian function renormalized to take account of the three-wave processes in second order of perturbation (as explained in Sec. 1.5.2).

It turns out that in the case of continuous excitation the above equation leads to the modulational [146,147] and decay [148] instability of monochromatic spin waves propagating in a film. In the case of pulsed excitation it leads to the formation of spin-wave envelope solitons [149-151]. We discuss some of these possibilities below.

We assume that a narrow packet of spin waves, having a central in-plane wave vector $\mathbf{Q}_0$ and a central frequency ω_0, is linearly excited in a film by a localized external pumping field of frequency $\omega_p \simeq \omega_0$. The conservation laws , Eq. (1.134), for the four-wave interaction process in this packet can be written as

$$2\omega_{n0} = \omega_{n1} + \omega_{n2} , \qquad 2\mathbf{Q}_0 = \mathbf{Q}_1 + \mathbf{Q}_2 , \tag{1.155}$$

where $\omega_{ni} \equiv \omega_{n\mathbf{Q}_i}$ with $i = 0,1,2,\ldots$. Also we assume that all the waves taking part in this process belong to the same dispersion branch (same n) and have the same distribution of variable magnetization along the film thickness described by the function $\Phi_n(\xi)$ [see Eq. (1.87)].

The equation of motion (1.154) for the spin-wave amplitudes b_{n0}, b_{n1}, and b_{n2} in this case can be written in the form [70,73,92]:

$$\frac{\partial b_{n0}}{\partial t} + i\tilde{\omega}_{n0} b_{n0} = 0 , \tag{1.156}$$

$$\frac{\partial b_{n1}}{\partial t} + i\tilde{\omega}_{n1} b_{n1} + iT_{12}(b_{n0})^2 b^*_{n2} = 0 , \tag{1.157}$$

$$\frac{\partial b^*_{n2}}{\partial t} - i\tilde{\omega}_{n2} b^*_{n2} - iT_{12}(b^*_{n0})^2 b_{n1} = 0 , \tag{1.158}$$

where

$$\tilde{\omega}_{n0} = \omega_{n0} + T_{00}|b_{n0}|^2 , \tag{1.159}$$

$$\tilde{\omega}_{n1} = \omega_{n1} + 2T_{10}|b_{n0}|^2 , \tag{1.160}$$

$$\tilde{\omega}_{n2} = \omega_{n2} + 2T_{20}|b_{n0}|^2 , \tag{1.161}$$

and the four-wave Hamiltonian coefficients are: $T_{00} = T_{n,\mathbf{Q}_0,\mathbf{Q}_0,\mathbf{Q}_0,\mathbf{Q}_0}$, $T_{10} = T_{n,\mathbf{Q}_1,\mathbf{Q}_0,\mathbf{Q}_1,\mathbf{Q}_0}$, $T_{20} = T_{n,\mathbf{Q}_2,\mathbf{Q}_0,\mathbf{Q}_2,\mathbf{Q}_0}$, and $T_{12} = T_{n,\mathbf{Q}_0,\mathbf{Q}_0,\mathbf{Q}_1,\mathbf{Q}_2}$.

The instability increment ν for the process in Eq. (1.155) is given by [71,73]:

$$\nu^2 = |T_{12} b^2_{n0}|^2 - \frac{1}{4}\left[\tilde{\omega}_{n1} + \tilde{\omega}_{n2} - 2\tilde{\omega}_{n0}\right]^2 . \tag{1.162}$$

In a dissipative medium, where the spin-wave modes b_{n1} and b_{n2} have dissipative parameters Γ_1 and Γ_2, the threshold amplitude of the linearly excited spin wave mode b_{n0} for this process can be written in a form analogous to that of Eq. (1.135):

$$|b_{no}|^2_{th} = (\Gamma_1\Gamma_2)^{1/2}/T_{12} . \tag{1.163}$$

As pointed out in Sec. 1.6.1, the Hamiltonian coefficients T_{00}, T_{10}, T_{20}, and T_{12} are analytical functions. Therefore when $\mathbf{Q}_1$ and $\mathbf{Q}_2$ are close to $\mathbf{Q}_0$ (denoting $\mathbf{Q}_1 = \mathbf{Q}_0 + \vec{\kappa}$ and $\mathbf{Q}_2 = \mathbf{Q}_0 - \vec{\kappa}$, with $\kappa \ll Q_0$), we can neglect the wave-vector dependence in the coefficients by assuming that $T_{10} \simeq T_{20} \simeq T_{12} \simeq T_{00}$. Also, for simplicity, we assume that $\vec{\kappa}$ is collinear to $\mathbf{Q}_0$.

Next we expand the nonlinear spin-wave dispersion relations $\tilde{\omega}_{n\mathbf{Q}}$ in Eqs. (1.160) and (1.161) in a Taylor series near the point with $\mathbf{Q} = \mathbf{Q}_0$ and $\omega_{n\mathbf{Q}} = \omega_{n\mathbf{Q}_0}$:

$$\tilde{\omega}_{n1,2} = \omega_{no} \pm v_g\kappa + \frac{D}{2}\kappa^2 + \ \ldots \ + 2T_{00}|b_{no}|^2 , \tag{1.164}$$

where

$$v_g = \left(\frac{\partial\omega_{nQ}}{\partial Q}\right)_{Q=Q_0} , \qquad D = \left(\frac{\partial^2\omega_{nQ}}{\partial Q^2}\right)_{Q=Q_0} , \tag{1.165}$$

are the group velocity and dispersion, calculated at the center of the spin-wave packet. Substituting Eq. (1.164) for $\tilde{\omega}_{n1,2}$ in Eq. (1.162) we get the following expression for the increment ν of the four-wave parametric process:

$$\nu^2 = - D\kappa^2(D\kappa^2 + 4T_{00}|b_{no}|^2)/4 . \tag{1.166}$$

The four-wave instability , Eq. (1.155), takes place when $\nu^2 > 0$. From Eq. (1.166) it is clear that this occurs only when

$$T_{00}D < 0 , \tag{1.167}$$

so that the dispersion D and the nonlinearity T_{00} have opposite signs. Equation (1.167) is the *Lighthill criterion* [152]. Even when the Lighthill criterion is satisfied, the four-wave instability can occur only for small values of κ:

$$0 < \kappa < 2|T_{00}| \ |b_{no}|^2/|D| , \tag{1.168}$$

and the maximum instability increment $\nu = T_{00}|b_{no}|$ corresponds to

$$\kappa = \sqrt{2}\ |T_{00}|\ |b_{no}|^2/|D|\ . \tag{1.169}$$

The new spin wave modes b_{n1} and b_{n2} which appear as a result of the process in Eq. (1.155) have in-plane wave vectors which are consistent with $\kappa \ll Q_0$. The process can be interpreted as the amplitude and phase modulation of the linearly excited spin wave b_{no}. Therefore this type of instability is usually called *modulational instability* (see, e.g., [153,154]). The modulational instability of dipole-exchange spin waves, propagating in a YIG film was observed for the first time in [146]. Similar effects for magneto-elastic waves and hybrid electromagnetic / spin waves were observed in YIG films [155,156].

Since the spin-wave packet, Eq. (1.155), is spectrally narrow (with $\kappa \ll Q_0$) we can analyze the evolution of this packet by studying slow variations of its envelope. To obtain the equation for the envelope we expand the spin wave dispersion equation in the film [Eq. (1.159)] in a Taylor series near the point with $Q = Q_0$, substitute this expansion into Eq. (1.154), and introduce "slow" dimensionless variables

$$\varphi_n(\kappa, t) = \frac{1}{J}\, b_{n1}[(Q_0 + \kappa), t]\ \exp[-i\omega_{no}(Q_0)t]\ . \tag{1.170}$$

Here ζ is a coordinate along the direction of wave propagation (see Fig. 1.5) and $J = \sqrt{(2M_0/\gamma)}$ is a normalization coefficient having the same dimension as the spin-wave amplitude b_{n1}. Using the "slow" Fourier transformation

$$\varphi_n(\zeta, t) = \int \mathrm{d}\kappa\ \varphi_n(\kappa, t)\ e^{i\kappa\zeta}\ , \tag{1.171}$$

and the properties that $T_{12} \simeq T_{00} \equiv T/J^2$ (= constant) at $\kappa \ll Q_0$, we transform Eq. (1.154) from the wave-number (or κ) representation to the coordinate ζ representation. This yields the equation for the envelope $\varphi_n(\zeta, t)$ of the spin-wave packet [Eq. (1.155)] as

$$\left[i\left(\frac{\partial}{\partial t} + v_g \frac{\partial}{\partial \zeta}\right) + \frac{D}{2}\frac{\partial^2}{\partial \zeta^2} - T\ |\varphi_n|^2\right] \varphi_n(\zeta, t) = 0\ . \tag{1.172}$$

This is a well-known *nonlinear Schrödinger equation* which was solved exactly by means of the "inverse scattering problem" method in [157]. It describes the evolution of a wave packet in a nonlinear medium with dispersion. The term $v_g(\partial/\partial\zeta)$ describes the motion of the wave packet as a whole with group velocity v_g. The term $(D/2)(\partial^2/\partial\zeta^2)$ describes the spin wave dispersion, and the last term $T|\varphi|^2$ describes the nonlinear self-interaction in the packet.

When the Lighthill criterion ($TD < 0$) is observed, Eq. (1.172) has a single soliton solution of the form

$$\varphi(\zeta, t) = \varphi_0 \mathrm{sech}\left[\varphi_0 \; |T/D|^{1/2} (\zeta - V_0 t)\right] , \tag{1.173}$$

where φ_0 and V_0 are independent arbitrary constants denoting the amplitude and velocity of a soliton. *Envelope solitons* described by Eq. (1.173) are stable nonlinear excitations propagating in a nonlinear dispersive medium. After collisions with each other they retain their amplitudes and velocities. N-soliton solutions are also possible. As shown in [157] an arbitrary (but sufficiently smooth) initial envelope $\varphi(\zeta, 0)$ in the limit $t \longrightarrow \infty$ splits into a number of solitons, the parameters of which can be determined for a given initial envelope. There are also periodic solutions of Eq. (1.172) for $\varphi(\zeta, t)$ in the form of knoidal waves [158]. These solutions describe the effect of modulational instability of a monochromatic (constant envelope) wave.

It is worth noting that Eq. (1.172) can alternatively be derived directly from the spin-wave dispersion equation (1.144) by substituting Eq. (1.117) for M_0 and using the method of adiabatic approximation described in [159] (for details, see Chapter 9 of this book).

The new nonlinear excitations described above arise as a result of competition between the spreading effects of dispersion and the self-steepening effects of nonlinearity on the initial quasi-monochromatic spin-wave packet. The balance between dispersion and nonlinearity, which guarantees stability of envelope solitons and periodically modulated spin waves in films, exists only in a certain range of spin-wave amplitudes. With increasing amplitude, the spectrum of spin-wave modulation in films becomes more complicated and finally chaotic after a series of successive bifurcations [146,160]. This evolution of the spectrum of modulational instability of propagating spin waves in films is in many ways analogous to the evolution of the spectrum of spin wave auto-oscillations observed in ferromagnetic spheres (see Chapter 7).

The results of the experimental observations and the numerical modeling of the process of spin wave modulational instability in YIG films are presented in Chapter 9 of this book.

1.7. CONCLUSIONS

In this chapter we have given a summary of the basic properties of spin waves, starting with the well-established results for the bulk spin waves in effectively infinite media and then focusing on the bulk and localized spin waves in media with restricted geometries. The properties of (linear) spin waves in films and multilayers (including superlattices) were detailed in Secs. 1.4 and 1.5, respectively, for the different regimes of magnetic behavior: exchange-dominated, dipole-exchange, magnetostatic, and electromagnetic. The nonlinear aspects of spin-wave behavior were introduced in Secs. 1.5 and 1.6 and were applied to spin-wave parametric processes, spin-wave auto-oscillations, and magnetic solitons. Most of the above topics are explored in greater detail in the following chapters of this book.

We have restricted attention to the case of geometries with parallel planar surfaces and interfaces. An interesting generalization of the parallel-sided film is to the case where two planar surfaces meet at an angle θ to form a wedge-shaped magnetic sample. The surfaces are now such as to remove the translational symmetry in *two* different directions, with obvious consequences for the symmetry of the spin-wave excitations. Under appropriate conditions there may be localized excitations known as *wedge modes* (or as *edge modes* in the case of θ = 90°) that propagate in a wave-like fashion along the apex of the wedge but decay with increasing distance into the wedge from its apex and faces. For example, the spin waves localized at the 90° edge of a Heisenberg ferromagnet have been studied in [161,162]. There have also been calculations for spin waves in geometries involving curved surfaces, e.g., spheroidal and cylindrical samples. The results are relatively straightforward in the case of magnetostatic modes (see [49] for a general account).

We have not included in this introductory chapter a discussion of the experimental techniques for investigating spin waves in limited magnetic systems. These techniques are described in several of the later chapters including, in particular, Chapter 6. To date the main experimental methods that have been applied to localized spin waves in ferromagnetic and ferrimagnetic samples are Brillouin light scattering (BLS) and microwave spectroscopy (MS) and resonance. At present there are few direct measurements of surface spin waves in antiferromagnets (e.g., see [163]), but it is likely that improvements to far infrared techniques such as attenuated total reflection (ATR) will provide results in the future.

In their traditional form the above-mentioned methods of MS and BLS usually provide information about the spin waves in films averaged across the sample thickness (except in highly absorptive media and/or for large film thicknesses). However, the spin-wave processes in films very often have significant spatial dependences, as we have discussed in this chapter, and it is of interest to investigate these. The modification of traditional methods or/and the development of new localized probe methods of investigating spin-wave processes in films is therefore necessary to overcome this limitation of the conventional experimental techniques.

Modifications of the traditional BLS method in order to obtain information about the spatial distribution of spin waves in YIG films and thin slabs were made in [124,163-165]. By placing various diaphragms behind the collecting lens of their BLS spectrometer, the authors of [124,163] were able to select spin waves having a specific in-plane wave number and in-plane propagation angle This modification of the BLS method made possible the direct measurement of the directions of propagation of linearly and parametrically excited spin waves in tangentially magnetized YIG films. Another modification of the traditional BLS technique was made in [164]. Here the YIG sample was held to a XY translator driven by two stepper motors to allow its motion in two directions perpendicular to the laser beam. This scanning of the laser beam on the sample surface allowed the authors of [164] to measure spatial distributions (along the sample surface) of the variable magnetization of inhomogeneous magnetostatic resonant modes excited in the rectangular YIG sample.

The optical probing method to study the spatial distributions of the intensity in magnetostatic waves propagating in YIG films was developed in [165]. The laser beam scanning on the surface of the film was modulated due to the magneto-optical interactions with the variable magnetization of propagating magnetostatic waves, and the spatial distributions of the wave amplitude along the film surface were determined. This method is very similar to the method described in [164], but has lower sensitivity as it does not use the multi-pass Fabry-Perot interferometer.

One of the simplest and most efficient methods for the localized probing of spin waves in magnetic films is the method of magnetic-induction probe developed in [166-171]. This method is based on Faraday's law of electromagnetic induction. A loop of wire (or a micro-strip antenna [168]) with its plane perpendicular to the surface of the magnetic film is moved along the film surface. The components of the magnetic field of the magnetostatic wave perpendicular to the plane of the loop create the induction current in the loop which is proportional to the wave amplitude. By rotating the loop about the vertical axis (perpendicular to the film plane) it is possible to determine the direction of the in-plane wave vector **q** (the phase-propagation direction) of the MSW in the film. Thus a detailed picture of the amplitude distribution and phase propagation directions of a spin-wave process in the film can be obtained. This method was used in both linear and nonlinear regimes of spin-wave propagation to determine the spatial structure of surface and bulk magnetostatic waves propagating in tangentially magnetized YIG films [169,170]. A further modification of this method [171] made it applicable in the pulsed regime of magnetostatic-wave excitation in YIG films.

A photothermal method for the localized probing of spin waves in magnetic slabs was developed in [172,173]. This method is based on the effect of the probe laser beam deflection in the temperature-induced gradient of the optical refractive index in the air near the surface of a magnetic slab in which microwave spin waves are excited. The probe laser beam in this method is parallel to the sample surface for a distance which is of the order of the diffusion length in air. The registered deflection angle is proportional to the local intensity of the microwave magnetization under the probe beam, but is still a spatial average in the direction of this beam. The photothermal deflection method was used in [173] to study the spatial structure of microwave absorption in magnetostatic resonance modes of a rectangular YIG slab.

REFERENCES

1. W. Heisenberg, *Z. Phys.* **49** (1928) 619.
2. D.C. Mattis, *The Theory of Magnetism* (Harper and Row, New York, 1965).
3. D. Wagner, *Introduction to the Theory of Magnetism* (Pergamon, Oxford, 1972).
4. A.I. Akhiezer, V.G. Bar'yakhtar, and S.V. Peletminskii, *Spin Waves* (North-Holland, Amsterdam, 1968).
5. P.W. Anderson, in *Magnetism*, Vol. 1, ed. G.T. Rado and H. Suhl (Academic, New York, 1963), p. 25.
6. C. Herring, in *Magnetism*, Vol. 4, ed. G.T. Rado and H. Suhl (Academic, New York, 1966), p. 1.
7. F. Bloch, *Z. Phys.* **61** (1930) 206.
8. G. Heller and H.A. Kramers, *Proc. K. Akad. Wet.* **37** (1934) 378.
9. T. Holstein and H. Primakoff, *Phys. Rev.* **58** (1940) 1098.
10. F. Keffer, *Handbuch der Physik* **18**/2 (1966) 1.
11. W. Marshall and S.W. Lovesey, *Theory of Thermal Neutron Scattering* (Clarendon, Oxford, 1971).
12. M.G. Cottam and D.J. Lockwood, *Light Scattering in Magnetic Solids* (Wiley, New York, 1986).
13. T.G. Phillips and H.M. Rosenberg, *Rep. Prog. Phys.* **29** (1966) 285.
14. D. Bloor and G.M. Copland, *Rep. Prog. Phys.* **35** (1972) 1173.

15. C. Kittel, *Introduction to Solid State Physics*, 6th edn. (Wiley, New York, 1986).
16. M.G. Cottam and D.R. Tilley, *Introduction to Surface and Superlattice Excitations* (Cambridge University Press, Cambridge, 1989).
17. C. Kittel, *Quantum Theory of Solids*, 2nd edn. (Wiley, New York, 1987).
18. L.I. Schiff, *Quantum Mechanics* (McGraw-Hill, New York, 1955).
19. B.A. Auld, *J. Appl. Phys.* **31** (1960) 1642.
20. J.D. Jackson, *Classical Electrodynamics*, 2nd edn. (Wiley, New York, 1975).
21. L.D. Landau, E.M. Lifshitz, and L.P. Pitaevskii, *Electrodynamics of Continuous Media*, 2nd edn. (Pergamon, Oxford, 1984).
22. E.F. Sarmento and D.R. Tilley, in *Electromagnetic Surface Excitations*, ed. A.D. Boardman (Wiley, New York, 1982).
23. D.L. Mills, in *Surface Excitations*, eds. V.M. Agranovich and R. Loudon (North-Holland, Amsterdam, 1984).
24. G.T. Rado and J.R. Weertman, *J. Phys. Chem. Solids* **11** (1959) 315.
25. R.F. Soohoo, *Phys. Rev.* **131** (1963) 594.
26. G.T. Rado and R.J. Hicken, J. Appl. Phys. **63** (1988) 3885.
27. R.W. Damon and J. Eshbach, *J. Phys. Chem. Solids* **19** (1961) 308.
28. V.G. Baryakhtar and M.I.Kaganov, Ch. 7 in *Ferromagnetic Resonance*, ed. S.V. Vonsovsky (Fizmatgiz, Moscow, 1961) [in Russian].
29. R.W. Damon and H. van de Vaart, *J. Appl. Phys.* **36** (1965) 3453.
30. L.R. Walker, *Phys. Rev.* **105** (1957) 390.
31. F.R. Morgenthaler, *Proc. IEEE* **76** (1988) 138.
32. R.F. Soohoo, *Magnetic Thin Films* (Harper and Row, New York, 1965).
33. N.M. Salanskij and M.Sh. Erukhimov, *Physical Properties and Applications of Magnetic Films* (Nauka, Moscow, 1975) [in Russian].
34. C. Kittel, *Phys. Rev.* **110** (1958) 1295.
35. B.N. Filippov, *Sov. Phys. - Solid State* **9** (1967) 1048 [*Fiz. Tverd. Tela* **9** (1967) 1339].
36. B.A. Kalinikos and A.N. Slavin, *J. Phys. C* **19** (1986) 7013.
37. M.H. Seavey and P.E. Tannenwald, *Phys. Rev. Lett.* **1** (1958) 168.
38. V.V. Gann, *Sov.Phys. - Solid State* **8** (1967) 2537 [*Fiz. Tverd. Tela*

9 (1966) 3167].

39. L.V. Mikhailovskaya and R.G. Khlebopros, *Sov. Phys. - Solid State* **11** (1969) 1381 [*Fiz. Tverd. Tela* **11** (1969) 2854].

40. R.E. DeWames and T. Wolfram, *J. Appl. Phys.* **41** (1970) 987.

41. T. Wolfram and R.E. DeWames, *Phys. Rev. B* **4** (1971) 3125.

42. M. Sparks, *Phys. Rev. Lett.* **24** (1970) 1178.

43. M. Sparks, *Phys. Rev. B* **1** (1970) 3831.

44. B.N. Filippov, *Phys. Met. Metall.* **32** (1971) 911 [*Fiz. Met. Metalloved* **29** (1970) 1131)].

45. O.G. Vendik and D.N. Chartorizhskii, *Sov. Phys. - Solid State* **12** (1970) 2537 [*Fiz. Tverd. Tela* **12** (1970) 1538].

46. O.G. Vendik, B.A. Kalinikos, and D.N. Chartorizhskii, *Sov.Phys. - Solid State* **19** (1977) 222 [*Fiz. Tverd. Tela* **19** (1977) 387].

47. B.A. Kalinikos, *Sov. Phys. J.* **24** (1981) 719 [*Izvestiya VUZov - Fizika* **24** (1981) 42].

48. L.D. Landau and E.M. Lifshitz, *Quantum Mechanics Non-Relativistic Theory*, 2nd edn. (London, Pergamon, 1965).

49. T. Wolfram and R.E. DeWames, *Prog. Surf. Sci.* **2** (1972) 233.

50. M.G. Cottam and D. Kontos, *J. Phys. C* **13** (1980) 2945.

51. H. Puszkarskii, *Prog. Surf. Sci.* **9** (1979) 191.

52. D.L. Mills, in *Light Scattering in Solids V*, eds. M. Cardona and G. Güntherodt (Springer-Verlag, Heidelberg, 1989), p. 13.

53. P. Grünberg, in *Light Scattering in Solids V*, eds. M. Cardona and G. Güntherodt (Springer-Verlag, Heidelberg, 1989), p. 303.

54. P. Grünberg, *J. Appl. Phys.* **51** (1980) 4338.

55. P. Grünberg, R. Schreiber, Y. Pang, M.B. Brodsky, and H. Sowers, *Phys. Rev. Lett.* **57** (1986) 2442.

56. P. Grünberg, S. Demokritov, A. Fuss, M. Vohl, and J.A. Wolf, *J. Appl. Phys.* **69** (1991), 4789.

57. R.E. Camley, T.S. Rahman, and D.L. Mills, *Phys. Rev. B* **27** (1983) 261.

58. P. Grünberg and K. Mika, *Phys. Rev. B* **27** (1983) 2955.

59. R.E. Camley and M.G. Cottam, *Phys. Rev. B* **35** (1987) 189.

60. C.A. Ramos, D. Lederman, A.R. King, and V. Jaccarino, *Phys. Rev.*

Lett. **65** (1990) 2913.

61. C.F. Majkrzak, J.W. Cable, J. Kwo, M. Hong, D.B. McWhan, Y. Yafet, J.V. Waszczak, and C. Vettier, *Phys. Rev. Lett.* **56** (1986) 2700.
62. N. Raj and D.R. Tilley, *Phys. Rev. B* **36** (1987) 7003.
63. N. Bloembergen and R. Damon, *Phys. Rev.* **85** (1952) 699.
64. N. Bloembergen and S. Wang, *Phys.Rev.* **93** (1954) 72.
65. H. Suhl, *J. Phys. Chem. Solids* **1** (1957) 209.
66. E. Schlömann, J. Green, and V. Milano, *J. Appl. Phys.* **31** (1960) 3865.
67. F.R. Morgenthaler, *J. Appl. Phys.* **31** (1960) 955.
68. R.W. Damon, in *Magnetism: A Treatise on Modern Theory and Materials*, eds. G.T. Rado and H. Suhl (Academic Press, Mew York, 1963), **1**, 551.
69. E. Schlömann, *Phys. Rev.* **116** (1959) 829.
70. V.E. Zakharov, V.S. L'vov, and S.S. Starobinets, *Sov. Phys. - Solid State* **11** (1970) 1658 and 2368 [*Fiz. Tverd. Tela* **11** (1969) 2047 and 2922].
71. V.E. Zakharov, V.S. L'vov, and S.S. Starobinets. *Sov. Phys. - Usp.*, **17** (1975) 896 [*Usp. Fiz. Nauk* **114** (1974) 609].
72. T.S. Hartwick, E.R. Peressini, and M.T. Weiss, *J. Appl. Phys.* **32** (1961) 223.
73. V.S. L'vov, *Nonlinear Spin Waves*, (Nauka, Moscow, 1987) (in Russian).
74. V.S. L'vov, Ch. 5 in *Solitons*, ed. S.E. Trullinger, V.E. Zakharov, and V.L. Pokrovsky (Elsevier, Amsterdam, 1986).
75. H.G. Schuster, *Deterministic Chaos* (Physik-Verlag, Weinheim, 1984).
76. A.M. Kosevich, B.A. Ivanov, and A.S. Kovalev, *Magnetic Solitons: A New Type of Collective Excitation in Magnetically Ordered Crystals*, in Soviet Scientific Reviews, Section A, Physics Reviews **6** (1985) 161, ed. I.M. Khalatnikov.
77. A.M. Kosevich, B.A. Ivanov, and A.S. Kovalev, *Nonlinear Magnetization Waves: Dynamic and Topological Solitons* (Naukova Dumka, Kiev, 1988) (in Russian).
78. V.E. Zakharov, *Sov. Phys. - JETP* **60** (1971) 1286 [*Zh. Eksp. Teor. Fiz.* **60** (1971) 1717).

79. F.R. Morgenthaler, *J. Appl. Phys.* **31** (1960) 955.

80. V.V. Zautkin, V.E. Zakharov, V.S. L'vov, S.L. Musher, and S.S. Starobinets, *Sov. Phys. - JETP* **35** (1972) 926 [*Zh. Eksp. Teor. Fiz.* **62** (1972) 1782].

81. V.L. Grankin, V.S. L'vov, V.I. Motorin, and S.L. Musher, *Sov. Phys. - JETP* **54** (1981) 405 [*Zh. Eksp. Teor. Fiz.* **81** (1981) 757].

82. K. Nakamura, S. Ohta, and K. Kawasaki, *J. Phys. C* **15** (1982) L143.

83. S.M. Rezende, O.F. de Alcantra Bonfim, and F.M. de Aguiar, *Phys. Rev. B* **33** (1986) 5153.

84. F.M. de Aguiar and S.M. Rezende, *Phys. Rev. Lett.* **56** (1986) 1070.

85. C.D. Jeffries, P.H. Bryant, and K. Nakamura, *J. Appl. Phys.* **64** (1988) 5382.

86. G. Srinivasan, *J. Appl. Phys.* **69** (1991) 5.

87. C.E. Patton and W.Jantz, *J. Appl. Phys.* **50** (1979) 7082.

88. S.M. Rezende, F.M. de Aguiar, and A. Azevedo, *J. Appl. Phys.* **67** (1990) 5624.

89. S.M. Rezende, A. Azevedo, A. Cascon, and J. Koiller, *J. Appl. Phys.* **69** (1991) 5430.

90. R.D. McMichael and P.E. Wigen, *Phys. Rev. Lett.* **64** (1990) 64.

91. B.A. Kalinikos, *Sov. Tech. Phys. Lett.* **9** (1983) 325 [*Pis'ma Zh. Tekh. Fiz.* **9** (1983) 811].

92. A.N. Slavin and B.A. Kalinikos, *Sov. Phys. Tech. Phys.* **32** (1987) 1446.

93. A.D. Boardman and S.A. Nikitov, *Phys. Rev. B* **38** (1988) 11444.

94. P.H. Bryant, *Phys. Rev. B* **39** (1989) 4363.

95. R.D. McMichael and P.E. Wigen, *J. Appl. Phys.* **69** (1991) 5425.

96. M. Chen and C.E. Patton, *J. Appl. Phys.* **69** (1991) 5782.

97. N.N. Bogoliubov, *J. Phys. (USSR)* **11** (1947) 23.

98. F.J. Dyson, *Phys. Rev.* **102** (1956) 1230.

99. V.G. Vaks, A.I. Larkin, and S.A. Pikin, *Sov. Phys - JETP* **26** (1968) 647 [*Zh. Eksp. Teor. Fiz.* **53** (1967) 1089].

100. H.J. Spencer, *Phys. Rev.* **167** (1968) 434.

101. M.G. Cottam, *J. Phys. C* **4** (1971) 2658 and 2673.

102. P. Mazur and D.L. Mills, *Phys. Rev. B* **29** (1984) 5081.

103. D. Kontos and M.G. Cottam, *J. Phys. C* **19** (1986) 1189 and 1203.

104. V.V. Tarasenko and V.D. Kharitonov, *Sov. Phys. - Solid State* **16** (1974) 1031 [*Fiz. Tverd. Tela* **16** (1974) 1576].

105. A. Monosov, *Radiotekhnika i Electronika* **5** (1960) 85, 143.

106. A. Monosov, *Nonlinear Ferromagnetic Resonance* (Nauka, Moscow, 1971) (in Russian).

107. W. Jantz and J. Schneider, *Solid State Commun.* **9** (1971) 69.

108. G.P. Bryant, C. Jeffries, and K. Nakamura, *Phys. Rev. Lett.* **60** (1988) 1185; *Phys. Rev. A* **38** (1988) 4223.

109. G. Wiese, *Z. Physik B* **82** (1991) 453.

110. G. Wiese and H. Benner, *Z. Physik B* **79** (1990) 119.

111. G. Gibson and C. Jeffries, *Phys. Rev. A* **19** (1984) 811.

112. V.P. Lukomskii and A.V. Kuz'ko, *Sov. Phys. - Solid State* **11** (1970) 2390 [*Fiz. Tverd. Tela* **11** (1970) 2951].

113. A. Corciovei and D. Vamanu, *Thin Solid Films* **12** (1972) 51.

114. O.G. Vendik, B.A. Kalinikos, and D.N. Chartorizhskii, *Sov. Phys. - Solid State* **16** (1975) 1785 [*Fiz. Tverd. Tela* **16** (1974) 2757].

115. G.A. Melkov, *Sov. Phys. - Solid State* **30** (1988) 1458 [*Fiz. Tverd. Tela* **30** (1988) 2533).

116. B.A. Kalinikos and A.N. Slavin, *IEEE Trans. on Magnetics* **27** (1991) 5444.

117. J.B. Comly, T. Penney, and R.V. Jones, *J. Appl. Phys.* **34** (1963) 1145.

118. A. Berteaud and H. Pascard, *J. Appl. Phys.* **36** (1965) 970.

119. B.A. Kalinikos, N.G. Kovshikov, and N.V. Kozhus', *Sov. Phys. - Solid State* **27** (1985) 1621 [*Fiz. Tverd. Tela* **27** (1985) 2794].

120. L.M. Gorbunov, *Sov. Phys. - JETP* **40** (1974) 689 [*Zh. Eksp. Teor. Fiz.* **67** (1974) 1386].

121. V.S. L'vov and A.M. Rubenchik, *Preprint #31* (Institute of Automation and Electrometry, Novosibirsk), (1976) 26 (in Russian).

122. G.A. Vugal'ter, *Sov. Phys. - JETP* **70** (1990) 1072 [*Zh. Eksp. Teor. Fiz.* **97** (1990) 1901].

123. J.D. Adam and S.N. Stitzer, *Appl. Phys. Lett.* **36** (1980) 485.

124. A.M.Mednikov, *Sov. Phys. - Solid State* **23** (1981) 136 [*Fiz. Tverd.

Tela **23** (1981) 242].

125. O.A. Chivileva, A.G. Gurevich, A.N. Anisimov, B.N. Gusev, G.A. Vugal'ter, and E.S. Sher, *Sov. Phys. - Solid State* **29** (1987) 1221 [*Fiz. Tverd. Tela* **29** (1987) 1774].
126. G. Srinivasan, C.E. Patton, and P.R. Emtage, *J. Appl. Phys.* **61** (1987) 2318.
127. G.A. Melkov and S.V. Sholom, *Sov. Phys. - Solid State* **29** (1987) 2967 [*Fiz. Tverd. Tela* **29** (1987) 3257].
128. G.A. Melkov and S.V. Sholom, *Sov. Phys. - JETP* **69** (1989) [*Zh. Eksp. Teor. Fiz.* **96** (1989) 712].
129. P.E. Zil'berman, N.S. Golubev, and A.G. Temiryazev, *Sov. Phys. - JETP* **70** (1990) 353 [*Zh. Eksp. Teor. Fiz.* **97** (1990) 634].
130. W.E. Curtney and J. Claricoats, *J. Electron. Contr.* **16** (1964) 1.
131. S. Wang, G. Thomas, and T.L. Hsu, *J. Appl. Phys.* **39** (1968) 2719.
132. J.J. Green and E. Schlömann, *J. Appl. Phys.* **33** (1962) 1358.
133. V.V. Zautkin and S.S. Starobinets, *Sov. Phys. - Solid State* **16** (1974) [*Fiz. Tverd. Tela* **16** (1974) 678].
134. V.S. L'vov, S.L. Musher, and S.S. Starobinets, *Sov. Phys. - JETP* **37** (1973) 546 [*Zh. Eksp. Teor. Fiz.* **64** (1973) 1084].
135. E.N.Lorenz, *J. Atmos. Sci.* **20** (2963) 130.
136. M.I. Rabinovich and D.I. Trubetskov, *Oscillations and Waves* (Kluwer, Boston, 1989).
137. D. Ruelle and F. Takens, *Commun. Math. Phys.* **20** (1971) 167.
138. M.J. Feigenbaum, *J. Stat. Phys.* **19** (1978) 25.
139. Y. Pomeau and P. Manneville, *Commun. Math. Phys.* **74** (1980) 189.
140. H. Yamazaki, *J. Phys. Soc. Jpn.* **53** (1984) 1155.
141. S.M. Rezende, F.M. de Aguiar, and O.F. de Alcantara Bonfim, *J. Magn. Magn. Mater.* **54-57** (1986) 1127.
142. X. Zhang and H. Suhl, *Phys. Rev. B* **38** (1988) 4893.
143. J. Guckenheimer and P. Holmes, *Nonlinear Oscillations, Dynamical Systems, and Bifurations of Vector Fields* (Springer-Verlag, New York, 1983) vol.42.
144. G. Srinivasan, M. Chen, and C.E. Patton, *J. Appl. Phys.* **64** (1988) 5480.

145. G. Srinivasan, *Solid State Commun.* **80** (1991) 5.

146. B.A. Kalinikos, N.G. Kovshikov, and A.N. Slavin, *Sov. Tech. Phys. Lett.* **10** (1984) 392 [*Pis'ma Zh. Tekh. Fiz.* **10** (1984) 936].

147. B.A. Kalinikos, N.G. Kovshikov, and A.N. Slavin, *J. Appl. Phys.* **67** (1990) 5633.

148. P.E. Zil'berman, S.A. Nikitov, and A.G. Temiryazev, *Sov. Phys. - JETP Lett.* **42** (1985) 110 [*Pis'ma Zh. Eksp. Teor. Fiz.* **42** (1985) 92].

149. B.A. Kalinikos, N.G. Kovshikov, and A.N. Slavin, *Sov. Phys. - JETP Lett.* **38** (1983) [*Pis'ma Zh. Eksp. Teor. Fiz.* **38** (1983) 413].

150. B.A. Kalinikos, N.G. Kovshikov, and A.N. Slavin, *Sov. Phys. - JETP* **67** (1988) 303 [*Zh. Eksp. Teor. Fiz.* **94** (1988) 159].

151. B.A. Kalinikos, N.G. Kovshikov, and A.N. Slavin, *Phys. Rev. B* **42** (1990) 8658.

152. M.J. Lighthill, *J. Inst. Math. Appl.* **1** (1965) 269.

153. T.V. Bengamen and I.E. Feir, *J. Fluid Mech.* **27** (1967) 417.

154. V.E. Zakharov, *Zh. Prikl. Mekh. Tekh. Fiz.* **2** (1968) 80.

155. P.E. Zil'berman, G.T. Kazakov, and V.V. Tikhonov, *Sov. Tech. Phys. Lett.* **11** (1985) 319 [*Pis'ma Zh. Tekh. Fiz.* **11** (1985) 769].

156. A.G. Temiryazev, *Sov. Phys. - JETP Lett.* **50** (1989) 228 [*Pis'ma Zh. Eksp. Teor. Fiz.* **50** (1989) 202].

157. V.E. Zakharov and A.B. Shabat, *Sov. Phys. - JETP* **34** (1972) 62 [*Zh. Eksp.Teor.Fiz.* **61** (1972) 118].

158. V.E. Zakharov, S.V. Manakov, S.P. Novikov, and L.P. Pitaevskii, *Theory of Solitons: Inverse Scattering Problem Method* (Nauka, Moscow, 1980) (in Russian).

159. V.I. Karpman, *Nonlinear Waves in Dispersive Media* (Pergamon, New York, 1973).

160. G.M. Dudko and A.N. Slavin, *Sov. Phys. - Solid State* **31** (1989) 979 [*Fiz. Tverd. Tela* **31** (1989) 114].

161. T.M. Sharon and A.A. Maradudin, *Solid State Commun.* **13** (1973) 187.

162. A.A Maradudin, S.L. Moss, and S.L. Cunningham, *Phys. Rev. B* **15** (1977) 4490.

163. W. Wettling, W.D. Wilber, P. Kabos, and C.E. Patton, *Phys. Rev. Lett.* **51** (1983) 1680.

164. A. Azevedo and S.M. Rezende, *J. Mag. Magn. Mat.* **104-107** (1992) 1039.
165. N.P. Vlannes, *J. Appl. Phys.* **62** (1987) 972.
166. M. Bini, L. Millanta, and N. Rubino, *J. Appl. Phys.* **46** (1975) 3175.
167. V. Smith, Jr., J. M. Owens, T. J. Mears, and N. D. Parikh, *IEEE Trans. Magn.* **MAG-15** (1979) 1738.
168. H. Collins and F. A. Pizzarello, *Int. J. Electron.* **34** (1973) 319.
169. N.P. Vlannes, *J. Appl. Phys.* **61** (1987) 416.
170. A.B. Valyavsky, A.V. Vashkovsky, A.V. Stalmakhov, and V.A. Tyulyukin, *Sov. Phys. - Tech. Phys.* **34** (1989) 616 [*Zh. Tekh. Fiz.* **59** (1989) 51].
171. D.A. Barishev, A.V. Vashkovsky, and A.V. Stalmakhov, *Pis'ma Zh. Tekh. Phys.* **15** (1989) 57.
172. U. Netzelmann and J. Pelzl, *Can. J. Phys.* **64** (1986) 1307.
173. O.V. Geisau, U. Netzelmann, S.M. Rezende, and J. Pelzl, *IEEE Trans. Mag.* **26** (1990) 1471.

CHAPTER 2

DIPOLE-EXCHANGE SPIN-WAVE SPECTRUM OF MAGNETIC FILMS

B.A. KALINIKOS

St. Petersburg Institute of Electrical Engineering,
197376 St. Petersburg, Russia

Contents

2.1. INTRODUCTION

An understanding of the full wave spectrum of magnetic films is the key to solving many fundamental and applied problems in this field. The theme of this chapter is the spectrum of dipole-exchange spin waves in magnetically-biased ferromagnetic films.

The published papers on theoretical investigations of the dispersion characteristics of high-frequency propagating magnetization waves in films may be divided into two groups: those which consider the spin-wave spectrum in the non-exchange limit (when only the dipole-dipole interactions are taken into account), and those which consider the spin-wave spectrum taking into account both the dipolar and exchange interactions. The dispersion characteristics of purely dipole (non-exchange) waves which are usually called *magnetostatic waves* (MSW) have been included in Chapter 1. Here we concentrate in more detail on the more general case when both dipole and exchange effects are important. Consequently we refer to such waves as *dipole-exchange spin waves.*

It is well known that the dispersion characteristics of spin waves in thin ferromagnetic films (with thicknesses of the order of exchange spin-wave wavelengths) differ, because of exchange effects, from those in comparatively thick films. The clearest manifestation of exchange effects are to be seen in the phenomenon of *spin wave resonance* (SWR) [e.g., see Sec. 1.3.3 of Chapter 1]. At the present time there is experimental evidence that the exchange effects can manifest themselves not only in thin ferromagnetic films but also in relatively thick (L = 5-10 μm) films. Microwave experiments show that the dispersion characteristics of spin waves, even with relatively long wavelengths (in-plane wave vector $Q \ll \alpha^{-1/2}$, where α is the exchange constant), are influenced not only by dipole-dipole interactions, but also by exchange interactions (see, e.g., [1-7]). Hence there are wide ranges of film thicknesses and wave vectors for which dipole-dipole and exchange interactions must *both* be included in the theoretical treatment.

2.1.1. Theoretical approaches

The theoretical work devoted to the dipole-exchange spin-wave spectrum of ferromagnetic films is rather extensive. The main results have been published in the papers [8-29]. The authors of these papers have used a macroscopic continuum technique. It is based on the simultaneous solution of Maxwell's equations (without retardation) and the linearized Landau-Lifshitz equation of motion for the magnetization, together with the usual electrodynamic boundary conditions and also with exchange boundary conditions (the surface spin pinning conditions), as described in Sec. 1.3.1 of Chapter 1.

A survey of the above references shows that two main approaches adopted for solving eigenvalue boundary problems may be distinguished in the phenomenological dipole-exchange theory.

The first approach is based on a representation of the problem variables inside the ferromagnetic film (for example, the variable magnetization, magnetic field or magnetic potential) as a sum (or linear combination) of partial wave contributions, i.e., plane waves [8-21]. (Below, for short, we shall refer to this as the "*partial wave* (PW) *approach*"). These plane waves are characterized by total wave vectors $\mathbf{q}$ consisting of two parts: the in-plane propagation vector $\mathbf{Q}$ and the quantized wave vectors $\mathbf{q}_n$ across the film. (Taking the example where x is perpendicular to the film surface, we have $\mathbf{Q} = (q_y, q_z)$ and $q^2 = Q^2 + q_{xn}^2$). The values of the quantized wave vectors q_{xn} vary with Q. They are found as the roots of a sixth-order characteristic equation which is actually the well-known Holstein-Primakoff [30] and Herring-Kittel [31] dispersion relation as in an infinite ferromagnet. The investigation of q_{xn} as the roots of the characteristic equation, and their dependences on the film parameters and on the direction and value of constant bias magnetic field, is one of the major achievements of the PW approach (see, e.g., [8,10,13,15,21]). Outside the film the problem variables are constructed from plane waves which are obtained as solutions of Laplace's equation. By imposing the electrodynamic and exchange boundary conditions on the problem variables, a system of homogeneous linear algebraic equations for the unknown coefficients is found. The condition for the vanishing of the determinant of this system leads to the dispersion relation for dipole-exchange spin waves in the film. In a general case, such determinant has a 8×8 form and therefore the dispersion law $\omega_n(Q)$ has to be found numerically [20].

It is worth noting that recently the PW approach was adopted for calculations of the spin-wave spectra of multilayer systems (see, e.g., [32-34]).

The second approach is based on a solution of the integro-differential equation for variable magnetization which follows from the initial system of equations consisting of the linerized equation of motion for the magnetization, the Maxwell equations of magnetostatics, and the electrodynamic boundary conditions [23-29]. To solve this integro-differential equation, an expansion of the variable magnetization in the infinite series of complete vector functions satisfying the exchange boundary conditions - spin-wave modes is used. (Below, for short, we shall call this the "*spin-wave mode* (SWM) *approach*"). The natural form of the exact dispersion relation for dipole-exchange spin waves in the framework of the SWM approach is expressible in terms of a vanishing infinite determinant and an infinite convergent series.

One of the useful features of the SWM approach is that it gives the possibility of deriving exact and approximate dispersion relations in various forms. It is important that one of the exact forms coincides completely with the dipole-exchange dispersion relation of the PW approach. Such a coincidence "constructs a bridge" between different theories. It becomes evident (as it must be) that the two different theoretical approaches give one and the same dispersion law ω_{nQ} for spin waves in ferromagnetic films.

The approximate dispersion equations of the SWM approach have an explicit form ω_{nQ}. This allows one to give a clear physical picture of the dipole-exchange spin-wave spectrum and its dependence on each of the parameters.

In this Chapter, for the detailed discussion of the behavior of the dipole-exchange spin waves in ferromagnetic films, we shall use the SWM approach. But in the beginning we shall summarize the main results of the PM approach obtained for spin-wave spectra.

2.2. TRIGONOMETRIC DISPERSION RELATIONS

Consider an isotropic ferromagnetic film of thickness L and of infinite dimensions in its plane. It is magnetized normally or tangentially to the surface along the z-axis. The effective field $\mathbf{H}^{eff}(\mathbf{r},t)$, being part of the equation of motion for the magnetization,

$$\frac{\partial \mathbf{M}(\mathbf{r},t)}{\partial t} = -\gamma\left[\mathbf{M}(\mathbf{r},t) \times \mathbf{H}^{eff}(\mathbf{r},t)\right], \tag{2.1}$$

is assumed to be the sum of the internal static magnetic field and the variable dipole and exchange fields, i.e.

$$\mathbf{H}^{eff}(\mathbf{r},t) = \hat{\mathbf{z}}H_i + \mathbf{h}_d(\mathbf{r},t) + 4\pi\alpha\nabla^2\mathbf{m}(\mathbf{r},t),$$

where α is the exchange constant, $\mathbf{m}(\mathbf{r},t) = \mathbf{M}(\mathbf{r},t) - \mathbf{M}_0$ is the variable magnetization, and $\mathbf{M}_0$ is the saturation magnetization. The dipole (or demagnetization) field is determined by Maxwell's equations of magnetostatics:

$$\nabla \times \mathbf{h}_d(\mathbf{r},t) = 0 , \tag{2.2}$$

$$\nabla \times \left[\mathbf{h}_d(\mathbf{r},t) + 4\pi\mathbf{m}(\mathbf{r},t)\right] = 0 . \tag{2.3}$$

To solve the eigenvalue problem we shall apply the usual electrodynamic boundary conditions (which require the continuity of the tangential components of variable magnetic field and the normal flux component), and the exchange boundary conditions. Here we restrict our attention to the two limiting sets of exchange boundary conditions:

$$\mathbf{m}(\xi) = 0 , \quad \text{when } \xi = \pm L/2 , \tag{2.4}$$

for pinned surface spins, and

$$\frac{\partial}{\partial \xi}\mathbf{m}(\xi) = 0, \quad \text{when } \xi = \pm L/2 , \tag{2.5}$$

for unpinned surface spins. In Eqs. (2.4) and (2.5) ξ is the coordinate normal to the film surfaces.

Following the papers [8-16], we introduce a scalar magnetic potential Ψ such that $\mathbf{h}_d = \nabla\Psi$. It allows one to rewrite the initial system of

Eqs. (2.1)-(2.3) after linearization as a sixth-order differential equation for Ψ:

$$\left[F^2 + F - (\omega/\omega_M)^2\right]\nabla^2\Psi - F\,\nabla_z^2\Psi = 0 \tag{2.6}$$

where the operator F is defined by

$$F = (\omega/\omega_M) - \alpha\,\nabla^2 \tag{2.7}$$

and other parameters are $\omega_H = \gamma H_i$, $\omega_M = \gamma 4\pi M_0$. A time dependence like $\exp(i\omega t)$ is assumed.

Physically, the last differential equation may be called the dipole-exchange equation [10]. In fact, it determines the magnetic potential in the ferromagnet medium when Zeeman, dipole, and exchange contributions to the spin-wave energy are taken into account. In the non-exchange limit $\alpha = 0$ Eq. (2.6) converts to the well-known Walker's equation for magnetostatic waves [see Eq. (1.64)].

Dipole-exchange spin-wave spectrum depends critically on the direction of the bias static magnetization of the film. With the help of Eq. (2.6) we now discuss two particular cases: *perpendicular* and *parallel magnetizations*.

2.2.1. Perpendicular magnetization

In the case of perpendicular magnetization we consider that the ferromagnetic film is magnetized to saturation normally to its surface along the z-axis, and spin waves can propagate in the film plane xy. To describe wave propagation, we introduce the $2D$ in-plane wave vector $\mathbf{Q} = (q_x, q_y)$ and the $2D$ in-plane position vector $\mathbf{R} = (x, y)$. Owing to the property of translational invariance in the xy plane we may seek a solution of Eq. (2.6) inside the film in a form

$$\Psi(\mathbf{r}) = \Psi_Q(z)\,\exp(-i\mathbf{QR})\ . \tag{2.8}$$

Solutions for the distribution of the magnetic potential across the film $\Psi_Q(z)$ must be found as a combination of three pairs of waves:

$$\Psi_Q(z) = \sum_{n=1}^{3} \left[A_n \exp(iq_{zn}z) + B_n \exp(-iq_{zn}z)\right] \tag{2.9}$$

whose wave numbers q_{zn} are the roots of the characteristic bicubic equation:

$$q^2\left[\left(\alpha q^2 + (\omega_H/\omega_M)\right)^2 - (\omega/\omega_M)^2\right] + Q^2\left(\alpha q^2 + (\omega_H/\omega_M)\right) = 0 \ . \tag{2.10}$$

Here the total wave vector $q^2 = Q^2+q_z^2$ is introduced. As was already mentioned, an investigation of the nature of the roots of the last equation is the significant aspect of the PW approach. The reader may find it in the papers [8,10,13,15,21]. We stress that the waves $\exp(\pm iq_{zn}z)$ each taken separately cannot be observed. Only appropriate combinations of these waves, which give the distributions $\Psi_Q(z)$, $\mathbf{h}_Q(z)$ or $\mathbf{m}_Q(z)$, could be observed and have a physical meaning in this sense.

We note also that Eq. (2.10), when rewritten for the spin-wave frequency $\omega(\mathbf{q})$, gives the well-known dispersion relation for the spin-wave manifold in an isotropic magnetic medium [30,31] (see also Eq. (1.45)).

Outside the film the scalar magnetic potential obeys the Laplace's equation:

$$\nabla^2\Psi(\mathbf{r}) = 0 \ . \tag{2.11}$$

In accordance with Eq. (2.8), solutions of Eq. (2.11) can be written as

$$\Psi(\mathbf{r}) = C \exp(-|Q|z) \exp(-i\mathbf{Q}.\mathbf{R}) \ , \quad z > L/2 \tag{2.12}$$

and

$$\Psi(\mathbf{r}) = D \exp(|Q|z) \exp(-i\mathbf{Q}.\mathbf{R}) \ , \quad z < -L/2 \ . \tag{2.13}$$

The eight unknown coefficients A_n, B_n, C, and D are determined by imposing the electrodynamic and exchange boundary conditions. These calculations lead to the dispersion relation in the form of a vanishing determinant.

If we consider that the film has symmetric surface spin pinning conditions (i.e., the surface spins are either pinned (2.4) or unpinned (2.5) at both film surfaces), the wave solutions of the problem split into two types: symmetric and antisymmetric (relative to the central plane $z = 0$). This simplifies the solution for the vanishing of the determinant. After some calculation it may be written as two dispersion relations.

In the case of pinned surface spins, as in Eq. (2.4), we obtain

$$\sum_{n=1}^{3} c_n \, (q_{zn}/Q) \tan(q_{zn}L/2) = -1 \tag{2.14}$$

for symmetric spin waves, and

$$\sum_{n=1}^{3} c_n \, (q_{zn}/Q) \cot(q_{zn}L/2) = 1 \tag{2.15}$$

for antisymmetric spin waves.

In the case of unpinned surface spins, as in Eq. (2.5), we obtain

$$\sum_{n=1}^{3} c_n \, (Q/q_{zn}) \cot(q_{zn}L/2) = -1 \tag{2.16}$$

for symmetric spin waves, and

$$\sum_{n=1}^{3} c_n \, (Q/q_{zn}) \tan(q_{zn}L/2) = 1 \tag{2.17}$$

for antisymmetric spin waves.

The coefficients c_n in Eqs. (2.14)-(2.17) are given by

$$c_n = \frac{Q^2 \, [\alpha q_n^2 + (\omega_H/\omega_M)]}{\alpha^2 q_n^2 \, (q_n^2 - q_{n'}^2)(q_n^2 - q_{n''}^2)} \tag{2.18}$$

where $q_n^2 = Q^2 + q_{zn}^2$; $n, n', n'' = 1, 2, 3$, and $n \neq n' \neq n''$.

In order to distinguish between the above dispersion relations of Eqs. (2.14)-(2.17) and other forms, we shall sometimes refer to them later as the "*trigonometric dispersion relations*".

It may be shown that in the non-exchange limit $\alpha = 0$ trigonometric dispersion relations transform to the corresponding dispersion relations for magnetostatic waves [see Eqs. (1.72) and (1.73)].

It may be concluded that, in the case of perpendicular magnetization, the dipole-exchange spin-wave spectrum ω_n versus Q has a rather complicated implicit form. To find the spin-wave dispersion branches $\omega_n(Q)$ it is necessary to solve numerically the trigonometric dispersion equations (2.14)-(2.17) together with the characteristic equation (2.10).

2.2.2. Parallel magnetization

In the case of parallel magnetization we consider that the ferromagnetic film is magnetized to saturation tangentially to its surface along the z-axis, and spin waves can propagate in the film plane yz. We seek a solution to Eq. (2.6) inside the film in a form

$$\Psi(\mathbf{r}) = \Psi_Q(x)\ \exp(-i\mathbf{Q}.\mathbf{R}) \tag{2.19}$$

where $\mathbf{Q} = (q_y, q_z)$ is the 2D in-plane wave vector and $\mathbf{R} = (y, z)$ is the 2D in-plane position vector. In accordance with the PW approach, solutions for the distribution of the magnetic potential across the film, $\Psi_Q(x)$, are found as a combination of three pairs of partial waves:

$$\Psi_Q(x) = \sum_{n=1}^{3} \left[A_n \exp(iq_{xn}x) + B_n \exp(-iq_{xn}x)\right] \tag{2.20}$$

whose wave numbers q_{xn} are the roots of the characteristic equation:

$$q^2\left[\left(\alpha q^2 + (\omega_H/\omega_M)\right)^2 - (\omega/\omega_M)^2\right] + q_x^2\left(\alpha q^2 + (\omega_H/\omega_M)\right) = 0\ . \tag{2.21}$$

Here the total wave vector $q^2 = q_x^2 + Q^2$ is introduced.

Similarly to the previous case of perpendicular magnetization, the last Eq. (2.21) can be rewritten for the spin-wave frequency $\omega(\mathbf{q})$. This gives the dispersion relation for the spin-wave manifold in the magnetic medium [see Eq. (1.45)] if the relation $\sin^2\theta_q = q_x^2/q^2$ is recognized.

Further calculations in the case under consideration are analogous to the previous case of perpendicular magnetization and we shall not repeat them here. We shall note the differences only.

In contrast to the perpendicular magnetization case, now the spin-wave solutions may be divided into two pure types of symmetry (relative to the central plane $x = 0$) only when the direction of spin-wave propagation is parallel to the direction of the static magnetization (i.e., the z-axis). Physically such a result is obvious. The form of the wave symmetry is distorted when the direction of spin-wave propagation deviates from the direction of the bias field. Therefore in a general case it is impossible to derive dispersion relations for purely symmetric or purely antisymmetric eigenwaves. Thus, when the direction of spin-wave propagation is arbitrary, the dipole-exchange spin-wave dispersion relation is written in the form of the vanishing of a 8×8 determinant.

In the case of transverse propagation, $\mathbf{Q} = (q_y, 0)$, some elements of the determinant are equal to zero, and it becomes feasible to solve the determinant to derive the corresponding dipole-exchange trigonometric dispersion relation [19]. This trigonometric dispersion relation is very bulky but it enables one to see, in the non-exchange limit, the reduction to the well-known Damon-Eshbach dispersion equation [see Eq. (1.79)] for surface magnetostatic waves.

In the case of longitudinal propagation, $\mathbf{Q} = (0, q_z)$, the symmetry properties simplify the solution for the vanishing determinant, and it yields dispersion relations for symmetric and antisymmetric waves. If the surface spins are pinned, as in Eq. (2.4), we obtain

$$\sum_{n=1}^{3} c_n \, (q_{xn}/q_z) \tan(q_{xn}L/2) = 1 \tag{2.22}$$

for symmetric spin waves, and

$$\sum_{n=1}^{3} c_n \, (q_{xn}/q_z) \cot(q_{xn}L/2) = -1 \tag{2.23}$$

for antisymmetric spin waves. If the surface spins are unpinned, as in Eq. (2.5), we obtain

$$\sum_{n=1}^{3} c_n \, (q_z/q_{xn}) \cot(q_{xn}L/2) = 1 \tag{2.24}$$

for symmetric spin waves, and

$$\sum_{n=1}^{3} c_n \, (q_z/q_{xn}) \, \tan(q_{xn}L/2) = -1 \tag{2.25}$$

for antisymmetric spin waves.

The coefficients c_n in Eqs. (2.22)-(2.25) are given by

$$c_n = \frac{q_z^2 \, [\alpha q_n^{\,2} + (\omega_H/\omega_M)]}{\alpha^2 q_n^2 \, (q_n^2 - q_{n'}^2)(q_n^2 - q_{n''}^2)} \tag{2.26}$$

where $q_n^2 = q_{xn}^2 + q_z^2$; $n, n', n'' = 1,2,3$, and $n \neq n' \neq n''$.

The above exact trigonometric dispersion relations were presumably utilized by the authors of original papers for their numerical calculations of the dipole-exchange spin-wave dispersion properties. However, they were not presented explicitly in the original papers or they were presented in reduced (approximate) forms. We consider it useful to give the trigonometric dispersion relations here because they represent the exact solutions of the corresponding boundary value problems and may be used for numerical calculations of dipole-exchange spin-wave spectra.

Regarding the PW approach, we come to conclusion that the dispersion properties of dipole-exchange spin waves are hidden in the transcendental dispersion relations. To make these properties more clear, we now turn to the other theoretical approach within the framework of which we discuss the dipole-exchange spin-wave spectra in detail.

2.3. DIPOLE-EXCHANGE INTEGRO-DIFFERENTIAL EQUATION

We present the spin-wave mode approach by analyzing the general case of an arbitrary direction of the static magnetization. We consider a ferromagnetic film of thickness L in the direction with the surfaces corresponding to the planes $\xi = L/2$ and $\xi = -L/2$ and with infinite dimensions in the $\rho\zeta$ plane (see Fig. 1.5 of Chapter 1). The film is magnetized to saturation by a static external magnetic field $\mathbf{H}_0$ in an arbitrary direction. The axis ζ is always assumed to be oriented along the direction of propagation of the spin wave:

$$\mathbf{m}(\xi,\zeta,t) = \mathbf{m}_{\mathbf{Q}}(\xi)\exp[i(\omega t - Q\zeta)]. \tag{2.27}$$

For the convenience of further analysis, we introduce yet another coordinate system xyz, in which axis z is parallel to the direction of the saturation magnetization $\mathbf{M}_0$ and the internal bias magnetic field $\mathbf{H}_i$. The magnitude of the internal field $\mathbf{H}_i$ and its direction can be found using the solution of the magnetostatic problem when the magnitude H_0 and the direction of the external magnetic field $\mathbf{H}_0$ are known

$$\left.\begin{aligned} H_i\cos\theta &= H_0\cos\theta_0 - 4\pi M_0\cos\theta, \\ 2\pi M_0\sin2\theta &= H_0\sin(\theta-\theta_0). \end{aligned}\right\} \tag{2.28}$$

The principal feature of the SWM approach is that, as the first step, Maxwell's equations of magnetostatics for a ferromagnetic film are solved. As a result, a connection between the variable magnetization of Eq. (2.27) and the dipole field,

$$\mathbf{h}(\xi,\zeta,t) = \mathbf{h}_Q(\xi)\ \exp[i(\omega t-Q\zeta)], \tag{2.29}$$

is found in the integral form:

$$\mathbf{h}_Q(\xi) = 4\pi \int_{-\infty}^{\infty} \hat{\mathbf{G}}_{\xi\rho\zeta}(\xi,\xi')\ \mathbf{m}_Q(\xi')\ d\xi' \ . \tag{2.30}$$

The tensorial Green function in Eq. (2.30), defined in the coordinate system $\xi\rho\zeta$, is (see, e.g., [25,28])

$$\hat{\mathbf{G}}_{\xi\rho\zeta}(\xi,\xi') = \begin{bmatrix} G^P - \delta(\xi-\xi') & 0 & iG^Q \\ 0 & 0 & 0 \\ iG^Q & 0 & -G^P \end{bmatrix}, \tag{2.31}$$

where the components inside the film are

$$G^P = (Q/2)\ \exp(-Q|\xi - \xi'|)\ , \qquad G^Q = G^P\ \mathrm{sgn}(\xi' - \xi)\ , \tag{2.32}$$

and the components outside the film are

$$G^P = (Q/2)\ \exp(-Q|\xi - \xi'|)\ , \qquad G^Q = -G^P\ \mathrm{sgn}(\xi)\ . \tag{2.33}$$

It is easy to verify that the Eqs. (2.30)-(2.33) give a solution of

Maxwell's equations of magnetostatics *satisfying electromagnetic boundary conditions*.

As the second step, the integral form for the dipole field, Eq. (2.30), is substituted into the linearized Landau-Lifshitz equation of motion (2.1) for the magnetization. It is convenient to write the equation of motion in the coordinate system *xyz* where, due to linearization, the vector of variable magnetization in Eq. (2.27) could be considered two-dimensional:

$$\mathbf{m}_Q(\xi) = \hat{\mathbf{x}}m^x(\xi) + \hat{\mathbf{y}}m^y(\xi) = \begin{pmatrix} m^x \\ m^y \end{pmatrix} . \tag{2.34}$$

These calculations lead to the integro-differential equation for the Fourier amplitude (2.34) of the spin wave (2.27), namely

$$\hat{\mathbf{F}}\, \mathbf{m}_Q(\xi) = \int_{-L/2}^{L/2} \hat{\mathbf{G}}_{xy}(\xi,\xi')\, \mathbf{m}_Q(\xi')\, d\xi' , \tag{2.35}$$

where $\hat{\mathbf{F}}$ is the linear matrix-differential operator. It has four components:

$$\hat{\mathbf{F}} = \begin{pmatrix} F^{xx} & F^{xy} \\ F^{yx} & F^{yy} \end{pmatrix} , \tag{2.36}$$

$$F^{xx} = F^{yy} = (\omega_H/\omega_M) + \alpha Q^2 - \alpha\nabla_\xi^2 , \qquad F^{yx} = -F^{xy} = i\omega/\omega_M .$$

$\hat{\mathbf{G}}_{xy}$ in Eq. (2.35) is a two-dimensional part of matrix $\hat{\mathbf{G}}_{xyz}$. The latter quantity is obtained from the matrix $\hat{\mathbf{G}}_{\xi\rho\zeta}$ in Eq. (2.31) by using orthogonal transformations for rotations through angles ϕ and $(\pi/2 - \theta)$ [28].

The integro-differential equation (2.35) is the *principal relation* of the SWM approach. It is equivalent to the initial system of equations consisting of Maxwell's equations of magnetostatics together with the electrodynamic boundary conditions and the linearized equation of motion for the magnetization with allowance for *Zeeman, dipole* and *exchange energies*. A solution of it, satisfying the appropriate exchange boundary conditions, gives the full spectrum of dipole-exchange spin waves in a ferromagnetic film. In the magnetostatic (non-exchange) approximation Eq. (2.35) defines the spectrum of magnetostatic waves.

An equation similar to (2.35) was used by Sparks [22], but he utilized an infinite-space Green function. By the way, if we solve Eq. (2.35) by a variational method, with the trial functions for rectangular films suggested by Sparks [22], we immediately obtain the equation for dipole-exchange spin-wave resonance frequencies of rectangular samples.

2.4. SPIN-WAVE MODES

Various ways could be suggested to solve the integro-differential equation (2.35) with the appropriate exchange boundary conditions to find the eigen-waves and their spectrum. In order to do this we shall apply the normal mode expansion technique.

There are several options to choose the normal modes for an expansion of the variable magnetization in Eq. (2.35). For example, in the papers [24-26] distributions of the variable magnetization corresponding to the spin wave resonance frequencies were utilized. They were found as eigenfunctions of the operator $\hat{\mathbf{F}}$ [see Eq. (2.36)] in the SWR regime of $Q = 0$. A similar choice of the normal modes for an expansion of $\mathbf{m}_Q(\xi)$ may be convenient for a specific geometry (specific direction of the bias magnetization) since we utilize the exact solution of Eq. (2.35) in the important regime of standing spin waves $Q = 0$. However, in the case of an arbitrary direction of bias magnetization when we want to find the dispersion equation describing the angular dependence of the spin-wave spectrum, it is convenient to use as the normal modes the eigenfunctions of the pure exchange operator. Here, following our paper [28], as a set of functions for expanding $\mathbf{m}_Q(\xi)$ in Eq. (2.35) we shall use the eigenfunctions of the diagonal part $\hat{\mathbf{F}}_{\mathbf{diag}}$ of the matrix-differential operator $\hat{\mathbf{F}}$ [see Eq. (2.36)].

As a set of exchange boundary conditions at $\xi = \pm L/2$ we shall use mixed exchange boundary conditions in the form [35,36] :

$$\frac{dm^x(\xi)}{d\xi} + \eta_{1,2}\cos 2\theta \; m^x(\xi) = 0 \; , \qquad \frac{dm^y(\xi)}{d\xi} + \eta_{1,2}\cos^2\theta \; m^y(\xi) = 0 \; , \qquad (2.37)$$

where η_1 and η_2 are the pinning parameters on the upper ($\xi = L/2$) and the lower ($\xi = -L/2$) surfaces of the film.

The problem of determining the eigenfunctions of the operator $\hat{\mathbf{F}}_{diag}$, i.e.

$$\hat{\mathbf{F}}_{diag}\mathbf{S}(\xi) = F\,\mathbf{S}(\xi) \qquad (2.38)$$

with the boundary conditions (2.37), is a particular case of a general Sturm-Lionville problem. It is described in the work [28]. The obtained vector eigenfunctions are denoted by $\mathbf{S}^p_n(\xi)$ (where $p = x,y$) and are named *spin wave modes* (SWM). In the coordinate system *xyz* these functions are given by

$$\mathbf{S}^x_n(\xi) = \begin{pmatrix} \Phi^x_n(\xi) \\ 0 \end{pmatrix}, \qquad \mathbf{S}^y_n(\xi) = \begin{pmatrix} 0 \\ \Phi^y_n(\xi) \end{pmatrix}, \qquad (2.39)$$

where

$$\Phi^p_n(\xi) = A^p_n\left[\cos\{\kappa^p_n(\xi + L/2)\} + (\eta^p_2/\kappa^p_n)\sin\{\kappa^p_n(\xi + L/2)\}\right]. \qquad (2.40)$$

The quantity κ^p_n, which may be interpreted as the SWM transverse wave number, is determined by the equation

$$\left[(\kappa^p_n)^2 - \eta^p_1\eta^p_2\right]\tan(\kappa^p_n L) = \kappa^p_n(\eta^p_1 + \eta^p_2) \qquad (2.41)$$

where

$$\eta^x_{1,2} = \eta_{1,2}\cos 2\theta, \qquad \eta^y_{1,2} = \eta_{1,2}\cos^2\theta. \qquad (2.42)$$

The constant A^p_n may be obtained from the normalisation condition for the SWM. The eigenvalues F^p_n of the boundary problem (2.38) are given by

$$F^p_n = (\omega_H/\omega_M) + \alpha(q^p_n)^2 \qquad (2.43)$$

where q^p_n may be interpreted as a "full" wave vector satisfying

$$(q^p_n)^2 = Q^2 + (\kappa^p_n)^2. \qquad (2.44)$$

The SWM in Eq. (2.39), being the eigenfunctions of the Sturm-Liouville problem, form a set of complete orthogonal functions over the interval $-L/2 \le \xi \le L/2$. They satisfy over this interval the condition of orthonormality:

$$\left(\mathbf{S}_n^p(\xi), \mathbf{S}_{n'}^{p'}(\xi) \right) = L\, \delta_{nn'} \delta_{pp'} , \qquad (2.45)$$

where δ is the Kronecker's symbol.

The solutions of the transcendental Eq. (2.41) have been extensively discussed in the literature, especially in connection with the phenomenon of SWR (e.g., see [36-38]). Physically the solutions of Eq. (2.41) describe the distributions of magnetization of standing spin waves $Q = 0$ across the film thickness. These solutions are particularly simple in the extreme cases of the exchange boundary conditions, namely in the case of pinned surface spins ($\eta_1 = \eta_2 \rightarrow \infty$):

$$\Phi_n = \sqrt{2}\, \sin[\kappa_n(\xi + L/2)] \; , \quad \kappa_n = n\pi/L, \quad n = 1,2,3\ldots \qquad (2.46)$$

and in the case of unpinned surface spins ($\eta_1 = \eta_2 = 0$):

$$\Phi_n = \sqrt{2}(1+\delta_{0n})^{1/2} \cos[\kappa_n(\xi + L/2)], \quad \kappa_n = n\pi/L, \quad n = 0,1,2,\ldots \qquad (2.47)$$

2.5. EQUATIONS FOR SPIN-WAVE MODE AMPLITUDES

Expanding the transverse distribution of the variable magnetization corresponding to the spin wave (2.27) in the infinite series of the SWM, we get

$$\mathbf{m}_Q(\xi) = M_0 \sum_{n,p} m_n^p\, \mathbf{S}_n^p(\xi) = M_0 \sum_{n,p} \left(m_n^x \mathbf{S}_n^x + m_n^y \mathbf{S}_n^y \right) . \qquad (2.48)$$

Substituting Eq. (2.48) into the integro-differential Eq. (2.35), using the orthonormality condition (2.45), and introducing two-dimensional SWM vector amplitudes $\mathbf{m}_n$:

$$\mathbf{m}_n = \begin{pmatrix} m_n^x \\ m_n^y \end{pmatrix} , \qquad (2.49)$$

we obtain after some simple calculations an infinite system of algebraic equations for $\mathbf{m}_n$:

$$\hat{\mathbf{D}}_{nn}\mathbf{m}_n + \sum_{n' \neq n} \hat{\mathbf{R}}_{nn'}\mathbf{m}_{n'} = 0 . \tag{2.50}$$

Here the square matrices $\hat{\mathbf{D}}_{nn'}$ and $\hat{\mathbf{R}}_{nn'}$, and their elements $P^{pp'}_{nn'}$, $R^{pp'}_{nn'}$, and $T^{pp'}_{nn'}$ are determined by the following expressions:

$$\hat{\mathbf{D}}_{nn} = \begin{bmatrix} F^{x}_{n} + \sin^2\theta + AP^{xx}_{nn} & CP^{xy}_{nn} - i\left(\frac{\omega}{\omega_M}T^{xy}_{nn} - DQ^{xy}_{nn}\right) \\ CP^{xy}_{nn} + i\left(\frac{\omega}{\omega_M}T^{yx}_{nn} + DQ^{yx}_{nn}\right) & F^{y}_{n} + EP^{yy}_{nn} \end{bmatrix} , \tag{2.51}$$

$$\hat{\mathbf{R}}_{nn'} = \begin{bmatrix} AP^{xx}_{nn'} + iBQ^{xx}_{nn'} & CP^{xy}_{nn'} - i\left(\frac{\omega}{\omega_M}T^{xy}_{nn'} - DQ^{xy}_{nn'}\right) \\ CP^{yx}_{nn'} + i\left(\frac{\omega}{\omega_M} T^{yx}_{nn'} + DQ^{yx}_{nn'}\right) & EP^{yy}_{nn'} \end{bmatrix} , \tag{2.52}$$

$$P^{pp'}_{nn'}(Q) = P^{p'p}_{n'n}(Q) = \frac{1}{L} \int_{-L/2}^{L/2}\int \hat{\mathbf{G}}^{P}(\xi,\xi';Q)\Phi^{p}_{n}(\xi)\Phi^{p'}_{n'}(\xi')d\xi d\xi' , \tag{2.53}$$

$$Q^{pp'}_{nn'}(Q) = -Q^{p'p}_{n'n}(Q) = \frac{1}{2L} \int_{-L/2}^{L/2}\int \hat{\mathbf{G}}^{Q}(\xi,\xi';Q)\Phi^{p}_{n}(\xi)\Phi^{p'}_{n'}(\xi')d\xi d\xi' , \tag{2.54}$$

$$T^{pp'}_{nn'} = T^{p'p}_{n'n} = \frac{1}{L} \int_{-L/2}^{L/2} \Phi^{p}_{n}(\xi)\Phi^{p'}_{n'}(\xi')d\xi , \tag{2.55}$$

$$\left.\begin{aligned} &A = \cos^2\phi - \sin^2\theta(1+\cos^2\phi) , \qquad B = 2\cos\phi\sin\phi , \\ &C = -\cos\theta\sin\phi\cos\phi , \quad D = -2\sin\theta\sin\phi , \quad E = \sin^2\phi . \end{aligned}\right\} \tag{2.56}$$

It should be noted that the system of homogeneneous equations for the SWM amplitudes, Eq. (2.50), was derived with simultaneous regard for both the dipole and exchange interactions, as well as electromagnetic and exchange boundary conditions. This system rigorously describes the wave process in the magnetically biased ferromagnetic films. It may be used for precise analysis of the dipole-exchange (or purely dipole in the limit $\alpha = 0$) spin-wave spectra.

The physical meaning of the matrix elements $P_{nn'}^{pp'}$ and $Q_{nn'}^{pp'}$, given by Eqs (2.53) and (2.54), should be noted. These elements describe the dipole-dipole interactions of the SWM and therefore we call them *dipole matrix elements*. When the exchange boundary conditions are symmetrical, the elements $P_{nn'}^{pp'}$ describe dipole interaction of the SWM having one and the same type of symmetry, and the elements $Q_{nn'}^{pp'}$ describe dipole interaction of the SWM having opposite types of symmetry.

We note that the Zeeman and exchange energies determine the elements of the main diagonal of the matrix of Eq. (2.50) and do not influence the non-diagonal elements. It means that the solution of the system (2.50) in a diagonal approximation takes into account Zeeman and exchange contributions completely and the essential part of dipole interaction also. Such an approximation greatly simplifies the analysis of (2.50) and we shall discuss it below.

At first sight the system of equations for the SWM amplitudes, Eq. (2.50), looks complicated and fairly difficult to grasp. But owing to the symmetry properties in particular cases(for example, perpendicularly and tangentially magnetized films with pinned or unpinned surface spins) the appropriate solutions, both exact and approximate, could be found in comparatively simple forms.

2.6 EXACT DIPOLE-EXCHANGE DISPERSION RELATIONS

The main purpose of this section is to present exact dispersion equations following from Eq. (2.50) for some particular cases of interest. These equations allow one to calculate the dispersion characteristics of spin waves. Besides they may be directly compared with the corresponding equations of the PW approach.

2.6.1. Dispersion relation in the form of an infinite determinant

The condition for the vanishing of the determinant of the infinite system of homogeneous equations (2.50) yields the exact dispersion relation for dipole-exchange spin waves in a ferromagnetic film. We may say that it is the natural form of the dispersion relation for the SWM approach. The zeroes of this determinant give the eigenfrequencies ω_{nQ} of the film. In other words, the problem of determining the exact spin-wave spectrum of the film ω_{nQ} is reduced to the problem of calculating the eigenvalues of the block matrix of the infinite system (2.50). The calculations could be fulfilled by means of the well-known *QR* algorithm (see, e.g., Wilkinson and Reinsch [39]).

The results of such calculations are presented in our papers [28,29] and in the paper [40] by Vayhinger and Kronmuller. (They carried out calculations in connection with a problem of dipole-exchange spin waves in multilayers). We shall discuss numerical results describing the spin-wave spectra later.

2.6.2. Dispersion relations in the form of an infinite series

For the first time the dispersion relation in the form of an infinite series was derived by Vendik and Chartorizhskii [24]. They considered the case of a perpendicularly magnetized film with pinned surface spins [see Eq. (2.4)] when purely symmetrical and purely antisymmetrical spin waves could exist in the film. Later the method was extended to the waves with mixed type of symmetry [27].

To introduce the method, let us examine the case of perpendicular magnetization. For simplicity we consider the film with symmetrical boundary conditions, in particular with *pinned surface spins* (PSS), as in Eq. (2.4), and *unpinned surface spins* (USS), as in Eq. (2.5).

Introducing new circular variables $m^{\pm}_{n} = m^{x}_{n} \pm i m^{y}_{n}$ [for convenieuce we use for $m^{\pm}_{n}$ the notation m^{p}_{n} where superscript p = (1,2) = (-,+)] and writing the system (2.50) for $\theta = 0$ in terms of the new SWM amplitudes m^{p}_{n}, we get two coupled infinite systems of equations [26]:

$$(\Omega_{nQ}+\omega)m^{1}_{n} + (\omega_{M}/2) \sum_{n'} (m^{1}_{n'} - m^{2}_{n'})P_{nn'} = 0 , \tag{2.57}$$

$$(\Omega_{nQ}-\omega)m^{2}_{n} - (\omega_{M}/2) \sum_{n'} (m^{1}_{n'} - m^{2}_{n'})P_{nn'} = 0 , \tag{2.58}$$

where

$$\Omega_{nQ} = \omega_H + \omega_M\alpha(\kappa_n^2 + Q^2) . \quad (2.59)$$

In the case of pinned surface spins:

$$P_{nn'} = \frac{Q^2}{q_n^2}\delta_{nn'} + \frac{Q^2\kappa_n\kappa_{n'}}{q_n^2q_{n'}^2}V_n\left[1 + (-1)^{n+n'}\right] , \quad (2.60)$$

where $q_n^2 = Q^2 + \kappa_n^2$, $\kappa_n = n\pi/L$, $n=1,2,3,\ldots$. In the case of unpinned surface spins:

$$P_{nn'} = \frac{Q^2}{q_n^2}\delta_{nn'} - \frac{Q^4}{q_n^2q_{n'}^2}V_n\left[(1+\delta_{on})(1+\delta_{on'})\right]^{-1/2}\left[1 + (-1)^{n+n'}\right], \quad (2.61)$$

where $q_n^2 = Q^2 + \kappa_{n'}^2$, $\kappa_n = n\pi/L$, $n = 0,1,2,\ldots$. In Eqs. (2.60) and (2.61) the factor V_n is given by

$$V_n = [1 - (-1)^n\exp(-QL)]/QL \quad (2.62)$$

It is easy to see from Eqs. (2.57)-(2.61) that each of the coupled systems (2.57) and (2.58) *splits* into two independent system, each one corresponding to the even or to the odd integers n and n'. This has a clear physical explanation: the symmetric and antisymmetric spin waves in the perpendicularly magnetized film having symmetric boundary conditions are independent, i.e. there is no interaction between the even and odd spin-wave modes. The splitting of the systems enables us to fulfill the analysis of the symmetric spin waves (n and n' are odd in the case of PSS or n and n' are even in the case of USS) and of the antisymmetric spin waves (n and n' are even in the case of PSS or n and n' are odd in the case of USS) separately. Following the paper [24], we shall show that the solution of the coupled infinite systems (2.57) and (2.58) ("odd" or "even") can be written in a general form without investigating their determinant first. Adding together the rows of the systems which correspond to the same n we obtain

$$(\omega + \Omega_{nQ})m_n^1 = (\omega - \Omega_{nQ})m_n^2 . \quad (2.63)$$

The last relation enables one to rewrite the systems of equation for m_n^1 and m_n^2 [Eqs. (2.57) and (2.58)] into the system of equations for m_n^1 only or for m_n^2 only.

For example, let us eliminate the amplitudes m_n^1 from Eq. (2.57) with the aid of Eq. (2.63). The substitution of Eqs. (2.63) into (2.57) with allowance of the expression for matrix element $P_{nn'}$ for the case of PSS [Eq. (2.60)] gives

$$m_n^2 \left(\frac{\omega-\Omega_{nQ}}{\omega_{\mathrm{M}}} - \frac{\Omega_{nQ}}{\omega+\omega_{nQ}} \frac{Q^2}{q_n^2} \right) \frac{q_n^2}{Q\kappa_n} = 2V_n \sum_{n'} m_{n'}^2 \left(\frac{\Omega_{n'Q}}{\omega+\Omega_{n'Q}} \right) \frac{Q\kappa_n}{q_{n'}^2} . \tag{2.64}$$

The right-hand part of Eq. (2.64) is independent of the row number n because the factor V_n in Eq. (2.62) is now equal to $[1 \pm \exp(QL)]/QL$. This makes it possible to determine the ratio of the amplitudes of any SWM as

$$\frac{m_n^2}{m_{n'}^2} = \frac{\left(\dfrac{\omega-\Omega_{n'Q}}{\omega_{\mathrm{M}}} - \dfrac{\Omega_{n'Q}}{\omega+\omega_{n'Q}} \dfrac{Q^2}{q_{n'}^2} \right) q_{n'}^2 \kappa_n}{\left(\dfrac{\omega-\Omega_{nQ}}{\omega_{\mathrm{M}}} - \dfrac{\Omega_{nQ}}{\omega+\omega_{nQ}} \dfrac{Q^2}{q_n^2} \right) q_n^2 \kappa_{n'}} . \tag{2.65}$$

Now using Eqs. (2.63) and (2.65) we may obtain a solution of the system (2.57) and (2.58) in the form

$$m_n^1 = \frac{N\omega_{\mathrm{M}}(\omega-\Omega_{nQ})}{(q_n^2/Q\kappa_n)[\omega^2 - \Omega_{nQ}^2 - \Omega_{nQ}\omega_{\mathrm{M}}(Q^2/q_n^2)]} . \tag{2.66}$$

$$m_n^2 = \frac{N\omega_{\mathrm{M}}(\omega+\Omega_{nQ})}{(q_n^2/Q\kappa_n)[\omega^2 - \Omega_{nQ}^2 - \Omega_{nQ}\omega_{\mathrm{M}}(Q^2/q_n^2)]} . \tag{2.67}$$

Here N is a normalizing constant. It could be chosen, for example, so that the sum of the squares of the SWM amplitudes equals unity [24].

Substituting Eq. (2.65) into Eq. (2.64) we find the dispersion relation for dipole-exchange spin waves in the film with pinned surface spins:

$$\frac{2(1\pm e^{-QL})}{QL}\sum_{\substack{n_{odd}\\n_{even}}}\frac{Q^2\kappa_n^2\Omega_{nQ}\omega_M}{q_n^4[\omega^2-\Omega_{nQ}^2-\Omega_{nQ}\omega_M(Q^2/q_n^2)]}=1\ . \qquad (2.68)$$

In Eq. (2.68) n assumes either only odd or only even integer values. Thus Eq. (2.68) is equivalent to two independent dispersion relations for symmetric and antisymmetric normal spin waves.

These dispersion relations allow one to calculate the spin-wave branches $\omega_j(Q)$ constituting the dipole-exchange spectrum. Also, on the basis of the expansion (2.48) with the help of Eqs. (2.66) and (2.67) defining all unknown coefficients in this expansion, the distributions of the variable magnetization $\dot{\mathbf{m}}_{Qj}(\xi)$ of normal spin waves corresponding to the spin-wave branches $\omega_j(Q)$ could be computed.

Carrying out analogous calculations for the case of unpinned surface spins we get the following dispersion relations:

$$\frac{2(1\pm e^{-QL})}{QL}\sum_{\substack{n_{odd}\\n_{even}}}\frac{1}{1+\delta_{on}}\frac{Q^4\Omega_{nQ}\omega_M}{q_n^4[\omega^2-\Omega_{nQ}^2-\Omega_{nQ}\omega_M(Q^2/q_n^2)]}=-1\ . \qquad (2.69)$$

Applying the same method it is possible to find the spin-wave spectrum and solutions for the variable magnetizations of normal spin waves in a tangentially magnetized film ($\theta = 90^\circ$). In particular, for the case of longitudinal propagation ($\phi = 0$) the dipole-exchange dispersion relations are written in the forms:

$$\frac{2(1\pm e^{-QL})}{QL}\sum_{\substack{n_{odd}\\n_{even}}}\frac{Q^2\kappa_n\Omega_{nQ}\omega_M}{q_n^4\ [\omega^2-\Omega_{nQ}^2-\Omega_{nQ}\omega_M(\kappa_n^2/q_n^2)]}=-1\ . \qquad (2.70)$$

for the case of pinned surface spins (the upper and lower signs and summation indexes correspond to the symmetric and antisymmetric normal spin waves respectively), and

$$\frac{2(1\pm e^{-QL})}{QL}\sum_{\substack{n_{\text{odd}}\\ n_{\text{even}}}}\frac{1}{1+\delta_{on}}\;\frac{Q^4\Omega_{nQ}\omega_M}{q_n^4\,[\omega^2-\Omega_{nQ}^2-\Omega_{nQ}\omega_M(\kappa_n^2/q_n^2)]}=1\,. \tag{2.71}$$

for the case of unpinned surface spins (the upper and lower signs and summation indexes correspond to the antisymmetric and symmetric normal spin waves respectively).

As mentioned already, the infinite series in the above derived dispersion relations [Eqs. (2.68), (2.69), (2.70), and (2.71)] converge and may be used for numerical calculations as they stand.

It is very important and interesting to note that *after summation over* n *the dispersion relations in the form of series transform into the corresponding trigonometric dispersion relations.* In particular, the dispersion Eq. (2.68) transforms into Eqs. (2.14) and (2.15), Eq. (2.69) transforms into Eqs. (2.16) and (2.17), Eq. (2.70) transforms into Eqs. (2.22) and (2.23), and Eq. (2.71) transforms into Eqs. (2.24) and (2.25).

So we may conclude that, with the help of the SWM approach, one and the same *spin-wave dispersion law* ω_{nQ} *can be presented in different mathematical forms.* Bearing in mind that trigonometric dispersion relations are typical for the PW approach, we may also conclude that in the framework of the SWM approach a connection between these two different theoretical approaches is established.

It is a matter of taste to use one or another form of the exact dispersion relation for numerical calculation of the dipole-exchange spin-wave dispersion properties. But it seems that the form of a series is more convenient because it does not involve a solution of the sixth-order characteristic equation (2.10) or (2.21).

We note also that it is easy to obtain a series form of the dispersion relations for purely dipole spin waves (magnetostatic waves) by setting in Eqs. (2.68)-(2.71) the exchange constant $\alpha = 0$. It may be shown that after summation these dipole (non-exchange) dispersion relations transform into the corresponding trigonometric dispersion equations for perpendicularly [Eqs. (1.72) and (1.73)] and tangentially [Eqs.

(1.77) and (1.78)] magnetized films.

It is worth noting that one and the same final trigonometric dispersion equation for the MSW (specifically, Eq. (1.72) of Chapter 1) is obtained from both Eqs. (2.68) and (2.69), which are non-exchange dispersion relations originally obtained for different surface spin pinning conditions. This means that the exchange boundary conditions, which cause the mathematical difference between Eqs. (2.68) and (2.69) even in non-exchange limit $\alpha = 0$, do not really influence the MSW dispersion properties. In other words, at least two kinds of dispersion equations in the form of a series could be suggested to represent the dispersion law of purely dipole spin waves.

2.7. APPROXIMATE DIPOLE-EXCHANGE DISPERSION EQUATIONS

For practical purposes it is desirable to have an equation for the dispersion law ω_{nQ} in an *explicit* form. It is evident that an explicit form of the dispersion equation greatly simplifies a problem of calculating the spin-wave characteristics and their dependence on the parameters. It is possible to derive approximate but rather precise explicit dispersion equations for the dipole-exchange spin waves by applying perturbation theory methods to the analysis of the system of equations for the SWM amplitudes (2.50).

After some simple transformations we may rewrite the system of equations(2.50) in the form [28]:

$$\hat{\mathbf{H}}_{nn}\mathbf{m}_n + \sum_{n' \neq n} \hat{\mathbf{W}}_{nn'}\mathbf{m}_{n'} = 0 \qquad (2.72)$$

where

$$\hat{\mathbf{H}}_{nn} = \hat{\mathbf{I}} \det \hat{\mathbf{D}}_{nn} ,$$

$$\hat{\mathbf{W}}_{nn'} = \hat{\mathbf{D}}_{nn}^{-1} \hat{\mathbf{R}}_{nn'} \det \hat{\mathbf{D}}_{nn} \qquad (2.73)$$

and $\hat{\mathbf{I}}$ is the unit diagonal matrix.

Now we may represent the infinite system (2.72) in the form of a matrix equation

$$\hat{\mathbf{L}}\mathbf{m} = 0 \tag{2.74}$$

where $\hat{\mathbf{L}}$ is the infinite block matrix whose elements are the square matrices (2.73) and **m** is the infinite column vector consisting of the SWM vector amplitudes (2.49).

The dispersion equation for the dipole-exchange spin waves in this notation may be expressed as

$$\det \hat{\mathbf{L}} = 0 \quad . \tag{2.75}$$

It is clear from Eqs. (2.74) and (2.75) that the problem of evaluating the dispersion relations, corresponding to the separate normal spin waves with a given number n, may be reduced to the problem of determining the matrix "eigenvalues" $\hat{\lambda}_n$ for the infinite block matrix $\hat{\mathbf{L}}$. In other words, it is the problem of transforming $\hat{\mathbf{L}}$ to the block-diagonal form

$$\hat{\mathbf{L}} = \begin{bmatrix} \hat{\lambda}_0 & & & & 0 \\ & \hat{\lambda}_1 & & & \\ & & \ddots & & \\ & & & \hat{\lambda}_n & \\ 0 & & & & \ddots \end{bmatrix} \tag{2.76}$$

where the $\hat{\lambda}_j$ (j = 1,2,3...n...) are square matrix blocks. If the diagonal blocks $\hat{\lambda}_n$ are found, the determinant of Eq. (2.76) may be written in a form

$$\det \hat{\mathbf{L}} = \prod_{n=0}^{\infty} \det \hat{\lambda}_n \quad . \tag{2.77}$$

Then the dispersion equation for the spin wave of a number *n* would be of the form:

$$\det \hat{\lambda}_n = 0 \quad . \tag{2.78}$$

In the general case of arbitrary surface anisotropy and arbitrary direction of static magnetization, the problem of diagonalisation of the block matrix $\hat{\mathbf{L}}$ (and hence the problem of evaluating the SW dispersion relations) may be solved approximately using classical perturbation theory (see, e.g., [41]).

We represent the block matrix $\hat{\mathbf{L}}$ as a sum of its diagonal and off-diagonal parts:

$$\hat{\mathbf{L}} = \begin{bmatrix} \hat{\mathbf{H}}_{00} & & & & 0 \\ & \hat{\mathbf{H}}_{00} & . & & \\ & & . & & \\ & & & \hat{\mathbf{H}}_{nn} & \\ 0 & & & & . \end{bmatrix} + \begin{bmatrix} 0 & & & W_{nn'} & \\ & 0 & . & & \\ & & . & & \\ & W_{n'n} & & . \; 0 & \\ & & & & . \end{bmatrix} . \qquad (2.79)$$

We may now consider the diagonal part of Eq. (2.79) as the unperturbed operator and the off-diagonal part as the operator of perturbation. The physical interpretation of the perturbation operator is as follows. This operator describes the interaction between spin wave modes of different types $n \neq n'$ which is caused by the off-diagonal part of the magnetic dipole-dipole interaction. We note that the exchange interaction and the main (diagonal) part of the dipole-dipole interaction are described by the unperturbed diagonal operator in Eq. (2.79).

We may express the matrix "eigenvalues" of $\hat{\mathbf{L}}$ in a form of the usual series from the perturbation theory:

$$\hat{\lambda}_n = {}^{(0)}\hat{\lambda}_n + {}^{(1)}\hat{\lambda}_n + {}^{(2)}\hat{\lambda}_n + \dots \quad . \qquad (2.80)$$

Then in the non-degenerate case ($\det \hat{\mathbf{D}}_{nn} \neq \det \hat{\mathbf{D}}_{n'n'}$), the zero-order ($\hat{\mathbf{W}}_{nn'} = \hat{\mathbf{W}}_{n'n} = 0$) approximation for the dispersion equation of the nth spin wave has a form:

$$\det {}^{(0)}\hat{\lambda}_n = \det \hat{\mathbf{D}}_{nn} = 0 \quad . \qquad (2.81)$$

The zero-order approximation for the nth SWM vector amplitude is

$${}^{(0)}\mathbf{m}_n = \begin{pmatrix} 1 \\ 1 \end{pmatrix} \qquad (2.82)$$

and the zero-order approximation for the magnetization distribution of the nth spin-wave eigenwave, valid for arbitrary values of Q, may be found as follows:

$$\mathbf{m}_n(\xi) = M_0[\hat{x}\ \Phi_n^x(\xi) + \hat{y}\ \Phi_n^y(\xi)] \quad . \tag{2.83}$$

Substituting Eq. (2.51) for $\hat{\mathbf{D}}_{nn}$ in Eq. (2.81) we obtain the zero-order approximation for the dispersion equation $\omega_{nQ} = f(Q)$ of the nth propagating spin wave in the explicit form

$$\left(\frac{\omega_{nQ}}{\omega_M} T_{nn}^{yx} + DQ_{nn}^{yx}\right)^2 = \left(\frac{\Omega_{nQ}^x}{\omega_M} + \sin^2\theta + AP_{nn}^{xx}\right)\left(\frac{\Omega_{nQ}^y}{\omega_M} + EP_{nn}^{yy}\right) - C^2(P_{nn}^{yx})^2 \tag{2.84}$$

where

$$\Omega_{nQ}^{x,y} = \omega_M F_n^{x,y} = \omega_H + \omega_M \alpha[(\kappa_n^{x,y})^2 + Q^2] \ . \tag{2.85}$$

Dispersion Eq. (2.84) is valid for arbitrary direction of the internal magnetic field and for arbitrary exchange boundary conditions (2.37). In the case when both components of the magnetization vector (2.34) are pinned uniformly ($\eta_1^x = \eta_1^y = \eta_1$, $\eta_2^x = \eta_2^y = \eta_2$), but still differently on the different sides of the film ($\eta_1 \neq \eta_2$), we may rewrite Eq. (2.84) in a form which is similar to the Herring-Kittel form of the well known dispersion equation for spin waves in the bulk medium [31]:

$$\omega_{nQ}^2 = (\omega_H + \alpha\omega_M q_n^2)(\omega_H + \alpha\omega_M q_n^2 + \omega_M F_{nn}) \tag{2.86}$$

where

$$F_{nn} = P_{nn} + \sin^2\theta\left[1 - P_{nn}(1 + \cos^2\phi) + \omega_M \frac{P_{nn}(1 - P_{nn})\sin^2\phi}{(\omega_H + \alpha\omega_M^2 q_n)}\right] , \tag{2.87}$$

$q_n^2 = Q^2 + \kappa_n^2$, and κ_n can be determined from Eq. (2.41).

We note that the matrix element P_{nn} as a function of the dimensionless in-plane wave vector QL always varies in the range $0 \le P_{nn} < 1$ when $0 \le QL < \infty$. It will be recalled that the dipole matrix element in the cases of pinned and unpinned surface spins has the explicit form $P_{nn} = f(QL)$ (see Eqs. (2.60) and (2.61)). In the long-wavelength limit ($QL \ll 1$) very simple expressions for P_{nn} may be written. For example, in the case of unpinned surface spins the expression for the P_{nn} has the form

$$P_{nn} = QL/2 \qquad \text{when } n = 0 \ , \tag{2.88}$$

$$P_{nn} = (QL)^2/n^2\pi^2 \quad \text{when } n \neq 0 \quad . \tag{2.89}$$

In the case of pinned surface spins,

$$P_{nn} = 4QL/n^2\pi^2 \qquad \text{when } n = 1,3,5,\ldots \ , \tag{2.90}$$

$$P_{nn} = 3(QL)^2/n^2\pi^2 \quad \text{when } n = 2,4,6,\ldots \ . \tag{2.91}$$

Discussion of the dispersion Eq. (2.86) was already given in Section 1.3.4 of Chapter 1.

Equation (2.86) can be easily used for practical calculations of the dispersion characteristics of either dipole-exchange or magnetostatic waves ($\alpha = 0$) in ferromagnetic films. In particular, numerical calculations show that, in the long-wavelength part of the spectrum ($QL \ll 1$) in the case of unpinned surface spins, Eq. (2.86) gives the results which coincide very accurately with the results obtained from the non-exchange dispersion equations (see, e.g., (1.70) and (1.79)).

It is clear from the form of the perturbation operator in Eq. (2.79) that the first-order correction to the nth matrix "eigenvalue" ${}^{(1)}\hat{\lambda}_n$ is equal to zero (see Landau and Lifshitz [41]). Therefore the dispersion equations (2.84) and (2.86) remain in force in the first-order approximation of the perturbation theory.

The first non-trivial correction to the dispersion law (2.84) in the non-degenerate case appears in the second-order approximation of the perturbation theory which leads to the dispersion equation

$$\det\left[{}^{(0)}\hat{\lambda}_n + {}^{(2)}\hat{\lambda}_n\right] = \det\left[\hat{\mathbf{I}}\,\det\hat{\mathbf{D}}_{nn} + \sum_{n'\neq n} \frac{\hat{\mathbf{W}}_{nn'}\hat{\mathbf{W}}_{n'n}}{\det\hat{\mathbf{D}}_{nn} - \det\hat{\mathbf{D}}_{n'n'}}\right] = 0 \ . \tag{2.92}$$

The analysis shows that the dispersion curves (2.84) corresponding to the different numbers n may cross each other, i.e., there may be frequency degeneracy in some points of the SW spectrum ($\det\hat{\mathbf{D}}_{nn} = \det\hat{\mathbf{D}}_{n'n'}$). The secular equation which removes this degeneracy by taking into account the dipole-dipole hybridisation of the "interacting" dispersion branches in the first-order approximation of perturbation theory may be written as follows:

$$\det\begin{bmatrix} \hat{\mathbf{P}}_{nn} & \hat{\mathbf{R}}_{nn'} \\ \hat{\mathbf{R}}_{n'n} & \hat{\mathbf{D}}_{n'n'} \end{bmatrix} = 0 \quad . \tag{2.93}$$

For the particular cases of perpendicularly and tangentially magnetized ferromagnetic films with pinned and unpinned surface spins the explicit secular dispersion equations, which are deduced from Eq. (2.93), will be given below.

Having summarized the main theoretical results concerning the dipole-exchange spin-wave dispersion relations, we now turn to the detailed discusion and illustration of spin-wave dispersion properties for some particular cases of interest.

2.8. DIPOLE-EXCHANGE SPECTRUM: PERPENDICULAR MAGNETIZATION

To describe the spin-wave spectrum of perpendicularly magnetized ferromagnetic films we shall use both exact and approximate dispersion equations. For simplicity, in the following discussion we shall mention only the cases of pinned and unpinned surface spins. As was shown above, the generalization for the mixed surface spin pinning conditions is straightforward.

The approximate dispersion equations following from Eqs. (2.86) and (2.93) for the case of perpendicular magnetization are

$$\omega_{nQ}^2 = (\omega_H + \alpha\omega_M q_n^2)(\omega_H + \alpha\omega_M q_n^2 + \omega_M P_{nn}) \tag{2.94}$$

and

$$(\omega^2 - \omega_{nQ}^2)(\omega^2 - \omega_{n'Q}^2) = \omega_M^2 \Omega_{nQ} \Omega_{n'Q} P_{nn'}^2 \quad . \tag{2.95}$$

We recall that here the factor Ω_{nQ} is given by Eq. (2.59).

The last secular equation (2.95) describes dipole "*repulsion*" between two crossing dispersion branches, as in Eq. (2.94), ω_{nQ} and $\omega_{n'Q}$ of one and the same type of symmetry. It yields two non-crossing dispersion branches

$$\omega_{1,2}^2 = \frac{1}{2}\left\{\omega_{nQ}^2 + \omega_{n'Q}^2 \pm \left[(\omega_{nQ}^2 - \omega_{n'Q}^2)^2 + 4\omega_M^2 \Omega_{nQ}\Omega_{n'Q} P_{nn'}^2\right]^{1/2}\right\} . \tag{2.96}$$

Equation (2.94) demonstrates explicitly the contributions of the Zeeman (the term ω_H), the dipole-dipole (the term $\omega_M P_{nn}$) and the exchange (the term $\omega_M \alpha q_n^2$) interactions to the spin-wave dispersion.

It is easy to see that the eigenfrequencies of the spin waves are characterized by the in-plane propagation wave vector **Q** and by the transverse wave vector κ_n. [We recall that the full wave vector q_n in Eq. (2.94) is determined by the expression $q_n^2 = Q^2 + \kappa_n^2$]. The wave vector κ_n may be called the "*standing wave vector*". It designates the number of half-wavelengths across the film, i.e., it describes distributions of variable magnetization across the film thickness (in this connection see Eqs. (2.46) and (2.47)). The role of κ_n (in the term $\alpha\kappa_n^2$ in Eq. (2.94)) is essential when the film thickness is rather small, in other words, when the film thickness is comparable with the length of exchange spin wave. We may refer to the dependence of the spin-wave frequency on κ_n (due to the term $\alpha\kappa_n^2$ in Eq. (2.94)) as *exchange* dispersion.

The in-plane propagation vector Q is responsible for both *dipole* and *exchange* dispersion. For small Q the dependence ω_{nQ} is principally due to dipole effects; for big Q the exchange energy dominates the dipole energy. For rather large Q, a quadratic Q dependence in dispersion law ω_{nQ} (usual for exchange spin waves) takes over.

The advantage of the approximate dispersion, represented by Eqs. (2.94) and (2.95), is that they have a simple form which enables one easily to see the transformation of the dipole-exchange or purely dipole (in the limit $\alpha = 0$) spin-wave spectrum when any of the film parameters or the magnetizing field are changed. As follows from Eq. (2.94), the spin-wave spectrum of a film is discrete with each mode having its own characteristic dispersion. Each of the spin-wave dispersion branches begins from a particular point $\omega_{nQ}(Q = 0)$ corresponding to one of the SWR frequencies:

$$\Omega_n = \omega_H + \omega_M \alpha\kappa_n^2 . \tag{2.97}$$

The initial slope of the dispersion branches depends strongly on the surface spin pinning conditions. It is clear from Eq. (2.94), and from

(2.88)-(2.89) or (2.90)-(2.91), that all odd and even branches n = 1,2,3,... besides the main one $n = 0$ in the case of USS and all even branches $n = 2,4,6...$ in the case of PSS have zero initial slope. The main branches ($n = 0$ and $n = 1$ in the cases of USS and PSS, respectively) have the biggest initial slope. In other words, the main mode spin waves have the biggest group velocity $V_g = \partial\omega_{nQ}/\partial Q$. It is worth mentioning that for the case of USS the initial slopes (group velocities) at $Q = 0$ of all branches (spin waves) coincide completely with those obtained from the MSW theory (see, e.g., [42]).

The dispersion relation (2.94) (together with Eq. (2.95) if crossings between dispersion branches (2.94) appear) allows one to describe clearly a process of dipole-exchange spectrum formation starting from the well-known purely dipole spectrum (MSW spectrum). Let us consider at first the MSW spectrum. As follows from Eq. (2.94) in the non-exchange limit $\alpha = 0$, this spectrum consists of the dispersion branches beginning at $\omega = \omega_H$ (see Fig. 2.1.). When the exchange interaction is incorporated, it shifts up the higher branches with $n > 0$. The value of the frequency shift increases with increasing eigen-wave number n and with decreasing film thickness L.

This frequency shift may cause the crossings of the dispersion

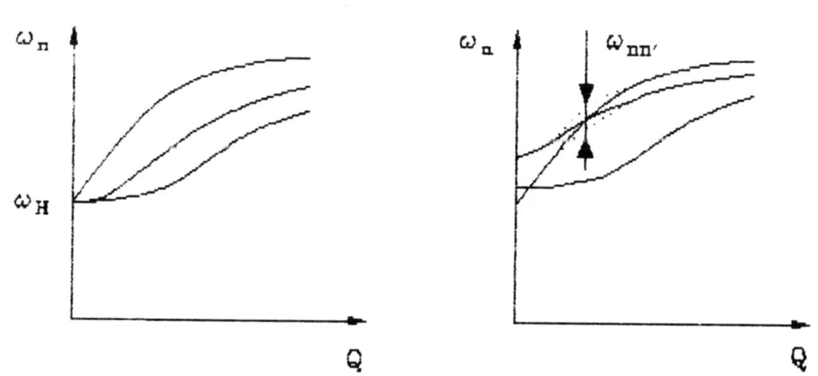

Fig. 2.1. Schematic effect of exchange interaction in the formation of the long-wavelength part of spin-wave spectrum: a) no exchange; b) the effect of exchange.

curves corresponding to the higher spin waves with the dispersion curve of the main spin wave (either $n = 0$ or $n = 1$). It is shown qualitatively in Fig. 2.1. (In the general case crossings of the different dispersion branches are possible). In the crossing points of the dispersion branches corresponding to spin waves of the same type of symmetry, the *dipole-dipole branch repulsion* takes place and dipole "gaps" in the spin-wave spectrum are formed.

The secular dispersion equation (2.96) enables us to calculate the width of the dipole "gap" in the spin-wave spectrum (i.e., the frequency distance between the repulsed branches ω_{1Q} and ω_{2Q}). The approximate expression for the width of the dipole gap can be written in the form

$$\Delta\omega_{nn'} = (\Omega_{nQ}\Omega_{n'Q})^{1/2}(\omega_M/\omega_o)\,|P_{nn'}| \tag{2.98}$$

where ω_0 is a frequency corresponding to the crossing point of the curves ω_{nQ} and $\omega_{n'Q}$. The dipole matrix element $P_{nn'}$ is to be calculated at the same point.

Analysis of Eq. (2.98) shows that the values of the dipole gaps depend heavily on film thickness and surface spin pinning conditions. These dependences are reflected by the matrix element $P_{nn'}$ which describes in the case in question the dipole interaction between different ($n \neq n'$) spin waves. To make the dependences clear we write down the expressions for $P_{nn'}$ for the long-wavelength part $QL \ll 1$ of spectrum. In the case of pinned surface spins,

$$P_{nn'} \approx 4QL/\pi^2 n, \qquad n' = 1, \quad n = 1,3,5,\dots . \tag{2.99}$$

In the case of unpinned surface spins,

$$P_{nn'} \approx \sqrt{2}(QL)^2/\pi^2 n, \qquad n' = 0, \quad n = 2,4,6,\dots . \tag{2.100}$$

Evidently, the widths of the gaps in the films with unpinned surface spins in the region $QL < 1$ are much smaller than in the films with pinned surface spins.

For example, in the YIG film of thickness $L = 5$ μm with pinned surface spins for $\omega_M/2\pi = 4900$ MHz, $\omega_H/2\pi = 4500$ MHz, $n = 1$, and $n' =$

25, we obtain from Eq. (2.98) $\Delta\omega_{nn'}/2\pi = 28$ MHz. Calculations for the film with unpinned surface spins, with the same parameters and $n = 0$, $n' = 24$ give $\Delta\omega_{nn'}/2\pi = 0.09$ MHz. From the experimental point of view, the dipole gaps could be observed if their widths are large compared with the relaxation frequency $\omega_r = \gamma\Delta H$. For high-quality YIG films the relaxation frequency equals approximately 0.5 MHz. It means that for the parameters of the above given example, dipole gaps could be observed only in the YIG film with pinned surface spins. In the film with unpinned surface spins the dipole gaps will be closed by dissipation.

The expression (2.98) is applicable only for the films with thickness L greater than a certain critical thickness L_c. The existence of the critical thickness L_c is connected with the fact that the decreasing of L leads to the increasing of the frequency distance between the dispersion branches and at the same time to the decreasing of the slopes of the dispersion curves. So, at $L = L_c$ even the lowest branch has such a low slope that it does not give any crossings in the dipole-exchange area of the spectrum. For the YIG films in the X-band this critical film thickness equals approximately 0.2 μm.

As follows from the preceeding discussion, it is impossible to make a universal picture for the dipole-exchange spin-wave spectrum ω_{nQ} versus Q even qualitatively (as it is possible to do for the purely dipole spectrum). The shape of the dipole-exchange spectrum depends critically on the film thickness. Depending on the film thickness there may be only one, several or many crossings (repulsions) between spin-wave branches, especially between the main and higher-order branches. In the case of a *thin* film $L < L_c$ crossings do not exist at all.

Fig. 2.2 illustrates the spectrum of dipole-exchange spin waves in a perpendicularly magnetized thin YIG film of thickness $L = 0.1$ μm with pinned and unpinned surface spins. For the case of PSS we note the exchange shift of the main branch $n = 1$ from the bottom of the spin-wave manifold ω_H.

To compare the exact and the approximate dispersion relations in the case of comparatively thick ferromagnetic film we present in Fig. 2.3 the results of numerical calculations of dipole-exchange spectra for

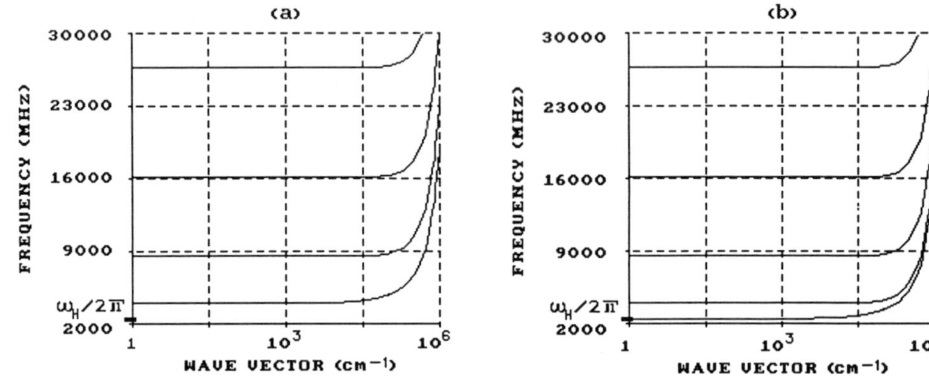

Fig. 2.2. Spin-wave spectra of a film L = 0.1 μm thick with pinned (a) and unpinned (b) surface spins. The parameters used are $\omega_H/2\pi$ = 2.8 GHz, $\omega_M/2\pi$ = 4.9 GHz, $\alpha = 3.1 \times 10^{-12}$ cm^2.

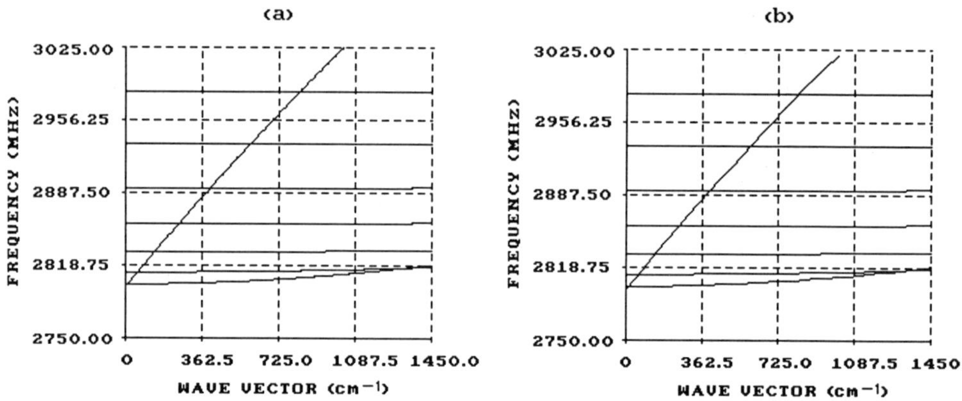

Fig. 2.3. Exact (a) and approximate (b) spin-wave spectra of a film L = 2 μm thick with unpinned surface spins. The parameters used are $\omega_H/2\pi$ = 2.8 GHz, $\omega_M/2\pi$ = 4.9 GHz, $\alpha = 3.1 \times 10^{-12}$ cm^2.

the YIG film of the thickness L = 27 μm having unpinned surface spins. It is seen that in spite of the dipole repulsion the main dispersion branch clearly survives.

The spectra presented in Fig. 2.3 are calculated with the help of the approximate equation (2.94) and the exact dispersion relation in the form of an infinite determinant (see Eqs. (2.57) and (2.58)). To the accuracy of a graphical error the exactly and approximately calculated dispersion branches coincide. We note, that the scale of the Fig. 2.3 a does not allow to see the dipole gaps which appear as a result of the exact calculations of the spectrum near the crossing points of dispersion branches corresponding to the spin waves of the same symmetry.

To elucidate the difference between the cases of unpinned and pinned surface spins, in Fig. 2.4 the exact dipole-exchange spin-wave spectrum is presented calculated for a YIG film of thickness L = 2 μm having pinned surface spins. The strong repulsions of the odd-numbered dispersion branches corresponding to the symmetrical spin waves are obvious. It is evident that the dipole gaps are rather wider in comparison with those in Fig. 2.3.

The distributions of the y-component of variable magnetization are presented in Fig. 2.5. The calculations of $m^y_{Qj}(\xi)$ are made at Q = 250 cm^{-1} for four lower normal spin waves (the corresponding points are marked with letters A, B, C, D in the spectrum of Fig. 2.4). As was already mentioned, the shape of the magnetization of standing spin waves Q = 0 is described accurately by the spin-wave modes $\Phi_n(\xi)$. Near the dipole gaps the distributions of $m^y_{Qj}(\xi)$ are admixtures of different states and deviate from a purely sinusoidal form considerably. Far from the dipole gaps with the increase of in-plane wave vector Q owing to the

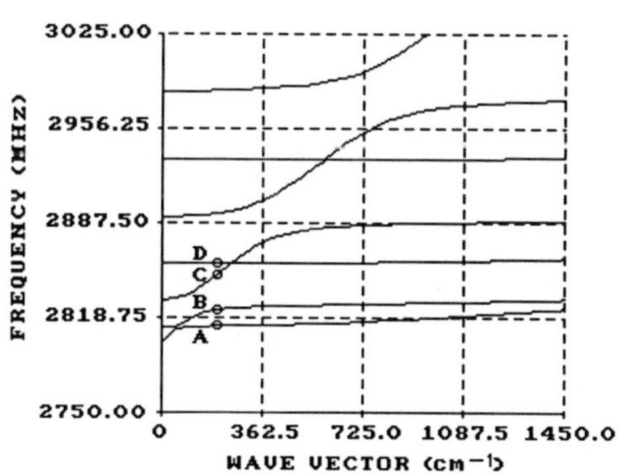

Fig. 2.4. Spin-wave spectrum of a film L = 2 μm thick with pinned surface spins. The parameters used are $\omega_H/2\pi$ = 2.8 GHz, $\omega_M/2\pi$ = 4.9 GHz, $\alpha = 3.1 \times 10^{-12}$ cm^2.

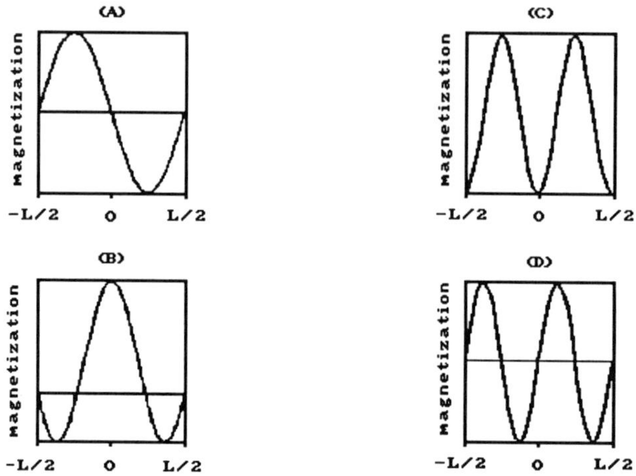

Fig. 2.5. Distributions of the variable magnetization of normal spin waves. The parameters used are the same as in Fig. 2.4.

weakening of dipole-dipole interaction the shapes of the magnetization tend to a sinusoidal (cosinusoidal) form again.

2.9. DIPOLE-EXCHANGE SPECTRUM: PARALLEL MAGNETIZATION

For simplicity we shall describe in this section the spin-wave spectrum for two extreme cases of exchange boundary conditions - PSS and USS. We shall consider the spectrum which is defined by the dispersion equation following from Eqs. (2.86) and (2.87) for the case in question:

$$\omega_{nQ}^2 = (\Omega_{nQ} + \omega_M - \omega_M P_{nn})(\Omega_{nQ} + \omega_M P_{nn} \sin^2\phi) \ . \tag{2.101}$$

From the last equation it is clear that the dispersion properties of propagating $Q \neq 0$ spin waves depend on the direction of their propagation relative to the direction of the static bias field. (We recall that ϕ is the angle between **Q** and $\mathbf{H}_i$). Below we shall consider two usually treated cases of longindinal $\mathbf{Q} \parallel \mathbf{H}_i$ and transverse $\mathbf{Q} \perp \mathbf{H}_i$ propagation. Consequently we shall refer to these cases as *longitudinal spin waves* and *transverse spin waves*.

Similarly to the perpendicular magnetization, the dispersion equation (2.101) demonstrates evidently the contributions of Zeeman, dipole-dipole and exchange interactions to the spin-wave dispersion and we shall not repeat again relevant conclusions here.

The dispersion branches for both longitudial and transverse spin waves begin from the points of the SWR frequencies:

$$\omega_{nQ}^2 = (\omega_H + \omega_M \alpha \kappa_n^2)(\omega_H + \omega_M \alpha \kappa_n^2 + \omega_M) \ . \tag{2.102}$$

These frequencies are situated above the upper boundary of the bulk spin-wave band [see Eq. (1.45)].

It follows from Eq. (2.102) that in the non-exchange approximation $\alpha = 0$ all dispersion branches converge to $\omega = [\omega_H(\omega_H + \omega_M)]^{1/2}$, which is the FMR frequency of the usual uniform mode for tangentially magnetized plate.

In contrast to the case of perpendicular magnetization, in the case under consideration only the longitudinal spin waves have a "pure" type of symmetry $\mathbf{m}_{Qj}(\xi)$: they are either purely symmetric or purely antisymmetric. Calculations show that the form of the wave symmetry is distorted when the direction of spin-wave propagation deviates from the direction of the static field, so we can speak of quasi-symmetric and quasi-antisymmetric spin waves. The most pronounced distortion from the "pure" type of symmetry is exhibited by transverse spin waves, the lowest type of which has a surface-like character of $\mathbf{m}_{Qj}(\xi)$. This "mixed" symmetry is a distinguishing characteristic of spin waves in a tangentially magnetized film. All the waves with a mixed form of symmetry demonstrate *field-displacement nonreciprocity*, i.e., their distributions $\mathbf{m}_{Qj}(\xi)$ are reversed from one film surface to another with the change of the bias field direction to the opposite one.

Let us consider now the spectrum of *longitudinal spin waves* in more detail. We obtain from Eq. (2.101) setting $\phi = 0$:

$$\omega_{nQ}^2 = \Omega_{nQ}(\Omega_{nQ} + \omega_M - \omega_M P_{nn}) \ . \tag{2.103}$$

Numerical calculations show that, in the non-exchange limit $\alpha = 0$ in the case of USS, Eq. (2.103) gives results which coincide very accurately with the results obtained from the non-exchange dispersion relations for symmetric (1.77) and antisymmetric (1.78) modes. In the case of PSS the initial slopes of the dispersion curves differ from those in the case of USS. This may be seen from a comparison of Eqs. (2.88)-(2.89) with Eqs. (2.90)-(2.91).

The dipole-exchange spectrum of longitudinal spin waves is always described by non-crossing dispersion branches. To illustrate the influence of film thickness two spin-wave spectra are given in Fig. 2.6. The broken curves represent the results of calculations of the main dispersion branch $n = 0$ in the non-exchange approximation $\alpha = 0$. A distinctive feature of the dipole-exchange spectrum is the presence of two zones corresponding to spin waves with negative and positive dispersion. It is reflected by the "sag" of the dispersion curves. It is clear from the dispersion relation (2.103) that this effect is caused by competition between dipole (demagnetization) and exchange energies. We note that with decreasing film thickness the minimum value of the spin-wave eigenfrequency at the sag point increases. In contrast to the non-exchange theory (MSW theory) this frequency may lie well above the lower boundary of the spin-wave manifold ω_H.

Let us turn now to the spectrum of *transverse spin waves*. We

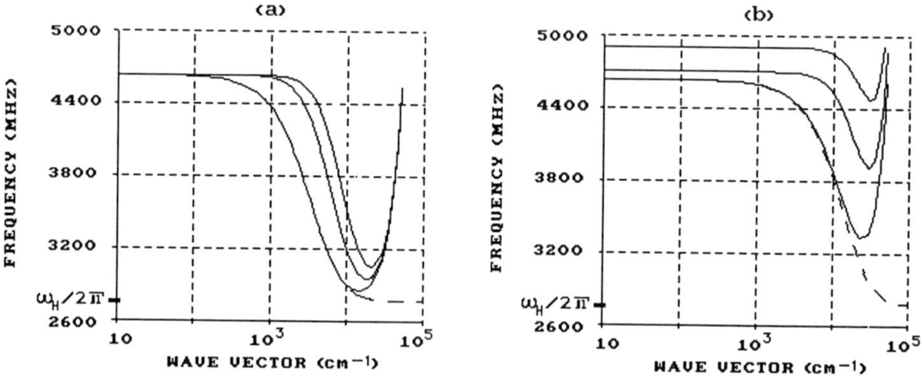

Fig. 2.6. Spin-wave spectra of longitudinal waves in films of thicknesses (a) $L = 4$ μm and (b) $L = 0.5$ μm with pinned surface spins. The parameters used are $\omega_H/2\pi = 2.8$ GHz, $\omega_M/2\pi = 4.9$ GHz, $\alpha = 3.1\times10^{-12}$ cm^2.

obtain from Eq. (2.101) considering $\phi = \pm 90^\circ$:

$$\omega_{nQ}^2 = (\Omega_{nQ} + \omega_M - \omega_M P_{nn})(\Omega_{nQ} + \omega_M P_{nn}) \; . \tag{2.104}$$

An analysis of the last equation shows that the dispersion branches ω_{nQ} may cross each other. Due to the dipole interaction the crossing branches repel each other and dipole "gaps" in the spin-wave spectrum are formed. Unlike the case of perpendicular magnetization the branch repulsions take place between all modes because of their mixed type of symmetry. Similarly to the case of perpendicular magnetization, in comparatively thick films the lowest mode branch ($n = 0$ or $n = 1$ for USS or PSS respectively) gives many repulsions with higher branches which appear in the long-wavelength part of the spin-wave spectrum. In comparatively thin films the repulsions may not exist at all.

Dipole hybridization of the dispersion branches ω_{nQ} [Eq. (2.104)] is described approximately by the secular equation

$$(\omega^2 - \omega_{nQ}^2)(\omega^2 - \omega_{n'Q}^2) = \omega_M^2[(\Omega_{nQ} + \omega_M/2)(\Omega_{n'Q} + \omega_M/2) - \omega^2]2(P_{nn'}^2 + 4Q_{nn'}^2)$$

$$+ \omega_M^4(P_{nn} - 1/2)(P_{n'n'} - 1/2)2(P_{nn'}^2 - 4Q_{nn'}^2) - \omega_M^4(P_{nn'}^2 - 4Q_{nn'}^2)^2 \; , \tag{2.105}$$

which follows from Eq. (2.93). From the properties of the dipole matrix elements $P_{nn'}$ $(n \neq n')$ and $Q_{nn'}$ $(n \neq n')$ it turns out that Eq. (2.105) could be written down as two equations: when the numbers of interacting modes n and n' are simultaneously even or odd, and when n is even (odd) and n' is odd (even). The widths of the corresponding dipole gaps calculated on the basis of Eq. (2.105) with the assumption $\Omega_{nQ} \simeq \Omega_{n'Q} \simeq \omega_H$ are

$$\Delta\omega_{nn'} = (\omega_M^2/\omega_o)\,|P_{nn'}| \tag{2.106}$$

and

$$\Delta\omega_{nn'} = 2(\omega_M^2/\omega_o)(P_{nn} + P_{n'n'})^{1/2}|Q_{nn'}| \; . \tag{2.107}$$

To estimate the values of the gaps in the long-wavelength $QL \ll 1$ part of the spectrum the expressions (2.88)-(2.91) for P_{nn} and the expressions (2.99)-(2.100) for $P_{nn'}$ may be applied. The required expressions for $Q_{nn'}$ are

$$Q_{nn'} \simeq 2nQL/\pi^2(n^2-1) , \qquad n' = 1, \quad n = 2,4,6,... \tag{2.108}$$

for the case of pinned surface spins, and

$$Q_{nn'} \simeq \sqrt{2}\ QL/\pi^2 n^2 , \qquad n' = 0, \quad n = 1,3,5,... \tag{2.109}$$

for the case of unpinned surface spins.

As in the case of perpendicular magnetization it is impossible to make a universal picture for the dipole-exchange spin-wave spectrum ω_{nQ} versus Q. But it is possible to choose spectra for two limiting cases of comparatively thin and thick films. In the case of a thin film there are no repulsions of the dispersion branches; in the case of a thick film there may be a lot of repulsions in the long-wavelength part of the spin-wave spectrum. These cases are illustrated by the next figures. The calculations are fulfilled with the help of the QR algorithm [39] on the basis of the dispersion relation in the form of an infinite determinant.

Fig. 2.7 shows the spectra of transverse spin waves in thin films $L = 0.1$ μm thick having pinned and unpinned surface spins. These spectra are essentially discrete.

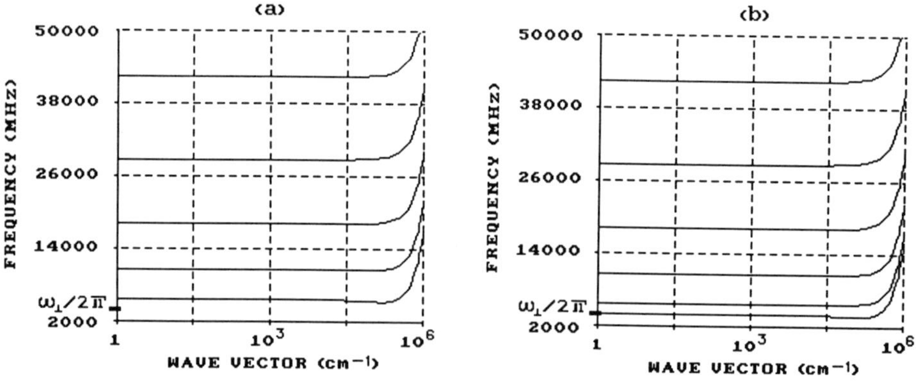

Fig. 2.7. Spin-wave spectra of transverse waves for a film $L = 0.1$ μm thick with pinned (a) and unpinned (b) surface spins. The parameters used are $\omega_H/2\pi = 2.8$ GHz, $\omega_M/2\pi = 4.9$ GHz, $\alpha = 3.1 \times 10^{-12}$ cm^2.

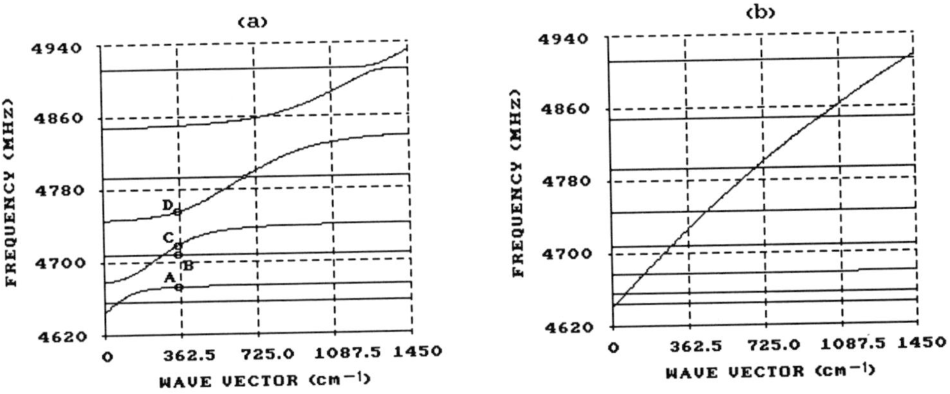

Fig. 2.8. Spin-wave spectra of transverse waves for a film L = 2 μm thick with pinned (a) and unpinned (b) surface spins. The parameters used are $\omega_H/2\pi$ = 2.8 GHz, $\omega_M/2\pi$ = 4.9 GHz, $\alpha = 3.1 \times 10^{-12}$ cm^2.

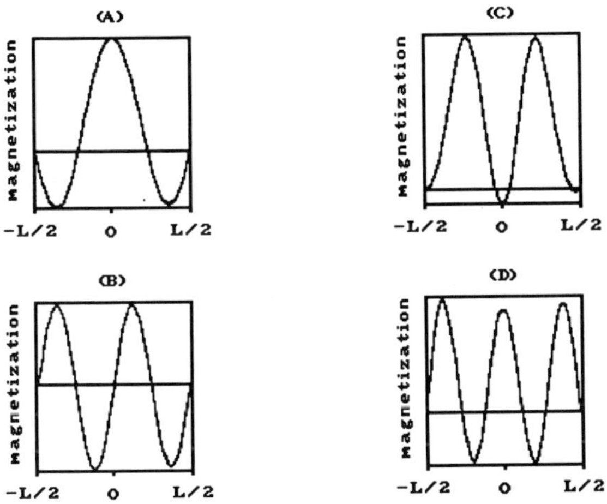

Fig. 2.9. Distributions of the variable magnetization of normal spin waves. The parameters used are the same as in Fig. 2.8.

Fig. 2.8 illustrates the spectra of transverse spin waves in films L = 2 μm thick having pinned and unpinned surface spins. We note the differences in the widths of dipole gaps. Fig. 2.9 presents the distributions of the x-component of variable magnetization $\mathbf{m}_{Qj}(\xi)$ for the eigen-waves. The distributions are calculated at $Q = 350\ \text{cm}^{-1}$ for four lower spin waves for the parameters of Fig. 2.9. (The corresponding points are indicated with the letters A,B,C,D). It is easy to see the surface-like character of the given distributions, especially for the lowest spin waves. In accordance with physical expectations, the numerical calculations show that in the short-wavelength part of the spin-wave spectrum $QL \gg 1$ the distributions $\mathbf{m}_{Qj}(\xi)$ lose their surface features because of the decreasing effectof the dipole contributions with the increase of in-plane wave vector $\mathbf{Q}$.

We have described the main features of spin waves in tangentially magnetized films for two typical cases of longitudinal $\mathbf{Q} \parallel \mathbf{H}_i$ and transverse $\mathbf{Q} \perp \mathbf{H}_i$ propagation. It could be computed, with the help of the above given dispersion relations, that the full dipole-exchange spectrum is described by a family of continuous dispersion sheets $\omega_{nQ}(\phi)$. This will be discussed in the next section taking the general example of an anisotropic ferromagnetic film.

2.10. EFFECTS OF MAGNETOCRYSTALLINE ANISOTROPY

So far in this chapter, only isotropic ferromagnetic films have been considered. However, the effects associated with magnetocrystalline anisotropy cannot be completely neglected even in the films having very weak anisotropy. In this Section we give a description of the influence of magnetocrystalline anisotropy on the dispersion properties of spin waves in films. This is not intended to be a full account of such a vast subject. Our main purpose is to give dispersion relations with the help of which the dispersion characteristics of dipole-exchange spin waves in anisotropic films could be found.

2.10.1 Formulation of the problem

The existing theory of spin-wave spectra in films considers the influence of crystallographic anisotropy on spin-wave dispersion mainly in the non-exchange limit (see, e.g., [43-52]) although papers where crystallographic anisotropy was taken into account in the dipole-exchange spectrum have been published also [20,29,53-57]. To derive dispersion relations the PW approach was utilized in a good many of the published papers [20,43-54]. Here we shall describe the role of the anisotropy on the basis of the SWM approach [29,55-57]. An arbitrary orientation of the external bias magnetic field with respect to the orientation of crystallographic axes and the film normal will be assumed.

The geometry of the problem is given in Fig. 2.10. It is similar to the case of an isotropic ferromagnetic film (see Fig. 1.5) but we note that now, because of the presence of magnetic anisotropy, in order to define the directions of the static external field $\mathbf{H}_0$ and the saturation magnetization $\mathbf{M}_0$ two different angles φ and φ_0 are needed. The film is magnetized to saturation by a bias external field $\mathbf{H}_0$ of an arbitrary direction which is determined by the angles θ_0 and φ_0. The direction of the static internal magnetic field $\mathbf{H}_i$ coincides with the direction of the saturation magnetization $\mathbf{M}_0 = \hat{\mathbf{z}}M_0$. This direction is determined by the angles θ and φ. In addition, to characterize the position of vector $\mathbf{M}_0$, the angle $\varphi_M = \varphi + \varphi_Q$ (the angle between the projection of $\mathbf{M}_0$ on the film plane and the axis $\hat{\mathbf{a}}$ lying in the film

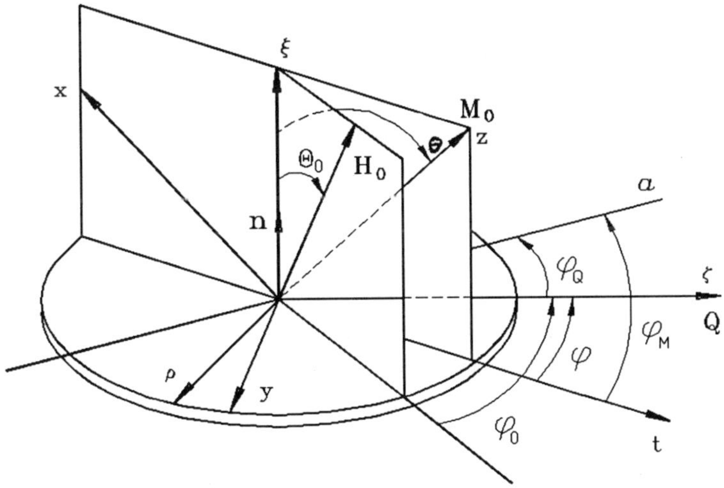

Fig. 2.10. The geometry and orientation of the coordinate axes.

plane) is introduced. The angles characterizing the *equilibrium orientation* of the saturation magnetization $\mathbf{M}_o$ are determined from the minimization procedure for the density of magnetic energy of the given anisotropic ferromagnetic film (for given angles θ_o, φ_o and φ_Q) [29,58].

The total effective magnetic field in the Landau-Lifshitz equation of motion for magnetization (2.1) in the case under consideration can be written in the form:

$$\mathbf{H}_{eff}(\mathbf{r},t) = \mathbf{H}_i + \mathbf{h}_d(\mathbf{r},t) + \mathbf{h}_{ex}(\mathbf{r},t) + \mathbf{h}^a(\mathbf{r},t) , \tag{2.110}$$

where $\mathbf{H}_i = \mathbf{H}_o + \mathbf{H}_{do} + \mathbf{H}_o^a$ is a static internal magnetic field consisting of an external bias field $\mathbf{H}_o$, a constant dipole (or demagnetization) field $\mathbf{H}_{do}$, and a constant field of the crystallographic anisotropy $\mathbf{H}_o^a$. All the other components of the effective magnetic field (2.110) are variables: $\mathbf{h}_d(\mathbf{r},t)$ is a variable dipole field, $\mathbf{h}_{ex}(\mathbf{r},t) = 4\pi\alpha\nabla^2\mathbf{m}(\mathbf{r},t)$ is a variable exchange field, and $\mathbf{h}^a(\mathbf{r},t)$ is a variable field of crystallographic anisotropy.

The effective fields of crystallographic anisotropy can be calculated if the expressions for the densities of anisotropy energy are known. For linear problems, when the condition $\mathbf{m}(\mathbf{r},t) \ll \mathbf{M}_o$ applies, there is, however, an easier way to calculate the effective magnetic fields associated with the magnetocrystalline anisotropy - namely the method of *effective demagnetization factors of anisotropy*. In the framework of this method the effective magnetic field of anisotropy $\mathbf{H}^a(\mathbf{r},t)$ is represented in the form

$$\mathbf{H}^a(\mathbf{r},t) = \mathbf{H}_o^a + \mathbf{h}^a(\mathbf{r},t) = - \hat{\mathbf{N}}^a\mathbf{M}(\mathbf{r},t) = - \hat{\mathbf{N}}^a\mathbf{M}_o - \hat{\mathbf{N}}^a\mathbf{m}(\mathbf{r},t) \tag{2.111}$$

where $\hat{\mathbf{N}}^a$ is the tensor of effective demagnetization factors of magnetocrystalline anisotropy. The components of the tensor $\hat{\mathbf{N}}^a$ can be demagnetization factors of any type of anisotropy or the sum of demagnetization factors of different types of anisotropy taken into account simultaneously.

For example, in the case of a cubic monocrystalline film having induced uniaxial anisotropy, the components of the tensor $\hat{\mathbf{N}}^a$ can be represented as a sum of demagnetization factors of cubic anisotropy N^A and uniaxial anisotropy N^u.

We underline that the calculation of the components of the tensor $\hat{\mathbf{N}}^a$ for each type of anisotropy may be considered in this context as a separate problem. The solution of such a problem for the case of a ferromagnetic ellipsoid having cubic and uniaxial anisotropy and the arbitrary direction of the crystallographic axes with respect to its geometrical axes is presented in the papers [59,60]. We shall use this solution below to determine the effective magnetic fields for different types of crystallographic anisotropy in ferromagnetic films. It is worth mentioning that an advantage of the method of effective demagnetization factors of anisotropy is that the effect of one or several types of anisotropy in the ferromagnet can be taken into account by substituting the appropriate expressions for the components of the tensor $\hat{\mathbf{N}}^a$ into the final dispersion equations.

Now, with the effective magnetic field (2.110) being substituted into the equation of motion, we apply the SWM approach to solve the boundary value problem for an anisotropic film. Following this approach, as was described for the case of an isotropic film in Secs. 2.3-2.5, for the case of anisotropic film with mixed exchange boundary conditions (2.37) we obtain the infinite system of algebraic equations for SWM vector amplitudes (2.49) in the form [29]:

$$\hat{\mathbf{D}}_{nn}\mathbf{m}_n + \sum_{n'\neq n} \hat{\mathbf{R}}_{nn'}\mathbf{m}_{n'} = 0 \tag{2.112}$$

where matrices $\hat{\mathbf{D}}$ and $\hat{\mathbf{R}}$ are determined by the following expressions:

$$\hat{\mathbf{D}}_{nn} = \begin{bmatrix} F_n^{\mathrm{x}} + \sin^2\theta + AP_{nn}^{\mathrm{xx}} + N_{\mathrm{xx}}^{\mathrm{a}} & CP_{nn}^{\mathrm{xy}} + N_{\mathrm{xy}}^{\mathrm{a}}T_{nn}^{\mathrm{xy}} - i\left(\dfrac{\omega}{\omega_{\mathrm{M}}}T_{nn}^{\mathrm{xy}} - DQ_{nn}^{\mathrm{xy}}\right) \\ CP_{nn}^{\mathrm{yx}} + N_{\mathrm{xy}}^{\mathrm{a}}T_{nn}^{\mathrm{yx}} + i\left(\dfrac{\omega}{\omega_{\mathrm{M}}}T_{nn}^{\mathrm{yx}} + DQ_{nn}^{\mathrm{yx}}\right) & F_n^{\mathrm{y}} + EP_{nn}^{\mathrm{yy}} + N_{nn}^{\mathrm{a}} \end{bmatrix} \tag{2.113}$$

$$\hat{\mathbf{R}}_{nn'} = \begin{bmatrix} AP^{xx}_{nn'} + iBQ^{xx}_{nn'} & CP^{xy}_{nn'} + N^{a}_{xy}T^{xy}_{nn'} - i\left(\frac{\omega}{\omega_M}T^{xy}_{nn'} - DQ^{xy}_{nn}\right) \\ CP^{yx}_{nn'} + N^{a}_{xy}T^{yx}_{nn'} + i\left(\frac{\omega}{\omega_M}T^{yx}_{nn'} + DQ^{yx}_{nn}\right) & EP^{yy}_{nn'} \end{bmatrix} \tag{2.114}$$

The expressions for matrix elements $P^{pp'}_{nn'}$, $Q^{pp'}_{nn'}$, $T^{pp'}_{nn'}$ and for the angle functions A, B, C, D, E are given by Eqs. (2.53)-(2.56). The expressions for the components of $\hat{\mathbf{N}}^{a}$ are given in Sec. 2.10.3 for some particular cases of interest.

In contrast to the case of an isotropic film, the static internal magnetic field H_i (which determines the frequency $\omega_H = \gamma H_i$) is determined as

$$H_i = H_{oz} + H_{doz} - N^{a}_{zz}M_o \tag{2.115}$$

where H_{oz} is the Z-projection of the external bias magnetic field $\mathbf{H}_o$, and H_{doz} is the Z-projection of the constant demagnetization field $\mathbf{H}_{do}$.

It is easy to see that if anisotropy is neglected, Eqs. (2.112)-(2.114) coincide completely with Eqs. (2.50)-(2.52) valid for an isotropic film.

The infinite system (2.112) gives the exact description of the linear spin-wave processes in anisotropic ferromagnetic films and enables us to obtain the expressions for the spin-wave spectrum and for the distributions of the variable magnetization of eigenwaves across the film thickness.

2.10.2. Spin-wave spectrum

In a similar manner to the isotropic case treated above, by solving Eq. (2.112) the exact dispersion relations describing both dipolar and dipole-exchange spectra may be written in different forms, namely in the form of an infinite determinant, in the form of an infinite series, and in the form of trigonometric equations. Besides the exact dispersion relations describing the spin-wave spectra ω_{nQ} in implicit form, with the help of pertubation theory methods the approximate dispersion equations describing spin-wave spectra in an explicit form may be obtained

[29,56,57]. Below we shall write down only approximate dispersion equations bearing in mind that one of the main exact dispersion relations (in the form of vanishing infinite determinant) is actually given by the system (2.112).

The numerical analysis of the exact dispersion relations shows that it is not possible to draw a universal picture of the spectrum for dipole-exchange waves in an anisotropic film. But it is possible to depict the main features of the dipole-exchange spin-wave spectrum. Thus the dipole-exchange spin-wave spectrum of an anisotropic ferromagnetic film in the coordinates (ω, Q, φ) consists of the non-crossing surfaces $\omega_{nQ}(\varphi)$ with "centres" corresponding to the frequencies of the spin-wave resonances ω_{nQ} at $Q = 0$. In the cases of pinned and unpinned surface spins the distances between the SWR frequencies do not depend on magnetocrystalline anisotropy. The SWR frequencies are easily calculated for arbitrary direction of static magnetization, each frequency corresponding to its own dispersion surface. Therefore it is convenient to number the dispersion surfaces using the numbers of the corresponding modes of spin-wave resonance. The qualitative example of the dipole-exchange spectrum of a tangentially magnetized film is sketched in Fig. 2.11. While moving along the dispersion surface, either along the radius or along the azimuth we find that spin-wave dispersion and transverse distribution of the variable magnetization change considerably. We emphasize that the curvature of the dispersion surface and the form of the dispersion curve corresponding to the given direction of spin-wave propagation are determined by competition between the dipole-dipole interaction and the inhomogeneous exchange interaction and by the magnetocrystalline anisotropy. We emphasize also that, in comparison with the case of the dipole-exchange spectrum of an isotropic film, the occurrence of magnetic anisotropy does not change the number of dispersion surfaces, i.e., magnetocrystalline anisotropy does not produce new dispersion surfaces for dipole-exchange spin waves in anisotropic film (as happens in the case of purely dipole waves [49-52]).

When the tensor of effective demagnetization factor of anisotropy $\hat{\mathbf{N}}^a$ has such a form that

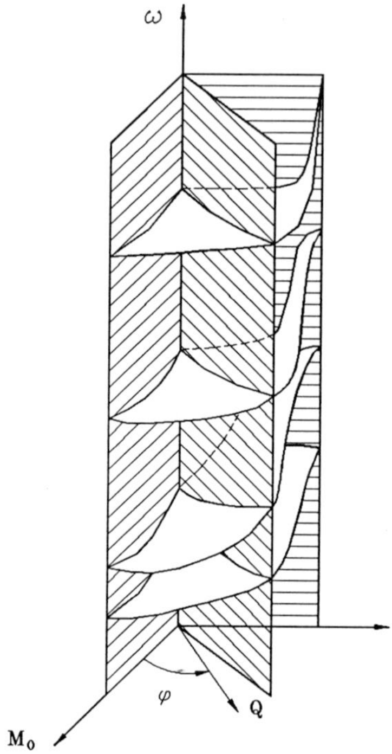

Fig. 2.11. Sketch of the dipole-exchange dispersion surfaces for a tangentially magnetized film.

$$N^{a}_{xx} = N^{a}_{yy} , \qquad N^{a}_{xy} = 0 , \tag{2.116}$$

the only effect of the magnetocrystalline anisotropy is a shift of the spin-wave spectrum. This shift may be accounted for a change of static internal magnetic field in the "isotropic" dispersion equation $H_i \longrightarrow \tilde{H}_i$ where

$$\tilde{H}_i = H_i + N^{a}_{xx} M_o . \tag{2.117}$$

Here H_i is determined by Eq. (2.115). Physically the conditions in Eq. (2.116) mean that anisotropy does not change the ellipticity of polarization of the propagating spin waves in the plane perpendicular to the direction of equilibrium magnetization $\mathbf{M}_o$. This conclusion is in accordance with the analogous conclusion for infinite magnetic media.

Now we proceed to consider the particular cases of *perpendicular* and *parallel* magnetization briefly. In the case of perpendicular magnetization the conditions (2.116) are fulfilled only for the (111) and (100) oriented films having cubic and/or normal uniaxial anisotropy. In the other cases the anisotropy leads to a dependence of the spin-wave eigenfrequencies on the direction of spin-wave propagation relative to the crystallographic axes in the film plane.

In the case of parallel magnetization, if the conditions (2.116) are not fulfilled, the influence of magnetic anisotropy on the spin-wave dispersion may be rather complicated. (We say "parallel magnetization" when the vector of the static magnetization $\mathbf{M}_o$ is parallel to the film plane). As an illustration, we shall consider transverse waves, φ = +90°. We start with the case of purely dipole waves (magnetostatic waves). An analysis of the exact dispersion relation derived from Eq. (2.112) for MSW (in the limit $\alpha = 0$) shows that if one (or both) of the conditions (2.116) are not satisfied anisotropy causes the appearance of a new set of dispersion branches of volume waves, in addition to the "*old*" branch of magnetostatic surface wave [the Damon-Eshbach wave of Eq. (1.79)] which is modified by anisotropy.

In the case when $N^a_{xx} \neq N^a_{yy}$ and $N^a_{xy} = 0$, the sign of the group velocity V_g of these volume waves coincides with the sign of the difference $(N^a_{xx} - N^a_{yy})$. We note that these new "anisotropic" magnetostatic volume waves may have both positive ($V_g > 0$) and negative ($V_g < 0$) dispersion. For example, in the (111) oriented ferromagnetic film with negative constant of cubic anisotropy ($K^A_1 < 0$) magnetized along the axis <121>, the "anisotropic" volume waves have positive dispersion, since we have $N^a_{xx} - N^a_{yy} = -H^A/M_o$ where $H^A = K^A_1/M_o$, $K^A_1 < 0$ [50]. In the ferromagnetic film with perpendicular uniaxial anisotropy of easy-axis type, the "anisotropic" volume waves have negative dispersion, since we have $N^a_{xx} - N^a_{yy} = -2H^u/M_o$ where $H^u = K^u_1/M_o$, $K^u_1 > 0$ [50].

In the case when $N^a_{xx} \neq N^a_{yy}$ and $N^a_{xy} \neq 0$, the anisotropy leads to the occurence of two new families of dispersion branches corresponding to positive and negative dispersion. For example, such a situation occurs in the in-plane magnetized film of orientation (111) when the direction of $\mathbf{M}_0$ does not coincide with the axis <121> [43,50,52].

Besides it is worth mentioning that two films of different orientation, (100) and (111), magnetized in-plane along one and the same axis <110>, have different spin-wave spectra. In the first case, in the dipolar spectrum the anisotropy causes magnetostatic volume waves having negative dispersion (when $K_1 < 0$) [54]; for the second case, the anisotropy causes magnetostatic volume waves having both negative and positive dispersion [43,50,52].

To clear up the roles of the magnetic anisotropy and exchange interaction in the process of forming the spin-wave spectrum, we consider this process gradually using the example of the transverse waves in a ferromagnetic film having uniaxial anisotropy $H^{\mathrm{u}} > 0$. The pictures in Fig. 2.12 illustrate the successive transformation of the spin-wave spectrum from the purely dipolar case for an isotropic film (picture (a)) to the dipole-exchange case for an anisotropic film (picture(d)). Obviously, it is possible to come to the resulting picture (d) of anisotropic-dipole-exchange spectrum step-by-step through the different ways: (a) $\longrightarrow$ (b) $\longrightarrow$ (d), (a) $\longrightarrow$ (c) $\longrightarrow$ (d), or directly (a) $\longrightarrow$ (d). At every step the influence of magnetic anisotropy and exchange interaction is clear.

The above given qualitative illustration of anisotropic effects in the formation of the spin-wave spectrum elucidates the general role of magnetic anisotropy (although it is based on the particular example of the transverse waves in the film with uniaxial anisotropy). As was already mentioned, the inclusion of magnetic anisotropy in the dipole-exchange spin-wave spectrum does not change the number of its dispersion surfaces but leads to the frequency shift of dispersion branches and to the modification of the dispersion character $\omega_{nQ}(\varphi)$.

The numerical calculations of the spin-wave spectra for the anisotropic films were done using the QR algorithm. The accuracy of the numerical calculations was determined by the order of reduction of the system (2.112) (i.e., by the dimension of the finite determinant which

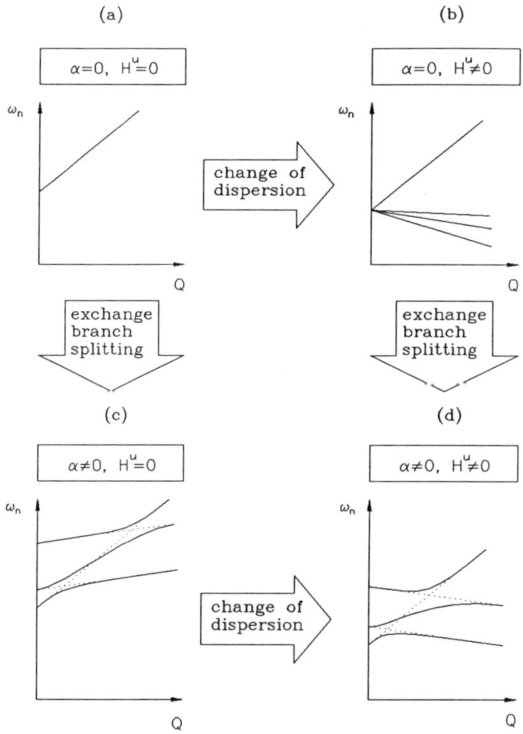

Fig. 2.12. Schematic illustration of how the inclusion of magnetic anisotropy modifies spin-wave spectrum.

was used in calculations instead of the full infinite determinant of the system (2.112)) and was not worse than 0.01%.

As an example of such calculations, Figs. 2.13 and 2.14 show spin-wave spectra of in-plane magnetized YIG films of thickness L = 2 μm with uniaxial anisotropy and pinned surface spins computed for purely dipole and dipole-exchange transverse waves respectively. From a comparison of these two spectra the influence of the exchange effects is evident. In comparison with the isotropic case (see Fig. 2.8(a)) the uniaxial anisotropy causes the frequency shift and change of group velocity of dipole-exchange spin waves, especially of the lowest branches.

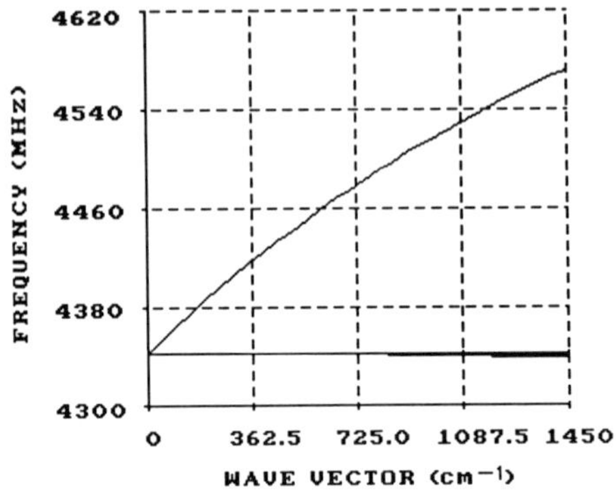

Fig. 2.13. Spin-wave spectrum of transverse non-exchange waves for a tangentially magnetized film $L = 2$ μm thick having uniaxial anisotropy $H^u = 170$ Oe. The parameters used are $\omega_H/2\pi = 2.8$ GHz, $\omega_M/2\pi = 4.9$ GHz.

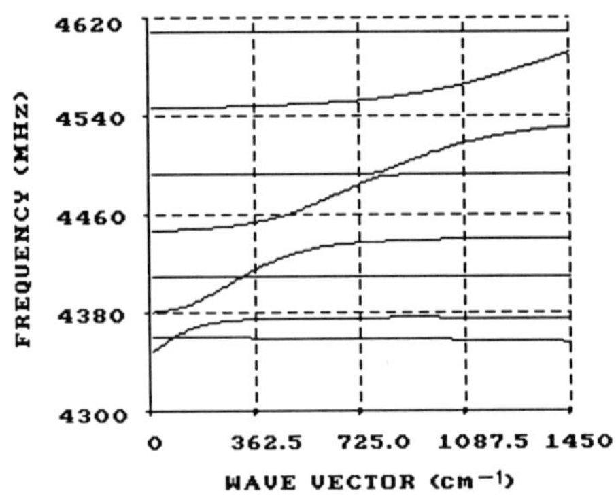

Fig. 2.14. Spin-wave spectrum of transverse dipole-exchange waves for a tangentially magnetized film $L = 2$ μm thick having uniaxial anisotropy $H^u = 170$ Oe and pinned surface spins. The parameters used are $\omega_H/2\pi = 2.8$ GHz, $\omega_M/2\pi = 4.9$ GHz, $\alpha = 3.1 \times 10^{-12}$ cm^2.

2.10.3. Approximate dispersion equations

As in the case of an isotropic film, to find approximate solutions of the system (2.112) we use classical pertubation theory [29]. In the first-order of the pertubation theory, when frequency degeneracy of different spin-wave modes is not taken into account, we obtain the approximate dispersion equation for dipole-exchange spin waves in an anisotropic film:

$$(\omega_{nQ}T^{yx}_{nn}/\omega_{M} + DQ^{yx}_{nn})^2 = (F^{x}_{n} + \sin^2\theta + AP^{xx}_{nn} + N^{a}_{xx})(F^{y}_{n} + EP^{yy}_{nn} + N^{a}_{yy})$$
$$- (CP^{yx}_{nn} + N^{a}_{yx}T^{yx}_{nn})^2. \qquad (2.118)$$

This equation is valid for arbitrary direction of the bias magnetic field and for mixed exchange boundary conditions (2.37). We note that in the isotropic limit Eq. (2.118) reduces to the dispersion equation for spin waves in an isotropic film (2.84).

In practice, as a rule, it is interesting to have the approximate dispersion equations for the particular cases of definite orientations of crystallographic axes and bias magnetic field in the anisotropic ferromagnetic film. Let us consider the dipole-exchange spin-wave spectrum in anisotropic films for the cases of perpendicular and parallel magnetization and for the limiting cases of exchange boundary conditions: pinned surface spins and unpinned surface spins.

In the case of perpendicular magnetization of the anisotropic film having PSS or USS the approximate dispersion equation has the form

$$\omega^2_{nQ} = (\Omega_{nQ} + \omega_{M}N^{a}_{xx} + \omega_{M}P_{nn}\cos^2\varphi)(\Omega_{nQ} + \omega_{M}N^{a}_{yy} + \omega_{M}P_{nn}\sin^2\varphi)$$
$$- \omega^2_{M}(P_{nn}\sin\varphi\,\cos\varphi - N^{a}_{xy})^2. \qquad (2.119)$$

As is clear from Eq. (2.119), anisotropy in a perpendicularly magnetized film can lead to the dependence of the spin-wave eigenfrequency on the angle φ, i.e., on the direction of wave propagation in the film plane. This effect takes place if the conditions (2.116) are not fulfilled.

For the particular cases of crystallographic orientations (111), (100), and (110), the demagnetization factors of anisotropy have the following values [58]:

(i) orientation (111):

$$N^{a}_{xx} = N^{a}_{yy} = -2H^{A}/M_{o}, \qquad N^{a}_{zz} = -2H^{A}/3M_{o}, \qquad N^{a}_{xy} = 0\,; \qquad (2.120)$$

(ii) orientation (100):

$$N^a_{xx} = N^a_{yy} = N^a_{xy} = 0 , \qquad N^a_{zz} = -2H^A/M_o ; \tag{2.121}$$

(iii) orientation (110):

$$N^a_{xx} = -3H^A(1 - \cos 2\varphi_M)/2M_o , \qquad N^a_{yy} = -3H^A(1 + \cos 2\varphi_M)/2M_o ,$$

$$N^a_{zz} = -H^A/M_o , \qquad N^a_{xy} = -3H^A \sin 2\varphi_M/2M_o \tag{2.122}$$

where φ_M is the angle between the axis t (see Fig. 2.10) and the crystallographic axis <100> in the film plane.

It follows from Eqs. (2.119)-(2.122) that the angular dependence of the spin-wave eigenfrequency exists only in the case of the orientation (110). In films with the orientations (111) and (100) the influence of the cubic anisotropy on the spectrum is equivalent to the change in magnitude of the internal magnetic field $H_i \longrightarrow \tilde{H}_i$, and leads to the frequency shift of the dispersion curves. The spin-wave dispersion equation for these orientations has the form of the dispersion equation (2.94) in an isotropic film where $\omega_H = \gamma\tilde{H}_i$ and $\tilde{H}_i$ has the following values: $\tilde{H}_i = H_{oz} - M_o - 4H^A/3$ for the orientation (111), and $\tilde{H}_i = H_{oz} - M_o + 2H^A$ for the orientation (100).

The influence of uniaxial normal anisotropy on the spin-wave spectrum in the perpendicularly magnetized film is also equivalent to the change in the internal magnetic field to the value $2H^u$. So the spin-wave spectrum in this case is also described by Eq. (2.94) where $\omega_H = \gamma\tilde{H}_i$ and $\tilde{H}_i = H_{oz} - M_o + 2H^u$.

We now proceed to the case of parallel magnetization of the anisotropic film having PSS or USS. In this case Eq. (2.118) can be written in the form

$$\omega^2_{nQ} = (\Omega_{nQ} + \omega_M + \omega_M N^a_{xx} - \omega_M P_{nn})(\Omega_{nQ} + \omega_M N^a_{yy} + \omega_M P_{nn}\sin^2\varphi) - (\omega_M N^a_{xy})^2. \tag{2.123}$$

Let us consider some particular cases of the film crystallographic orientation. In the case of a tangentially magnetized cubic ferromagnetic film, the demagnetization factors of anisotropy have the following values [58]:

(i) orientation (111):

$$N^{a}_{xx} = -2H^{A}/M_{o}\ ,\quad N^{a}_{yy} = N^{a}_{zz} = -H^{A}/M_{o}\ ,\quad N^{a}_{xy} = \sqrt{2}\ H^{A}\sin 3\varphi_{M}/M_{o}\ ; \qquad (2.124)$$

(ii) orientation (100):

$$N^{a}_{xx} = N^{a}_{xy} = 0\ ,\quad N^{a}_{zz} = -H^{A}(\cos 4\varphi_{M} + 3)/2M_{o}\ ,$$

$$N^{a}_{yy} = 3H^{A}(\cos 4\varphi_{M} - 1)/2M_{o}\ ; \qquad (2.125)$$

(iii) orientation (110):

$$N^{a}_{xx} = -3H^{A}\sin^{2}\varphi_{M}/M_{o}\ ,\quad N^{a}_{yy} = -9H^{A}\sin^{2}2\varphi_{M}/4M_{o}\ ,$$

$$N^{a}_{zz} = -H^{A}(\sin^{4}\varphi_{M} + 2\cos^{4}\varphi_{M})/M_{o}\ ,\quad N^{a}_{xy} = 0\ . \qquad (2.126)$$

Here for the case (i) the angle φ_M is measured from the axis <121>; for the cases (ii) and (iii) the angle φ_M is measured from the axis <100>.

In the case of perpendicular uniaxial anisotropy in tangentially magnetized film the demagnetization factors of anisotropy are:

$$N^{a}_{xx} = -2H^{u}/M_{o}\ ,\qquad N^{a}_{yy} = N^{a}_{zz} = N^{a}_{xy} = 0\ . \qquad (2.127)$$

In many experimental situations the group velocity of spin waves is the parameter of interest. The explicit form of the dispersion laws ω_{nQ} [Eqs. (2.118),(2.119) and (2.123)] easily permits one to find a relation for the group velocities of spin waves $V_{gn} = \partial\omega_{nQ}/\partial Q$, describing explicitly the dependence of V_{gn} on all the factors.

Finally we refer briefly to the calculations of the dipole-dipole hybridization in the spin-wave spectrum where the anisotropy effects are treated explicitly. Such calculations are given in the paper [29]. The explicit equations describing the hybridization of different spin-wave branches, and also the "anisotropic" dependence of dipole "gaps" in the dipole-exchange spectrum for perpendicularly and tangentially magnetized ferromagnetic films are found.

2.11. COMPARISON BETWEEN THEORY AND EXPERIMENT

This section is concerned with the experimental investigation of the dipole-exchange spectrum and also some dispersion properties of spin waves (such as their group velocity) in ferromagnetic films. Two experimental techniques are mainly used to measure parameters of spin waves in films: light scattering spectroscopy and microwave spectroscopy. The light scattering experimental results are reviewed in detail in Chapter 6, and here we shall describe some experimental results obtained with the help of *microwave spectroscopy*.

To study the spin-wave properties of films, linear and nonlinear (parametric) excitation of spin waves could be used. This means that two kinds of magnetic techniques in microwave spectroscopy, namely *linear* and *nonlinear* techniques can be distinguished.

For ferromagnets linear microwave spectroscopy is related to FMR spectroscopy. It is well-known that with the help of traditional FMR spectrometers a spectrum of magnetic *resonance oscillations* or, in other words, a spectrum of *standing waves* (modes) of an experimental sample can be measured. It is worth pointing out that within the last two decades, in addition to the traditional FMR and SWR techniques, novel microwave magnetic techniques aimed at studying films have been developed. These techniques are based on the microwave excitation (and reception) of *propagating* spin waves. These techniques form a new trend in microwave magnetic spectroscopy which we call *propagationg spin wave* (PSW) *spectroscopy*. We shall return to PSW spectroscopy below.

When FMR spectroscopy is used for film measurements, in order to make a comparison between experimental and theoretical results, strictly speaking we need a rigorous theory describing magnetic oscillations in a *finite* film sample. In other words, there is a need to have an exact theory for the magnetic resonance frequencies of a film sample of the realistic form used in the experiments. (The relevant example is Walker's theory for the spectrum of magnetostatic modes in a ferromagnetic spheroid [61]). In practice, however, an approximate theory is widely used. The usually used approximation is justified by the fact that the in-plane dimensions of a film sample are typically large compared with its thickness.

To obtain an equation for the resonance frequencies of a finite film sample, it is straightforward to modify the above derived dispersion relations (especially the approximate ones) for dipole-exchange spin waves. For example, let us consider a rectangular film sample with in-plane dimensions *a* and *b*. In such a sample the spin waves cannot propagate in any direction. This leads to the geometric *quantization* of the in-plane wavevector **Q** such that

$$Q^2 = (n_a\pi/a)^2 + (n_b\pi/b)^2 \tag{2.128}$$

where n_a, n_b = 1,2,3,... are the mode numbers. In accordance with Eq. (2.128), the value of $\sin^2\varphi$ which is a part of the dispersion equations (e.g., see Eq. (2.101)) is also quantized. Thus with the help of the substitution (2.128) in the above given dispersion relations, we obtain the spectrum of resonance frequencies of dipole-exchange spin-wave oscillations. These frequencies are characterized by three numbers: *n*, n_a, and n_b, each one corresponding to the certain distribution of variable magnetizationalong the given dimension *L*, *a*, and *b*.

The approach of a geometrical quantization like Eq. (2.128) was discussed in many papers (see, e.g., [22,62-66]) and was widely applied to interpret experiments, especially with *magnetostatic modes* in YIG samples. In many cases a very good coincidence between theoretical and experimental results was shown [63-66].

In [67] the *dipole-exchange spectrum* was studied by the FMR technique using perpendicularly magnetized finite circular thin YIG films - disks of radius *r*. For comparatively small wave vectors *Q* (experimental values with $Q < 10^4$ cm^{-1}), which were quantized as the roots of the zero-order Bessel function $J_o(Qr) = 0$, good agreement was observed between theory and experiment for both pinned and unpinned surface spins.

Now we turn to the consideration of the PSW spectroscopy data. First of all we describe the technique in more detail. The basic part of the PSW spectrometer is an arrangement which serves to excite and to receive *propagating* spin waves. Experimental information is being obtained from the reflected and/or transmitted microwave signals (pulsed or continuous) and their frequency dependences. In these arrangements in the capacity of launching and receiving antennas, microstrip transducers are used very often (see Fig. 9.3). With respect to the spin-wave wave vectors, these transducers may be wide-bandwidth or selective. For example, single microstrip transducers have broad-band wave-vector characteristics [68]; meander-line or grating microstrip transducers have selective wave-vector characteristics [69]. Moving microstrip antennas or induction probes (see, e.g., [70,71]) enable one to make a *direct* measurement of wavelengths of spin waves propagating in the experimental film sample and give the dispersion relations ω_{nQ}. The phase velocity of spin waves (and hence the spin-wave dispersion) is also obtainable from the interference patterns which occur when the delayed signal (transmitted through a ferromagnetic film between the input and output antennas) is superimposed on an undelayed signal [70,72]. The group velocity of the spin waves is calculated from the pulse delay time measured as a function of the distance l between the antennas (see, e.g., [70,73]).

There is a vast literature devoted to the experimental measurements of the spectrum and group velocity of spin waves (magnetostatic waves) in films. Below we consider the experiments which were carried out mainly with monocrystalline YIG films by PSW technique. The published papers on the subject may be divided into two groups. In the papers of the first group smooth (monotonic) dependences of ω_{nQ} versus Q and V_g versus ω were observed (e.g., see [54,72-77]). In the papers of the second group non-monotonic dependences were registered (e.g., see [1-5,70,78,79]). The obtained experimental results of these two groups seem to contradict each other, especially if we note that the experiments were carried out with the films having approximately the same thicknesses andsaturation magnetizations.

These contradictions can be removed if we suppose that, in the experiments of the first and second groups, film samples having *different* surface spin pinning conditions were used. (We underline that the authors of the papers of the first group *did not control* pinning conditions and explained their experiments in the framework of non-exchange theory). In particular, the smooth dependences of ω_{nQ} and $V_g(\omega)$ can be explained in the framework of dipole-exchange theory when we suppose that the experimental samples have unpinned surface spins. As regards the second group of experiments exibiting non-monotonic dependences of ω_{nQ} and $V_g(\omega)$ [1-5,70,78,79], the authors pointed out that film samples with pinned surface spins were chosen for the experiments.

Some results typical of the first group are presented in Fig. 2.15. It shows the dependences $\tau(\omega) = 1/V_g(\omega)$ obtained using monocrystalline YIG films with unpinned surface spins for three main experimental configurations: (a) perpendicular magnetization; (b) parallel magnetization, $\varphi = 0$; (c) parallel magnetization, $\varphi = 90^\circ$. Solid lines represent the theoretical curves for the main branch $n = 0$ calculated onthe basis of the dispersion Eqs. (2.94) and (2.101). As was already noted in the previous sections, the same theoretical curves are obtained from the corresponding non-exchange trigonometric dispersion relations. The good coincidence between theory and experiment is not unusual: it was found practically in all papers of the first group. When the magnetic anisotropy may be neglected, the value of the saturation magnetization is commonly used as a fitting parameter.

For the results of the second group the gaps in the dependences of ω_{nQ} and the non-monotonic behavior (oscillations) of the dependences of $\tau(\omega)$ are typical. The clearest manifestation of these effects are to be seen in the oscillatory form of the transmission-loss (amplitude-frequency) characteristic of an experimental arrangement [1-5]. These oscillations are due to dipole gaps of the dipole-exchange spin-wave spectrum (see, e.g., Fig. 9.2 d). Probably, for the first time the gaps in the spectrum ω_{nQ} were found in [70]. To measure the dispersion law ω_{nQ} of transverse spin waves propagating in the YIG film of thickness $L = 3.5\mu m$ the author of [70] used a moving probe. His experimental results for ω_{nQ} are in a good coincidence with dipole-exchange theory. Later similar gaps were observed in YIG films for both volume spin waves (perpendicular magnetization) [1] and transverse spin waves (parallel magnetization) [2] as the oscillations of transmission-loss characteristics. The frequency positions of the gaps are well-explained by dipole-exchange theory.

The influence of the surface spin pinning conditions on the trans-

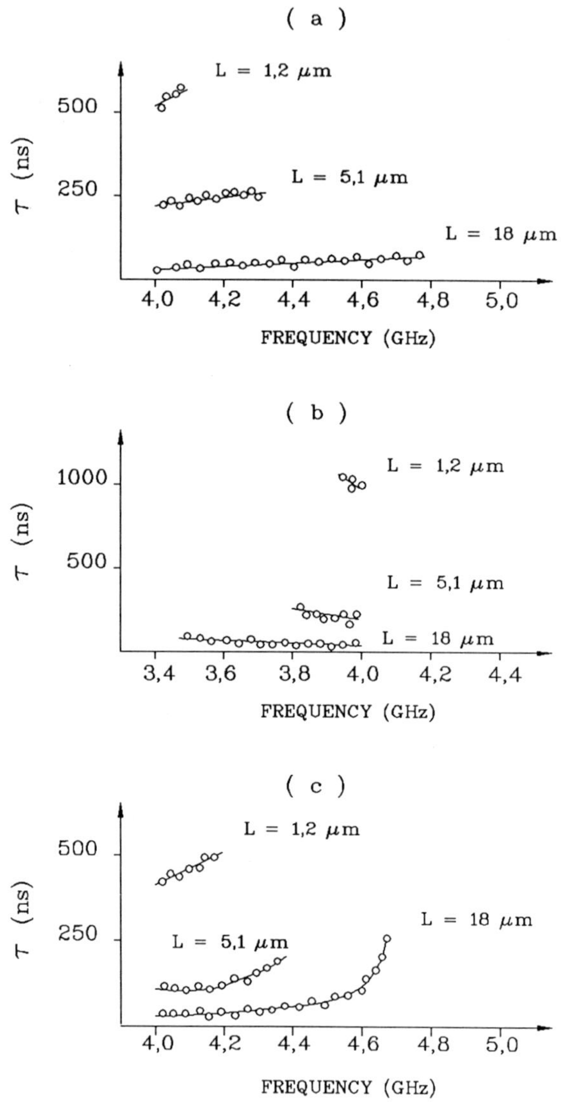

Fig. 2.15. Delay-time as a function of frequency for three main types of spin waves in YIG films.

mission-loss characteristics of transverse waves propagating in YIG films was experimentally studied in [78]. To vary the pinning conditions, ion implantation was used. It was shown that the gradual decrease of the pinning parameter leads to the narrowing of the dipole gaps right up to their full disappearance. In [79] the orientation dependence of the width of the dipole gaps in the spectrum of arbitrarily magnetized YIG films was studied. This dependence of the width of the gaps in the spectrum (or of the intensity of the oscillations of the transmission-loss characteristics) on the direction and magnitude of the external magnetic field $\mathbf{H}_o$ were observed for YIG film samples of thicknesses ranging from 3 to 15 μm, both for films implanted by He, Ar , and Ne ions with energies 150-175 keV and for nonimplanted films with an initial growth surface anisotropy. In accordance with predictions of the dipole-exchange theory [28], in most of samples the value of the angle φ corresponding to the smoothest transmission-loss characteristic depended on the implantation regime, but remained within the interval $20^\circ < \theta_o < 30^\circ$.

The adequacy of the dipole-exchange spin-wave theory is confirmed not only in the experiments with linear excitation of spin waves but in the experiments with nonlinear spin waves also. In particular, the experiments on modulational instability and envelope solitons of dipole-exchange spin waves in YIG films [7,80] show that the measured dispersion properties of propagating spin waves ω_{nQ} and $V_g(\omega)$ are in good agreement with the predictions of the dipole-exchange theory (in this connection see Chapter 9 also).

As was already mentioned, nonlinear microwave spectroscopy based on the parametric excitation of spin waves could be used to study the spin-wave characteristics. Thus parametric excitation of spin waves inthe regime of parallel pumping was used to measure the exchange constant α [82], the microwave thresholds of parametric excitation of spin waves with different numbers n, and the wave vector dependences of the relaxation parameter ΔH_Q for spin waves of different types [82,83]. In particular, in [83] the excitation of transverse and longitudinal spin-wave branches ω_{nQ} with $n = 0,1,2,3,4,5,6$ and wave vectors Q with values ranging from zero up to 10 cm^{-1} was observed in a YIG film of thickness L = 0.5 μm with unpinned surface spins. Similar efects were observed in YIG films with thicknesses in the range $0.3 < L < 4$ μm [82,83]. Such observations give *direct evidence* that in the spin-wave spectrum of thin films a *set* of transverse (quasisurface) spin waves exist.

The parametric excitation of spin waves in the regime of perpendicular pumping (at the frequency 3.3 GHz) was studied in [84]. In this work an impressive agreement between theoretically calculated (on the basis of the dispersion equation in the form of an infinite series (2.71)) and experimentally obtained dispersion ω_{nQ} for longitudinal spin waves was demonstrated. The results were obtained for tangentially magnetized YIG films of thicknesses 4.3 μm, 5.4 μm, 7.8 μm, and 12.4 μm. The dispersion branches of longitudinal spin waves $n = 0,1,2,3,4$ were measured in the frequency regions of a few hundred megahertz situated near the "sag" points of the curves ω_{nQ}.

So, the existing experimental data obtained with the help of parametric excitation of spin waves in YIG films prove the validity of the above described dipole-exchange theory.

2.12. CONCLUSION

The SWM approach described in the preceding section has been employed not only in the calculation of dipole-exchange spin-wave spectrum. It has been already used for solving various problems concerning spin waves in films. In particular, it has been used in the theory of dipole-exchange spin-wave spectrum of single ferromagnetic films [27] and multilayered ferromagnetic films structures [40,85], in the theory of response functions for dipole-exchange spin waves in ferromagnetic films [86,87] and layered structures [88], in the theory of impedance of spin-wave single-element and multi-element transducers [89,90], in the theory of parametric instability of spin waves, and in the theory of spin-waves envelope solitons [56,91,92]. Recently the SWM approach has been applied in the theory of Brillouin light scattering from dipole-exchange spin waves in ferromagnetic films [93]. The above references are not exhaustive: they reflect only the main results obtained with the help of SWM approach.

In our opinion, the SWM approach could be successfully used in further calculations, for example, for solving various problems on the linear and nonlinear dynamics of spin-wave and multi-wave processes in magnetic superlattices and other multilayer systems, ferromagnetic, antiferromagnetic, and magnetic semiconductor films, etc. More specifically, we may mention the problems of the dipole-exchange spin-wave spectrum and response functions for antiferromagnetic films, relaxation theory for thin ferromagnetic and antiferromagnetic films, the same problems for films with in-plane periodicity of their parameters, impurities and rough surfaces, and others.

Besides the topics of fundamental interest, it is relevant to mention here some applied problems. These are the application of spin waves (magnetostatic waves) propagating in ferromagnetic films and layered structures for analog signal processing directly at microwave frequencies, and also the application of optical guided mode interaction with propagating spin waves for analog signal processing (see, e.g., special issues of the journals [94,95]). A good many of the papers cited in the Sec. 2.11 were fulfilled in close connection with the problem of developing spin-wave (magnetostatic wave) devices. The emergence of this new technology based on the propagation of spin waves in magnetic films has become possible owing to the theoretical and experimental work on the subject. Further progress in the elaboration of these new microwave and magnetooptical devices demands a better understanding of linear and nonlinear spin-wave processes in films.

REFERENCES

1. J.D. Adam, T.W. O'Keefe, and R.W. Patterson, *J.Appl. Phys.* **50** (1979) 2446.
2. Yn.V. Gulyaev, A.S. Bugaev, P.E. Zul'berman, I.A. Ignat'ev, A.G. Konovalov, A.V. Lugovskoi, A.M. Mednikov, B.P. Nam and E.I. Nikolaev, *Sov. Phys.- JETP Lett.* **30** (1979) 565 [*Pis'ma Zh. Eksp. Teor. Fiz.* **30** (1979) 600].
3. Yu.F. Ogrin and A.V. Lugovskoi, *Sov. Tech. Phys. Lett.* **9** (1983) 421 [*Pis'ma Zh. Tekh. Fiz.* **9** (1983) 421].
4. A.S. Andreev, Yu.V. Gulyaev, P.E. Zil'berman, V.B. Kravchenko, A.V. Lugovskoi, Yu.F. Ogrin, A.G. Temiryazev, and L.M. Filimonova, *Sov. Phys.-JETP* **59** (1984) 586 [*Zh. Eksp. Teor. Fiz.* **59** (1984) 1005].
5. B.A. Kalinikos, N.G. Kovshikov, and A.N. Slavin, *Sov. Phys.-JETP Lett.* **38** (1983) 413 [*Pis'ma Zh. Eksp. Teor. Fiz.* **38** (1983) 343].
6. V.I. Dmitrijev, B.A. Kalinikos, and N.G. Kovshikov, *Sov. Phys. Tech. Phys.* **30** (1985) [*Zh. Tekh. Fiz.* **55** (1985) 2051].
7. B.A. Kalinikos, N.G. Kovshikov, and A.N. Slavin, *Sov. Phys.-JETP* **94** (1988) 303 [*Zh. Eksp. Teor. Fiz.* **94** (1988) 157].
8. V.V. Gann, *Sov. Phys.-Solid State* **8** (1967) 2537 [*Fiz. Tverd. Tela,* **9** (1966) 3167].
9. T. Wolfram and R.E. De Wames, *Solid State Commun.* **8** (1970) 191.
10. R.E. De Wames and T. Wolfram, *J. Appl. Phys.* **41** (1970) 987.
11. T. Wolfram and R.E. De Wames, *Phys. Rev.* B **4** (1971) 3125.
12. T. Wolfram and R.E. De Wames, *Prog. Surf. Sci.* **2** (1972) 233.
13. B.N. Filippov, *Phys. Met. Metall.* **32** (1971) [*Fiz. Met. Metalloved.* **32** (1971) 911].
14. B.N. Filippov and I.G. Titjakov, *Phys. Met. Metall.* **35** (1973) [*Fiz. Met. Metalloved.* **35** (1973) 28].
15. R.G. Khlebopros and L.V. Mikhailovskaya, Bull. *Acad. Sci. USSR Phys. Ser.* **36** (1972) [*Izv. AN SSSR - Ser. Fiz.* **36** (1972) 1522].
16. L.V. Mikhailovskaya and R.G. Khlebopros, *Sov. Phys.-Solid State* **16** (1974) 46 [*Fiz. Tverd. Tela* **16** (1974) 77].
17. Yu.V. Gulyaev, P.E. Zil'berman and A.V. Lugovskoi, *Sov. Phys.-Solid State* **23** (1981) 660 [*Fiz. Tverd. Tela* **23** (1981) 1136].

18. A.V. Lugovskoi and P.E. Zil'berman, *Sov. Phys.-Solid State* **24** (1982) 259 [*Fiz. Tverd. Tela* **24** (1982) 458].

19. P.E. Zil'berman and A.V. Lugovskoi, *Sov. Phys. Tech. Phys.* **57** (1987) [*Zh. Tekh. Fiz.* **57** (1987) 3].

20. G.T. Rado and R.J. Hicken, *J.Appl. Phys.* **63** (1988) 3885.

21. M. Sparks, *Phys. Rev. Lett.* **24** (1970) 1178.

22. M. Sparks, *Phys. Rev.* B **1** (1970) 3831.

23. O.G. Vendik and D.N. Chartorizhskii, *Sov. Phys.-Solid State* **11** (1970) 1957 [*Fiz. Tverd. Tela* **11** (1969) 2420].

24. O.G. Vendik and D.N. Chartorizhskii, *Sov. Phys.-Solid State* **12** (1970) 1209 [*Fiz. Tverd. Tela* **12** (1970) 1538].

25. O.G. Vendik, B.A. Kalinikos and D.N. Chartorizhskii, *Sov. Phys.-Solid State* **19** (1977) 222 [*Fiz. Tverd. Tela* **19** (1977) 387].

26. B.A. Kalinikos, *IEE Proc. Pt.* H **127** (1980) 4.

27. B.A. Kalinikos and S.I. Miteva, *Sov. Phys. Tech. Phys.* **51** (1981) [Zh. Tekh. Fiz. **51** (1981) 2213].

28. B.A. Kalinikos and A.N. Slavin, *J. Phys* C **19** (1986) 7013.

29. B.A. Kalinikos, M.P Kostylev, N.V. Kozhus', and A.N. Slavin, *J. Phys.: Condens. Matter* **2** (1990) 9861.

30. T. Holstein and H. Primakoff, *Phys. Rev.* **58** (1940) 1098.

31. C. Herring and C. Kittel, *Phys. Rev.* **81** (1951) 869.

32. B. Hillebrands, *Phys. Rev.* **B 41** (1990) 530.

33. J.F. Cochran and J.R. Dutcher *J. Magn. Magn. Mater.* **73** (1988) 299.

34. J. Barnas, *J. Phys.*C **21** (1988) 4097.

35. G.T. Rado and J.R Weertman, *J. Phys. Chem. Solids* **11** (1959) 315.

36. R.F. Soohoo, *Magnetic Thin Films* (Harper and Row, New York, 1965).

37. H. Puszkarski, *Prog. Surf. Sci.* **9** (1979) 191.

38. R.C. Moul and M.G. Cottam, *J. Phys.* C **16** (1983) 1307.

39. J. H. Wilkinson and C. Reinsch, *Handbook for Automation of Linear Algebra* (Springer, Berlin, 1976).

40. K. Vayhinger and H. Kronmuller, *J. Magn. Magn. Mater.* **62** (1986) 159.

41. L.D. Landau and E.M. Lifshitz, *Quantum Mechanics Non-Relativistic Theory*, 2nd edn. (Pergamon, London, 1965).

42. C.E. Patton, *Physics Reports* **103** (1984) 251.
43. B. Sneider, *Phys. Status Solidi* (b) **51** (1972) 325.
44. C. Vittoria and N.D. Wilsey, *J. Appl. Phys.* **45** (1974) 414.
45. S.N. Bajpai, I. Rattah and N.C. Srivastava, *J. Appl. Phys.* **50** (1979) 2887.
46. C.E. Patton, *Phys. Status Solidi* (b) **92** (1979) 211.
47. C.E. Patton, *Phys. Status Solidi* (b) **93** (1979) 63.
48. R.A. Lemous and B.A. Auld, *J. Appl. Phys.* **52** (1981) 7360.
49. A.S. Beregov, *Radioelectron. Commun. Syst.* **27** (1984) 7 [*Izv. Vyssh. Uchebn. Zaved., Radioelectron.* **27** (1984) 9].
50. A.S. Beregov and E.V. Kudinov, *Electron. Tech. ser. 1 Electronica SVCH* (1987) 8.
51. A.S. Beregov and E.V. Kudinov, *Sov. Phys. J.* **31** (1988) [*Izv. Vyssh. Uchebn. Zaved., Physica* **31** (1988) 106].
52. R. Gieniusz and L. Smoczynski, *J. Magn. Magn. Mater.* **66** (1987) 361.
53. V.B. Anfinogenov , L.A. Mitlina, A.F. Popkov, A.A. Sidorov, V.G. Sorokin, and V.V. Tihonov, *Sov. Phys. - Solid State* **30** (1988) [*Fiz. Tverd. Tela* **30** (1988) 2032].
54. G.T. Rado and L. Zhang, *Phys. Rev.* **63** (1986) 5081.
55. S. Miteva and T. Nurgaliev, *Bulgarian Phys. J.* **13** (1986) 358.
56. B.A. Kalinikos, and N.V. Kozhus', *Sov. Tech. Phys.* **34** (1989) 1105 [*Zh. Tekn. Fiz.* 59 (1989) 24].
57. S.V. Belyakov, B.A. Kalinikos, and N.V. Kozhus', *Electron Tech. ser 1 Electronica SVCH* (1989) 22.
58. S.V. Belyakov, *Electron. Tech. ser* 1 *Electronica SVCH* (1988) 39.
59. S.V. Beyakov, *Electron. Tech. ser.* 1 *Electronica SVCH* (1984) 35.
60. S.V. Belyakov, *Electron. Tech. Ser.* 1 *Electronica SVCH* (1986) 26.
61. L.R. Walker, *Phys. Rev.* **105** (1957) 390.
62. B.E. Storey, A.O. Tooke, A.P. Cracknell, and J.A. Przystawa, *J. Phys. C: Solid State Phys.* **10** (1977) 875.
63. A. Schonecker, S. Bornman, and W. Haubenreisser, *Phys. Stat. Sol.* (a) **22** (1974) 53.
64. H. Reimann, L.C. Hsia, and P.E. Wigen, *J. Appl. Phys.* **52** (1981) 2264.

65. L.C. Hsia, H. Reimann, and P.E. Wigen, *IEEE Trans. Mag.* **17** (1981) 2961.

66. A.P. Cracknell, B.E. Storey, and A.O. Tooke, *Solid State Comm.* **26** (1978) 377.

67. R. Henry, S.D. Brown, and P.E. Wigen, *Phys. Rev. Lett.* **28** (1972) 1272.

68. V.F. Dmitriev, B.A. Kalinikos, and N.G. Kovshikov, *Sov. Tech. Phys.* **31** (1986) 1300 [*Zh. Tekh. Fiz.* **56** (1986) 2169].

69. J.C. Sethares, and I.J. Weinberg, *Circuits, Systems, and Signal Processing* **4** (1985) 41.

70. W. Schilz, *Philips Res. Reports* **28** (1973) 50.

71. N.P. Vlannes, and F.R. Morgenthaler, *J. Appl. Phys.* **57** (1985) 3721.

72. W.L. Bongianni, *Microwave J.* **17** (1974) 49.

73. W.L. Bongianni, *J. Appl. Phys.* **43** (1972) 2541.

74. N.D.J. Miller, *Phys. stat. sol.* (a) **37** (1976) 83.

75. Z.M. Bardai, J.D. Adam, J.H. Collins, and J.P. Parekh, *AIP Conf. Proc.* **34** (1976) 268.

76. T.W. O'Keeffe, and R.W. Patterson, *J. Appl. Phys.* **49** (1978) 4886.

77. B.A. Kalinikos, N.G. Kovshikov, and N.V. Kozhus', *Sov. J. Radio-electronics and Communications Systems* **26** 65 [*Izv. Vyssh. Uchebn. Zaved., Radioelectronika* **26** (1983) 77].

78. Yu.M. Yakovlev, E.G. Rzhikhina, T.A. Krylova, et. al. *Sov. Phys. Solid State* **30** (1988) 360 [*Fiz. Tverd. Tela* **30** (1988) 622].

79. P.A. Kolodin, E.G. Rzhikhina, and A.N. Slavin, *Sov. Phys. Tech. Phys.* **35** (1990) 951 [*Zh. Tekn. Fiz.* **60** (1990) 132].

80. B.A. Kalinikos, N.G. Kovshikov, P.A. Kolodin, and A.N. Slavin, *Solid State Comm.* **74** (1990) 989.

81. I.L. Beresin, B.A. Kalinikos, N.G. Kovshikov, S.P. Orobinskii, and D.N. Chartorizhskii, *Sov. Phys. Solid State* **20** (1978) 1214 [*Fiz. Tverd. Tela* **20** (1978) 2101].

82. B.A. Kalinikos, N.G. Kovshikov, and N.V. Kozhus', *Sov. Phys. Solid State* **26** (1984) 1735 [*Fiz. Tverd. Tela* **26** (1984) 2867].

83. B.A. Kalinikos, N.G. Kovshikov, and N.V. Kozhus', *Sov. Phys. Solid State* **27** (1985) 1681 [*Fiz. Tverd. Tela* **27** (1985) 2794].

84. A.G. Temiryazev, *Sov. Phys. Solid State* **29** (1987) [*Fiz. Tverd. Tela* **29** (1987) 313].

85. B.A. Kalinikos, and P.A. Kolodin, *Sov. J. Radiaphysics and Quantum Electronics* **32** (1989) 957 [*Izv. Vyssh. Uchebn. Zaved., Radiofizika* **32** (1989) 1290].

86. B.A. Kalinikos, *Sov. Tech. Phys.* **29** (1984) 1083 [*Zh. Tekh. Fiz.* **54** (1983) 1846].

87. B.A. Kalinikos, and A.N. Slavin, *Asta Physi Polonica* **A75** (1989) 541.

88. B.A. Kalinikos, and P.A. Kolodin, *J. Magn. Magn. Mater.* **83** (1990) 103.

89. V.F. Dmitriev and B.A. Kalinikos, *Sov. Phys. J.* **31** (1988) 875 [*Izv. Vyssh. Uchebn. Zaved, Fizika* **31** (1988) 24].

90. V.F. Dmitriev and B.A. Kalinikos, *Sov. Phys. Tech. Phys.* (1989) [*Zh. Techn. Fiz.* **59** (1989) 197].

91. B.A. Kalinikos, *Sov. Tech. Phys. Lett.* **9** (1983) 349 [*Pis'ma Zh. Tekn. Fiz.* **9** (1983) 811].

92. B.A. Kalinikos and A.N. Slavin, *Sov. Phys. Solid State* **26** (1984) 2077 [*Fiz. Tverd. Tela* **26** (1984) 3456].

93. M.G. Cottam and A.N. Slavin, *IEEE Trans. Magn.* **27** (1992) 5489.

94. *Circuits, Systems, and Signal Processing* **4**, No. 1-2 (1985).

95. *IEEE Proc.* **76**, No. 2 (1988)

CHAPTER 3

SPIN WAVES IN MULTILAYERS

J. BARNAŚ

Institute of Physics, Technical University of Poznań,
60-965 Poznań, Poland

Contents

3.1 INTRODUCTION

Theoretical studies of spin-wave excitations in layered magnetic structures have been stimulated, at least to some extent, by recent achievements in the growth techniques. The spectrum of magnetic excitations in such artificial structures presents many unique features which are absent in bulk and single-film systems [1–4]. To describe them qualitatively and/or quantitatively, one should first discuss certain basic features of the spin-wave modes in a single-layer magnetic film. These were investigated extensively in the past two decades and several, more or less general techniques have been worked out for their theoretical description. Since the single-film problem is the subject of other chapters in this book as well as of many excellent reviews [5–9], we will not discuss it in detail but, where necessary, will refer to the appropriate publications.

An exact description of the spin-wave modes in the whole frequency-wavevector range might be obtained within the microscopic model with a Hamiltonian including all relevant interactions. For rare-earth structures one can work within the model of localized spins. For transition metals such an approach is rather oversimplified and more realistic models, based on the band magnetism theory, should be applied [10]. However, many interesting features of the spin-wave spectrum in transition metal structures, particularly in the long wavelength range, are well reproduced by the model of localized spins, which is much simpler for analytical and numerical analyses. Thus, we will restrict our discussion to the localized spin model—also because the spin-wave modes of layered structures within the band magnetism theory are in an early stage of investigation. Consequently, we consider excitations in layered structures which can be described by a Hamiltonian $\mathcal{H}$ consisting of the Heisenberg exchange coupling $\mathcal{H}^{exch}$, dipolar interactions $\mathcal{H}^{d}$, magnetocrystalline anisotropy $\mathcal{H}^{an}$ and Zeeman energy $\mathcal{H}^{z}$,

$$\mathcal{H} = \mathcal{H}^{exch} + \mathcal{H}^{d} + \mathcal{H}^{an} + \mathcal{H}^{z}. \tag{3.1}$$

In the general case, the full Hamiltonian (3.1) is complicated enough, so that further approximations and simplifications are required [4,11]. These are well known from appropriate considerations for bulk materials and single thin films. Suppose that a well-defined wave propagates in a magnetic material. If the angles α formed by adjacent spins differ considerably from those in the ground state configuration ($\alpha = 0$ for the ferromagnetic equilibrium state), then the exchange contribution to the spin-wave energy is dominant and one can neglect the dipolar coupling between the spins. In the case of *acoustic* spin-wave modes this condition is fulfilled for all wavevectors $\mathbf{q}$, except those for which the condition $qa \ll 1$ holds (a is here the lattice parameter). For *optic* modes it is usually true for the whole Brillouin zone. In many cases the anisotropy and Zeeman terms are also important in the exchange-dominated region and have to be included into the Hamiltonian. In the other limit, opposite to that discussed above, the angles α are very small, $\alpha \ll 1$. Consequently, the exchange contribution to the spin wave energy is negligible. This is true for *acoustic* modes if the condition $qa \ll 1$ is fulfilled, *i.e.*, in the long wavelength limit. In this so-called dipolar limit, the dipole-dipole interactions can be included approximately by way of a macroscopic microwave field $\mathbf{h}(\mathbf{r}, t)$. To find this field one usually applies magnetostatic equations and the corresponding excitations are referred to as magnetostatic modes. If the full set of Maxwell equations is employed to determine the magnetic field, one obtains so-called magnetic polaritons which are mixed states of magnons and photons [12,13]. Since these modes are the subject of the next chapter of this book, we will completely omit discussing the retardation effects. The region where the exchange and dipolar contributions are comparable is of particular interest from the theoretical point of view. A detailed analysis of this crossover region shows how the microscopic model of exchange-dominated modes and the continuum model of magnetostatic waves can be unified to give consistent results. The crossover region can be considered within the macroscopic continuum approximation. This follows from the fact that the corresponding wavelengths are long enough, allowing us to pass from the discrete to the continuum description and to include the dipole-dipole coupling by the macroscopic dipolar field (as for the magnetostatic modes) and the exchange coupling by a macroscopic effective exchange field.

We will restrict the following discussion to linear undamped modes, neglecting all relaxation processes originating in non-linear magnon-magnon interactions, eddy-currents, *etc.* These processes lead, in general, to a small shift of the excitation frequencies in thin films [14] and are important for the description of the line profiles in appropriate resonance and light scattering experiments [14,15].

3.2 EXCHANGE DOMINATED WAVES

As mentioned above, the exchange-dominated region is well described by the Hamiltonian (3.1) with the dipolar term omitted. To determine the spin-wave spectrum of any layered structure one might basically apply the same methods as for bulk materials. The difference resides in the crystallographic elementary cell which, in bulk materials, usually contains several magnetic atoms. In superlattices their number can be quite large. However, the problem can be simplified by taking into account the internal periodicity of each sublayer. A few different techniques have been developed for the treatment of spin-wave excitations in layered structures [16–25]. One of them—the transfer matrix method—will be described here in detail. The others will be discussed but very briefly.

3.2.1 Transfer matrix formalism

To describe the transfer matrix method and basic features of the spin-wave spectrum in the exchange-dominated region we consider a model simple-cubic layered structure with ferromagnetic exchange coupling between nearest neighbours (nn) and elementary unit (elementary bilayer) composed of N_A atomic planes of a material A followed by N_B atomic planes of a material B. The elementary bilayer is repeated periodically along the stacking direction which we assume along the axis x of the coordinate system. We will consider structures which are either infinite, semi-infinite or finite along the axis x, but extend infinitely in the plane yz. For simplicity we will neglect the bulk and surface or interface anisotropy in the following discussion of this section. Let the axis z of the coordinate system be along an external static magnetic field H_0. The Hamiltonian of the system can now be written explicitly as

$$\mathcal{H} = -\frac{1}{2}\sum_l \sum_\rho \sum_{\delta_\parallel} J_l \mathbf{S}_{l,\rho} \cdot \mathbf{S}_{l,\rho+\delta_\parallel} - \sum_l \sum_\rho J_{l,l+1} \mathbf{S}_{l,\rho} \cdot \mathbf{S}_{l+1,\rho}$$

$$+ \mu_B H_0 \sum_l \sum_\rho g_l S^z_{l,\rho}\,, \tag{3.2}$$

where $\boldsymbol{\rho}$ determines the position of a spin in a given atomic plane, $\boldsymbol{\delta}_\parallel$ is the position vector to the in-plane nearest neighbours, J_l is the exchange constant for nn lying in the l-th atomic plane, $J_{l,l+1}$ is the exchange integral for nn lying in the l-th and $(l+1)$-st atomic planes, g_l is the Landé factor for the l-th atomic plane and μ_B denotes the Bohr magneton. For the heterostructures discussed here we assume; $J_l = J_A$ for $mN+1 \le l \le mN+N_A$, $J_l = J_B$ for $mN+N_A+1 \le l \le (m+1)N$, $J_{l,l+1} = J_A$ for $mN+1 \le l \le mN+N_A-1$, $J_{l,l+1} = J_B$ for $mN+N_A+1 \le l \le (m+1)N-1$ and $J_{l,l+1} = J_I$ for $l = mN$ and $l = mN+N_A$, where $N = N_A+N_B$, J_A and J_B are the bulk exchange constants for the materials A and B, J_I is the exchange constant between the two materials, and m labels the elementary bilayers in the stack. For simplicity, we assume the same g factors for both materials, $g_l = g$ for all l.

The dynamical properties of the system are determined by the quantum-mechanical equation of motion for the spin raising operators,

$$i\hbar \frac{\partial S^+_{l,\rho}(t)}{\partial t} = [S^+_{l,\rho}, \mathcal{H}]. \tag{3.3}$$

It is now convenient to introduce a time and two-dimensional space Fourier transformation,

$$S^+_{l,\rho}(t) = \frac{1}{(2\Pi)^3} \int dE \int d\mathbf{Q} S^+_l(\mathbf{Q}, E) \exp\left[i(\mathbf{Q}\cdot\boldsymbol{\rho} - \frac{Et}{\hbar})\right], \quad (3.4)$$

where $\mathbf{Q}$ is a two-dimensional wavevector in the plane yz. In the linear approximation and in the low-temperature limit Eq. (3.3) leads to the following equation for the spin-wave amplitudes $u_l = S^+_l(\mathbf{Q}, E) \mid 0\rangle$:

$$(E - E_l)u_l + S_l J_{l,l+1} u_{l+1} + S_l J_{l,l-1} u_{l-1} = 0, \quad (3.5)$$

where, for brevity, we have dropped the dependence of u_l on $\mathbf{Q}$ and E. In the above equation S_l is the spin number for magnetic atoms in the l-th atomic plane, *i.e.*, $S_l = S_A(S_B)$ for all atomic planes of the material $A(B)$, whereas $E_l = E_A(E_B)$ for all internal atomic planes of the material $A(B)$ and $E_l = E^i_A(E^i_B)$ for the interfacial atomic planes of the material $A(B)$, with

$$E_A = g\mu_B H_0 + 4\Lambda S_A J_A + 2 S_A J_A, \quad (3.6a)$$

$$E^i_A = g\mu_B H_0 + 4\Lambda S_A J_A + S_A J_A + S_B J_I, \quad (3.6b)$$

and E_B and E^i_B given, respectively, by Eq. (3.6a) and Eq. (3.6b) with the indices A and B interchanged. The parameter Λ in the above equations is defined as

$$\Lambda = 1 - \frac{1}{2}\left[\cos(Q_y a) + \cos(Q_z a)\right], \quad (3.7)$$

and $\Lambda \in \langle 0, 2\rangle$ for Q ranging throughout the whole two-dimensional Brillouin zone.

Equation (3.5) is a recurrence equation for the spin-wave amplitudes, which allows one to determine the amplitude for a given atomic plane provided the amplitudes at the two preceding atomic planes are available. This equation can be rewritten in a convenient matrix form as [19]

$$\begin{pmatrix} u_{l+1} \\ u_l \end{pmatrix} = \mathbf{M}_l \begin{pmatrix} u_l \\ u_{l-1} \end{pmatrix}, \tag{3.8}$$

where $\mathbf{M}_l = \mathbf{M}_A$ for $mN + 2 \leq l \leq mN + N_A - 1$, $\mathbf{M}_l = \mathbf{M}_B$ for $mN + N_A + 2 \leq l \leq (m+1)N - 1$, $\mathbf{M}_l = \mathbf{M}_A^{il}$ for $l = mN + 1$, $\mathbf{M}_l = \mathbf{M}_A^{ir}$ for $l = mN + N_A$, $\mathbf{M}_l = \mathbf{M}_B^{il}$ for $l = mN + N_A + 1$ and $\mathbf{M}_l = \mathbf{M}_B^{ir}$ for $l = (m+1)N$, with

$$\mathbf{M}_A = \begin{pmatrix} \frac{E_A - E}{S_A J_A} & -1 \\ 1 & 0 \end{pmatrix}, \qquad \mathbf{M}_A^{ir} = \begin{pmatrix} \frac{E_A^i - E}{S_A J_I} & -\frac{J_A}{J_I} \\ 1 & 0 \end{pmatrix},$$

$$\mathbf{M}_A^{il} = \begin{pmatrix} \frac{E_A^i - E}{S_A J_A} & -\frac{J_I}{J_A} \\ 1 & 0 \end{pmatrix}. \tag{3.9}$$

The matrices $\mathbf{M}_B$, $\mathbf{M}_B^{ir}$ and $\mathbf{M}_B^{il}$ are given by similar formulae, but with the index B instead of A.

An equivalent form of the transfer matrix formalism can be obtained by taking into consideration a vector with the components u_l and $u_l - u_{l-1}$, instead of u_l and u_{l-1} [20,21].

By applying Eq. (3.8) to all N atomic planes in the elementary unit one obtains the following relation:

$$\begin{pmatrix} u_{mN+1} \\ u_{mN} \end{pmatrix} = \mathbf{M} \begin{pmatrix} u_{(m-1)N+1} \\ u_{(m-1)N} \end{pmatrix}, \tag{3.10}$$

where the transfer matrix $\mathbf{M}$ for the whole elementary bilayer is equal to

$$\mathbf{M} = \mathbf{M}_B^{ir}(\mathbf{M}_B)^{N_B-2}\mathbf{M}_B^{il}\mathbf{M}_A^{ir}(\mathbf{M}_A)^{N_A-2}\mathbf{M}_A^{il}. \tag{3.11}$$

Owing to the translational symmetry along the superlattice direction, the matrix $\mathbf{M}$ is independent of the bilayer index m and obeys the condition

$$\det \mathbf{M} = 1. \tag{3.12}$$

In the general case, the eigenmodes of the system can be described by the respective eigenvalues and eigenvectors of the matrix **M**. One can distinguish the following possibilities:

(*i*) $\quad | M_{11} + M_{22} | \leq 2$

The eigenvalues λ_+ and λ_- are now complex and $\lambda_- = \lambda_+^*$, except for $M_{11}+M_{22} = 2$ when $\lambda_+ = \lambda_- = 1$ and for $M_{11}+M_{22} = -2$ when $\lambda_+ = \lambda_- = -1$. These eigenvalues correspond to bulk modes (eigenmodes of an infinite stack). It is convenient to write both eigenvalues in a form which obeys the Bloch theorem, *i.e.*,

$$\lambda_\pm = \exp(\pm i q_x L), \tag{3.13}$$

with q_x real and with $L = Na$ denoting the thickness of the elementary bilayer. The upper limit ($M_{11} + M_{22} = 2$) then corresponds to $q_x = 0$ whereas the lower limit ($M_{11} + M_{22} = -2$) to $q_x L = \pm\Pi$.

(*ii*) $\quad | M_{11} + M_{22} | > 2.$

The eigenvalues λ_+ and λ_- are now real and positive for $M_{11}+M_{22} > 2$ or negative for $M_{11} + M_{22} < -2$. One may write them as

$$\lambda_\pm = \exp(\pm \beta L), \tag{3.14}$$

where

$$\beta = \kappa \tag{3.15a}$$

for $M_{11} + M_{22} > 2$, whereas

$$\beta = i\Pi/L + \kappa \tag{3.15b}$$

for $M_{11} + M_{22} < -2$, with κ real and positive by definition. The condition $M_{11} + M_{22} > 2$ determines the region where *acoustic* surface modes can propagate, whereas the condition $M_{11} + M_{22} < -2$ correponds to the region where *optic* surface modes can exist. For *acoustic* modes the equivalent spins in adjacent bilayers precess always in phase, whereas for *optic* modes they precess in anti-phase.

The eigenvalues of the matrix $\mathbf{M}$ are degenerate only for $M_{11} + M_{22} = \pm 2$, *i.e.*, at the band edges of bulk modes in infinite superlattices.

Bulk modes

Consider now normal modes of infinite layered structures. They correspond to the eigenvalues $\lambda_{\pm} = \exp(\pm i q_x L)$. The appropriate dispersion equation can be easily obtained by taking into account the explicit forms of the eigenvalues $\lambda_{\pm}$. This leads to the following expression:

$$\cos(q_x L) = (M_{11} + M_{22})/2. \tag{3.16}$$

The wavevector component q_x along the stacking direction is restricted to the range $-\Pi/L \leq q_x \leq \Pi/L$. The explicit form of the dispersion equation is determined by the explicit form of the matrix $\mathbf{M}$, and this can be found from Eq. (3.11).

Surface modes

Semi-infinite structures can also support modes which are localized exponentially at the surface. The spectrum of such modes as well as the conditions of their existence strongly depend on some surface parameters which, in general, can differ from the corresponding ones inside the layered structure. Consider for example the semi-infinite structure with the material A at the surface. In the general case, the exchange constant J_A^s for nn lying in the surface atomic plane differs from J_A, $J_A^s \neq J_A$. To find the spectrum of surface modes we will employ Eq. (3.10). Let $m = 1$ correspond to the elementary bilayer at the surface. Equation (3.10), however, does not apply directly to $m = 1$, because the surface atomic plane is not equivalent to its counterparts inside the structure. To overcome this difficulty we will use a method originally applied to thin homogeneous films, *i.e.*, we replace all surface parameters by the corresponding bulk ones and add a fictitious atomic plane of the material B at the surface, which corresponds to $l = 0$. If the parameters describing the fictitious atomic plane are the same as the corresponding ones inside the structure, then Eq. (3.10) is also true for $m = 1$. Although the original system and that with the fictitious atomic plane are not physically equivalent, one can impose adequate boundary conditions on the solutions for the fictitious system to ensure that these solutions—restricted to $l > 0$—will be the same as for the original problem.

The surface solutions can exist in the energy region where $| M_{11} + M_{22} |> 2$. For any exponentially decaying function one can write

$$\begin{pmatrix} u_1 \\ u_0 \end{pmatrix} = A\mathbf{u}_-, \tag{3.17}$$

where A is a constant and $\mathbf{u}_-$ is the eigenvector of the matrix $\mathbf{M}$, which corresponds to the eigenvalue $\lambda_- = \exp(-\beta L)$. One can easily verify this by applying Eq. (3.10) to Eq. (3.17). The amplitudes u_0 and u_1 have to fulfil conditions which will ensure the equivalence of the solutions for systems with and without the fictitious atomic plane. These conditions follow from the equation of motion for spins in the surface atomic planes of both systems and can be written in the form

$$\begin{pmatrix} u_1 \\ u_0 \end{pmatrix} = C \begin{pmatrix} 1 \\ p \end{pmatrix}, \tag{3.18}$$

where C is a constant and p is a parameter defined as

$$p = \frac{S_B}{S_A} - 4\Lambda \frac{J_A^s - J_A}{J_I}. \tag{3.19}$$

Equations (3.17) and (3.18) have to be satisfied simultaneously, which, on taking into account the explicit expression for the eigenvector $\mathbf{u}_-$, gives the following dispersion equation for the surface modes:

$$\exp(-\beta L) = pM_{12} + M_{11}. \tag{3.20}$$

This equation can be solved numerically inside the energy region where $| M_{11} + M_{22} |> 2$. Note that β as well as the $\mathbf{M}$-matrix components are certain functions of E, which is not indicated explicitly for brevity. The solutions with real β describe the surface *acoustic* modes for which the equivalent spins in all bilayers precess in-phase. If $\mathrm{Im}\beta = \Pi/L$, the corresponding solution describes surface *optic* modes with equivalent spins in adjacent bilayers precessing in anti-phase.

Finite structures

Consider now structures finite along the stacking direction. Let them consist of n complete elementary units. For the sake of generality, we assume an additional incomplete unit at one of the surfaces, say the $(n+1)$-st incomplete unit at the right-hand surface. Let the incomplete unit consist of the first N_i ($N_i < N$) atomic planes of the elementary bilayer. The method applied to surface modes can be easily adapted to finite structures. For the same reasons as before one introduces two fictitious atomic planes, the one at the left- and the other at the right-hand surface, replacing simultaneously all the surface parameters by the corresponding ones inside the structure. Apart from this, the parameters assigned to the fictitious atomic planes are equal to those resulting from the periodicity of the structure along the growth direction. As previously, the equivalence of the solutions for the systems with and without fictitious atomic planes can be ensured by appropriate boundary conditions. Such a condition for the left-hand surface of complete ($N_i = 0$) as well as incomplete ($0 < N_i < N$) structures is given by Eq. (3.18) in which, in order to distinguish the surfaces, we replace C and p by C_L and p_L, respectively, with the parameter p_L given by Eq. (3.19).

Consider first the complete case. The boundary condition for the right-hand surface can be obtained in the same way as for the left-hand one and takes the form:

$$\begin{pmatrix} u_{nN+1} \\ u_{nN} \end{pmatrix} = C_R \begin{pmatrix} p_R \\ 1 \end{pmatrix}, \tag{3.21}$$

where C_R is a constant and

$$p_R = \frac{S_A}{S_B} - 4\Lambda \frac{J_B^s - J_B}{J_I}. \tag{3.22}$$

In the above equation J_B^s is the exchange constant between spins lying in the atomic plane at the right-hand surface.

For arbitrary energy, except those particular energy values at which the eigenvalues of the matrix $\mathbf{M}$ are degenerate (*i.e.*, with the exception of the values corresponding to the band edges of the bulk modes in infinite structures, which must be considered separately), one can write

$$\begin{pmatrix} u_1 \\ u_0 \end{pmatrix} = A\mathbf{u}_- + B\mathbf{u}_+, \tag{3.23}$$

where A and B are constants. Applying Eq. (3.10) to Eq. (3.23) one now obtains

$$\begin{pmatrix} u_{nN+1} \\ u_{nN} \end{pmatrix} = A(\lambda_-)^n\mathbf{u}_- + B(\lambda_+)^n\mathbf{u}_+. \tag{3.24}$$

It is convenient to write all the eigenvalues of the matrix $\mathbf{M}$ in a unified form as $\lambda_\pm = \exp(\pm ik_\perp L)$, where $k_\perp$ is real ($k_\perp = q_x$) for $| M_{11} + M_{22} |< 2$ and $k_\perp = -i\beta$ for $| M_{11} + M_{22} |> 2$. The requirement that Eqs (3.18), (3.21), (3.23) and (3.24) shall have a non-trivial solution leads to the following dispersion equation:

$$\begin{aligned} &p_L p_R \left[\sin\left(k_\perp L(n-1)\right) - M_{22}\sin\left(k_\perp Ln\right)\right] \\ &\quad + \left(p_L M_{12} - p_R M_{21}\right)\sin\left(k_\perp Ln\right) \\ &\quad + \sin\left(k_\perp L(n+1)\right) - M_{22}\sin\left(k_\perp Ln\right) = 0. \end{aligned} \tag{3.25}$$

A similar analysis for structures with an additional incomplete unit leads to the dispersion equation

$$\begin{aligned} &\left(p_R^i M_{22}^i - M_{12}^i\right) \times \{p_L \left[\sin\left(k_\perp L(n-1)\right) - M_{22}\sin\left(k_\perp Ln\right)\right] \\ &- M_{21}\sin\left(k_\perp Ln\right)\} + \left(-p_R^i M_{21}^i + M_{11}^i\right) \times \{\sin\left(k_\perp L(n+1)\right) \\ &- M_{22}\sin\left(k_\perp Ln\right) + p_L M_{12}\sin\left(k_\perp Ln\right)\} = 0, \end{aligned} \tag{3.26}$$

where the parameter p_R^i differs, in general, from p_R and is dependent on the internal structure of the incomplete unit, whereas $\mathbf{M}^i$ is the total transfer matrix for the latter.

Equations (3.16), (3.20), (3.25) and (3.26) basically determine the spin-wave spectrum of infinite, semi-infinite and finite stacks of two alternating magnetic materials. They can be easily solved numerically to give the appropriate dispersion curves explicitly. Some numerical results and basic features of the spin-wave spectrum will be discussed later. First, however, we shall discuss briefly some other methods of the theoretical analysis of spin-wave excitations in magnetic layered structures.

3.2.2 Other formalisms

Owing to the translational symmetry in the plane yz the eigenvalue problem for layered structures reduces effectively to a one-dimensional problem. The method described above is, thus, the transfer matrix formalism in *site* representation. It is also possible to develop a transfer matrix formalism in plane-wave-like (complex wavevectors) representation. In this formulation one explicitly takes into account the exact solutions of the appropriate equation of motion inside each constituent material by writing the general solution for the spin-wave amplitudes inside films of the material A as [18]

$$\begin{aligned} u_l &= A_m^+ \exp\{ik^{(A)}[la-(m-1)L]\} \\ &\quad + A_m^- \exp\{-ik^{(A)}[la-(m-1)L]\} \end{aligned} \tag{3.27}$$

for $mN+1 \le l \le mN+N_A$. A_m^+ and A_m^- are here constants (amplitudes) and $k^{(A)}$ is a complex wavevector, which can be determined from the relation,

$$\cos(k^{(A)}a) = 1 + 2\Lambda - \frac{E - g\mu_B H_0}{2J_A S_A}. \tag{3.28}$$

The above equation follows directly from the well known dispersion relation for spin-wave modes in bulk materials. An expansion similar to (3.27) holds also for films of the material B, albeit with new constants and with $k^{(A)}$ replaced by $k^{(B)}$, where $k^{(B)}$ is given by Eq. (3.28) with the index B instead of A.

The equations of motion for the two atomic planes at each interface combine the amplitudes in adjacent layers leading finally to the relation

$$\begin{pmatrix} A^+_{m+1} \\ A^-_{m+1} \end{pmatrix} = \mathbf{T} \begin{pmatrix} A^+_m \\ A^-_m \end{pmatrix}, \tag{3.29}$$

where $\mathbf{T}$ is the appropriate transfer matrix. For the structures considered in Sect.3.2.1 it has the following explicit form:

$$\begin{aligned} T_{11(22)} = & \frac{\pm J_A^3 S_A^2 \exp(\pm i(N_A - 1)k^{(A)}a)}{4J_I^2 J_B S_B^2 \sin(k^{(A)}a)\sin(k^{(B)}a)} \\ & \times \Big\{ \Big[(J_I/J_A)\left(F_\pm^{(A)} + (S_B/S_A)F_-^{(B)}\right) + F_\pm^{(A)}F_-^{(B)} \Big]^2 \\ & \times \exp\left(-i(N_B - 1)k^{(B)}a\right) \\ & - \Big[(J_I/J_A)\left(F_\pm^{(A)} + (S_B/S_A)F_+^{(B)}\right) + F_\pm^{(A)}F_+^{(B)} \Big]^2 \\ & \times \exp\left(i(N_B - 1)k^{(B)}a\right) \Big\}, \end{aligned} \tag{3.30a}$$

$$\begin{aligned} T_{12(21)} = & \frac{\pm J_A^3 S_A^2 \exp\left(\mp i(N_A + 1)k^{(A)}a)\right)}{4J_I^2 J_B S_B^2 \sin(k^{(A)}a)\sin(k^{(B)}a)} \\ & \times \Big\{ \Big[(J_I/J_A)\left(F_\pm^{(A)} + (S_B/S_A)F_-^{(B)}\right) + F_\pm^{(A)}F_-^{(B)} \Big] \\ & \times \Big[(J_I/J_A)\left(F_\mp^{(A)} + (S_B/S_A)F_-^{(B)}\right) + F_\mp^{(A)}F_-^{(B)} \Big] \\ & \times \exp\left(-i(N_B - 1)k^{(B)}a\right) \\ & - \Big[(J_I/J_A)\left(F_\pm^{(A)} + (S_B/S_A)F_+^{(B)}\right) + F_\pm^{(A)}F_+^{(B)} \Big] \\ & \times \Big[(J_I/J_A)\left(F_\mp^{(A)} + (S_B/S_A)F_+^{(B)}\right) + F_\mp^{(A)}F_+^{(B)} \Big] \\ & \times \exp\left(i(N_B - 1)k^{(B)}a\right) \Big\}, \end{aligned} \tag{3.30b}$$

where $F_\pm^{(A)}$ and $F_\pm^{(B)}$ are defined as

$$F_\pm^{(A)} = \exp\left(\pm ik^{(A)}a\right) - 1, \tag{3.31a}$$

$$F_{\pm}^{(B)} = \frac{J_B S_B}{J_A S_A} \left[\exp\left(\pm ik^{(B)}a\right) - 1\right]. \tag{3.31b}$$

The eigenvalues of the matrix $\mathbf{T}$ describe the spectrum of excitations, similarly as the eigenvalues of the matrix $\mathbf{M}$. One can easily formulate the appropriate dispersion equations in terms of the $\mathbf{T}$-matrix components.

Another method developed recently for finite layered systems is the so-called interface rescaling method which consists in performing a series of rescalings of the respective boundary conditions and reducing the eigenvalue problem of the composite system to that of one of its individual constituents [22,23]. The rescailed boundary conditions, of course, contain all the information about the dynamical properties of the removed part of the system. To some extent this method is similar to the transfer matrix approach, in which the dynamical properties of the elementary unit as a whole are contained in the matrix $\mathbf{M}$ or $\mathbf{T}$. This reduces the whole composite system to some effective homogeneous system with new dynamical parameters. This can be concluded directly from Eq. (3.10) which gives the following relation:

$$u_{(m+1)N+1} + (M_{11} + M_{22})u_{mN+1} + u_{(m-1)N+1} = 0. \tag{3.32}$$

This equation has a form similar to that for a homogeneous system,

$$u_{l+1} + \left(\frac{E - E_0}{JS}\right) u_l + u_{l-1} = 0, \tag{3.33}$$

where $E_0 = (2 + 4\Lambda)JS + g\mu_B H_0$. Comparison of the above two equations shows that layered structures with N atomic planes in the elementary bilayers have been replaced by an effective homogeneous system with $M_{11} + M_{22}$ instead of $(E - E_0)/JS$.

Other methods developed for layered structures are based on Green's function calculations, the poles of which determine the approprite spectrum. They are known as *surface Green's function matching* [24] *and interface response theory* [25].

3.2.3 Spin-wave spectrum in multilayers with ferromagnetic inter-layer coupling

The general implicit dispersion equations discussed above allow us to examine basic features of the spin-wave modes in layered structures. Let us begin with the dependence of the spectrum on the inter-layer coupling. Consider first two identical ferromagnetic layers which are exchange-decoupled. The spin-wave spectrum of a single film in the exchange-dominated region is well known [6]. For a given in-plane wavevector $\mathbf{Q}$ (which is a good quantum number for a single film as well as for a layered structure) the eigenmodes of a single film form a set of discrete levels. Each energy level of the bilayer as a whole is then two-fold degenerate. The degeneracy is lifted by a nonzero inter-layer coupling. When the inter-layer exchange parameter is equal to the exchange constant inside the films, the spin-wave spectrum of the bilayer coincides with the spectrum of a single homogeneous film with doubled thickness. The dependence of the energy levels on the inter-layer coupling is shown in Fig. 3.1(a) for a given parameter Λ (which determines the in-plane wavevector $\mathbf{Q}$) and for the case when the ferromagnetic films consist of three atomic planes. Each energy level (doubly degenerate at $J_I = 0$) splits into two levels at $J_I \neq 0$. One of them corresponds to an even wavefunction and the second to an odd one. The even modes are independent of the inter-layer exchange parameter, because for these modes the spins on both atomic planes at the interface precess always in phase and, consequently, do not feel the coupling between the planes, contrary to the odd modes, the energy of which increases considerably with increasing inter-layer coupling. This is clearly visible in Fig. 3.1(a). If the number of magnetic films increases, a more complicated energy structure emerges. Finally, for an infinite stack of films one obtains a band structure like that shown in Fig. 3.1(b). To some extent, this structure resembles the electronic band structure in solids, where the electronic bands develop from single atomic levels as a result of inter-atomic interactions. Here, the band structure developes from single-film levels due to the interaction between the films. It is worth noting that the gaps be-

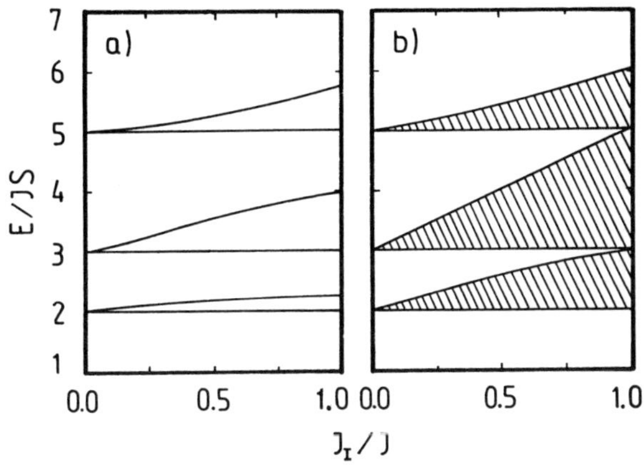

Fig. 3.1 (a) Spin-wave spectrum of two identical ($J_A = J_B = J$, $S_A = S_B = S$, $N_A = N_B = N$) exchange-coupled ferromagnetic films *versus* the inter-layer coupling for vanishing external magnetic field and for free surfaces ($J_A^s = J_B^s = J$). (b) Spin-wave spectrum of an infinite stack of such exchange-coupled films. The other parameters assumed in (a) and (b) are; $N = 3$ and $\Lambda = 0.5$.

tween the bands in Fig. 3.1(b) vanish for inter-layer coupling equal to that inside the films. At $J_I/J = 1$ the stack is simply equivalent to an infinite homogeneous system.

For a non-symmetrical double-layer there is no degeneracy at $J_I = 0$ (except an accidental one) and consequently no level splitting at $J_I \neq 0$. The modes are neither even nor odd. Consequently, all modes depend on the inter-layer coupling, contrary to the symmetrical bilayer discussed above, when the even modes were insensitive to the inter-layer coupling.

Spin-wave energy depends considerably on the in-plane wavevector $\mathbf{Q}$. This dependence is shown in Fig. 3.2 for a non-symmetrical bilayer (a) with $N_A = N_B = 3$ and with free surfaces, as well as for a trilayer (b) which, in comparison with the bilayer, has additional three atomic planes of the material A (structure $AAABBBAAA$). The spin-wave energy is plotted there against the parameter Λ which is related to $\mathbf{Q}$ by Eq. (3.7). It is worth noting that there are three additional modes in the trilayer in comparison with the bilayer. For the parameters

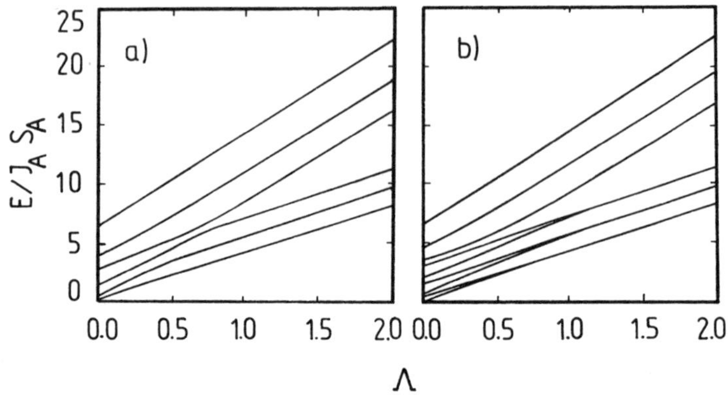

Fig. 3.2 Spin-wave spectrum of a bilayer (a) and a trilayer (b) for free surfaces ($J_A^s = J$, $J_B^s = J_B$). The relative spin-wave energy, E/J_AS_A, is shown as a function of the parameter Λ for the parameters $N_A = N_B = 3$, $J_B/J_A = 2$, $S_B/S_A = 1$, $J_I/J_A = 1.4$ and for vanishing external field, $H_0 = 0$.

assumed there these modes are degenerate for $\Lambda > 1$ with some other modes. The degeneracy is even more evident for structures composed of a large number of the elementary bilayers. In Fig. 3.3 we see the band structure (shaded regions) of bulk modes which can propagate in structures composed of an infinite number of the elementary bilayers. Each mode of Fig. 3.2(a) developes into the corresponding band in Fig. 3.3. For $\Lambda > 1$ all the bands are very narrow, indicating a high mode degeneracy. The upper band is very narrow for almost all Λ. The degeneracy is a consequence of the so-called mode confinement [16]. To understand this phenomenon let us have a look at the bands of bulk modes for both materials A and B. They are shown in Fig. 3.4. A mode of the stack, which corresponds to a point in the area where the bands overlap is of strongly collective nature. The appropriate wavefunction is of oscillatory type in both constituent materials. Consequently, the corresponding band in layered system is broad, as can be seen in Fig. 3.3. In the opposite limit a mode corresponds to a point which is far enough from either of the bands, say from the band B in Fig. 3.4. The appropriate wavefunction is then localized in a given film of the material A and decays exponentially

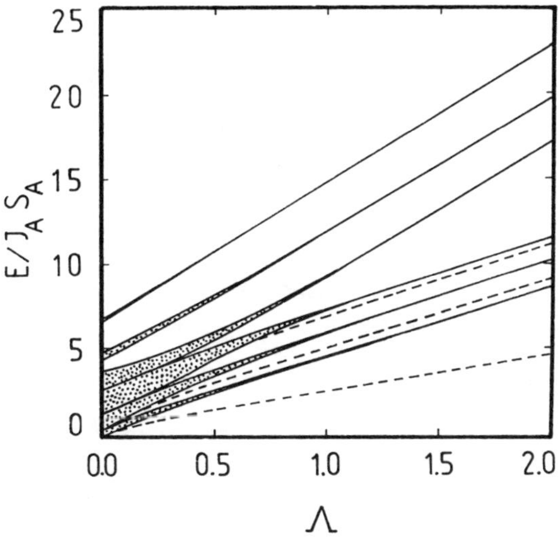

Fig. 3.3 Bulk (shaded regions limited by full lines) and surface (broken lines) modes in semi-infinite stacks for $J_A^S/J_A = 0.5$. The other parameters are the same as in Fig. 3.2.

inside the adjacent films of the material B. If the B-films are thick enough and/or the decay parameter is sufficiently large, the mode is well confined in a given film of the material A and the corresponding wavefunction does not overlap with the equivalent wavefunctions in the nearest films of the material A. Consequently, the corresponding band width vanishes, as is visible in Fig. 3.3 for $\Lambda > 1$. If the confinement is not perfect, the single-film modes delocalize and form a band of collective modes. The confinement is also visible in the wave propagation along the stacking direction, *i.e.*, for $\Lambda = 0$ ($Q = 0$) and $q_x \in < 0, \Pi/L >$. From Fig. 3.3 it follows that the gaps between the first four lower bands are very small at $\Lambda = 0$. For upper bands these gaps are quite significant (they are also called stop-bands) whereas the bands are relatively narrow. This is shown more explicitly in Fig. 3.5, where the spin-wave energy is plotted against q_x. The upper curve corresponds to modes which are well confined in films of the material B and, consequently, the curve is dispersionless.

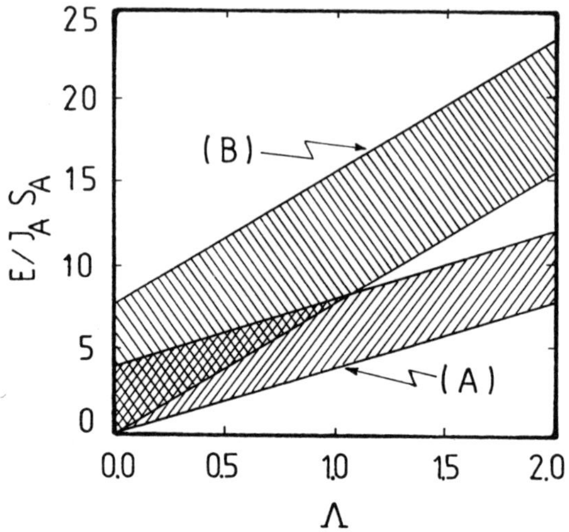

Fig. 3.4 Bands of the bulk modes in the materials A and B for the parameters used in Fig. 3.2 and Fig. 3.3.

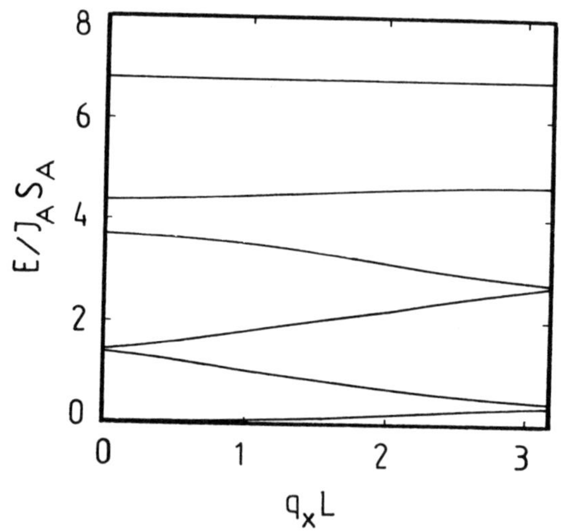

Fig. 3.5 Spin-wave spectrum *versus* q_x for infinite layered structures and for $\Lambda = 0$. The other parameters assumed here are the same as in Fig. 3.2.

Semi-infinite structures support additionally modes exponentially localized at the surface. If the spins in equivalent atomic planes in adjacent elementary bilayers precess in phase, the surface modes of the stack are of the *acoustic* type. If they precess in anti-phase, they are surface *optic* modes. In Fig. 3.3 these modes are shown (broken lines) for structures with the material A at the top and for weakened exchange coupling in the surface atomic plane. There are then three surface modes, one of the *optic* and two (the lower and upper broken curves) of the *acoustic* types. The spectrum of surface modes depends strongly on such parameters as the exchange constant in the surface atomic plane, the exchange constant between the two surface atomic planes, the thickness of the surface film *etc.* A characteristic feature of the superlattice surface modes is that they exist also for free surface, whereas no surface mode exists in the corresponding semi-infinite uniform material, *i.e.*, for semi-infinite Heisenberg model on a simple cubic lattice with nn-exchange coupling and with (100)-surface [17].

3.2.4 Other structures

In some layered structures ferromagnetic films couple antiferromagnetically when they are in direct contact, as in Fe/Gd systems, or when they are interspaced by a few atomic planes of a non-magnetic metal, as for example in Co/Cu multilayers. In the general case such structures have a complicated ground state configuration in an external magnetic field [26,27]. For example, in the case of Fe/Gd structures one finds a Fe-aligned configuration, Gd-aligned configuration and twisted phase. In the Fe-aligned phase the Fe-spins are parallel to the applied field and the Gd-spins are antiparallel, whereas in the Gd-aligned configuration the situation is reversed [26]. In some field and temperature ranges one finds a twisted phase in which neither Fe nor Gd spins are parallel to the field.

The dispersion equations discussed in Sect.3.2.1 are valid for antiferromagnetic coupling between the ferromagnetic films, but for a magnetic field strong enough to overcome the antiferromagnetic coupling and force parallel alignment of the different magnetizations.

One can also easily modify them to antiparallel configuration (aligned phases). For twisted phase, however, the problem is more complicated and one has to rely more on numerical calculations [28,29].

Another kind of structures investigated theoretically are ferromagnetic/antiferromagnetic layered structures [30,31]. As in ferromagnetic/ferromagnetic structures with antiferromagnetic inter-layer coupling, their ground state configuration is unstable with respect to application of an external magnetic field along the easy direction, and the system evolves from the ground state configuration to full alignment in a sufficiently strong field passing through a sequence of spin reorientation transitions. The spin-wave spectrum of such structures has been analysed numericaaly for all basic phases.

3.3 MAGNETOSTATIC MODES

Let us now proceed to the opposite limit, when the exchange coupling is negligible in comparison with the dipolar term in Eq. (3.1). Some general conditions for the validity of this approximation have been discussed in Sect.3.1 [32]. It will be helpful now to recall some basic features of the magnetostatic modes in a single ferromagnetic film with the static magnetization parallel to the film plane. For zero in-plane wavevector, $Q = 0$, there is a uniform mode which is degenerate with the standing modes. If Q increases, the uniform mode transforms, for some range of propagation directions, into the Damon-Eshbach (DE) surface mode, the amplitude of which varies exponentially across the film. This is accompanied by an increase in the mode frequency. Consequently, the DE-modes exist at frequencies which are well above the region where the bulk-type modes propagate. To describe properly the bulk-type modes in very thin films, one must take the exchange coupling into account. However, even for such very thin films the dipolar approximation is reasonable for the DE-modes. Of course, in a limited range of the in-plane wavevectors.

The DE-modes present very interesting and unique features. In a semi-infinite ferromagnet with the static magnetization parallel to the surface they can propagate only for a limited range of angles α,

$-\alpha_c < \alpha < \alpha_c$, where α is defined as the angle between the wavevector $\mathbf{Q}$ and a unit vector $\boldsymbol{\eta}$, with $\boldsymbol{\eta}$ parallel to the surface, normal to the static magnetization $\mathbf{M}_0$ and obeying the condition that $\mathbf{M}_0 \times \boldsymbol{\eta}$ points from the surface into the magnetic system. At a critical angle $\alpha = \pm\alpha_c$ the DE-mode merges into the spectrum of bulk-type modes. Consequently, the DE-mode in a semi-infinite ferromagnet is a non-reciprocal mode. Generally, a mode is non-reciprocal if two waves with opposite wavevectors differ in frequency, $\omega(-\mathbf{q}) \neq \omega(\mathbf{q})$ [9]. In a single film the spin-wave spectrum is an even function of the in-plane wavevector. This is a consequence of the existence of two surfaces. However, the DE-modes with opposite wavevectors are localized at opposite surfaces.

The dipolar modes are well described by the classical continuum model, in which the dipolar interactions are taken into account by way of a macroscopic dipolar field $\mathbf{h}(\mathbf{r}, t)$, whereas the spin variables are considered as a continuous classical spin variable $\mathbf{s}(\mathbf{r}, t)$. Instead of $\mathbf{s}(\mathbf{r}, t)$, one can equivalently use the magnetization $\mathbf{M}(\mathbf{r}, t)$. On neglecting eddy-currents and retardation corrections, the dipolar field $\mathbf{h}(\mathbf{r}, t)$ can be found from the magnetostatic equations,

$$\boldsymbol{\nabla} \times \mathbf{h}(\mathbf{r}, t) = 0, \tag{3.34a}$$

$$\boldsymbol{\nabla} \cdot [\mathbf{h}(\mathbf{r}, t) + 4\Pi\mathbf{m}(\mathbf{r}, t)] = 0, \tag{3.34b}$$

where $\mathbf{m}(\mathbf{r}, t)$ is the dynamical component of the magnetization, $\mathbf{M}(\mathbf{r}, t) = \mathbf{M}_0 + \mathbf{m}(\mathbf{r}, t)$. The magnetization $\mathbf{m}(\mathbf{r}, t)$ and the magnetic field $\mathbf{h}(\mathbf{r}, t)$ are additionally related through the Landau equation of motion which can be easily solved in terms of the dynamical susceptibility $\boldsymbol{\chi}(\omega)$ of the system or, equivalently, in terms of the dynamical magnetic permeability tensor $\boldsymbol{\mu}(\omega) = \mathbf{1} + 4\Pi\boldsymbol{\chi}(\omega)$, where $\mathbf{1}$ is the 3×3 unit matrix. For axial systems $\boldsymbol{\mu}(\omega)$ has the following general form; $\mu_{xx}(\omega) = \mu_{yy}(\omega) \equiv \mu_{\perp}(\omega)$, $\mu_{xy}(\omega) = -\mu_{yx}(\omega) \equiv i\mu_{\times}(\omega)$ and $\mu_{zz}(\omega) = 1$, with the other components vanishing. To solve the magnetostatic equations one usually introduces a scalar potential $\varphi(\mathbf{r}, t)$ by the definition $\mathbf{h}(\mathbf{r}, t) = -\boldsymbol{\nabla}\varphi(\mathbf{r}, t)$.

3.3.1 General dispersion equations for magnetic/non-magnetic structures with in-plane magnetization

The DE-modes in alternating magnetic/non-magnetic layered structures with static magnetization parallel to the film planes have been discussed theoreticaly for the first time by Camley *et al* [33] and by Grünberg and Mika [34]. Subsequently, many other theoretical papers have appeared, in which the magnetostatic modes were analysed for various compositions and geometries [35-40]. To present and interpret some basic features of the DE-modes in layered structures we restrict our discussion to wave propagation in Voigt geometry, *i.e.*, for $\mathbf{Q}$ perpendicular to the static magnetization; $\mathbf{Q} = Q_y\hat{\mathbf{y}}$, with $\hat{\mathbf{y}}$ being the unit vector along the axis y (see Fig. 3.6). For a general propagation geometry one finds behaviour similar to that in thin films, *i.e.*, the DE-modes vanish at some critical propagation direction.

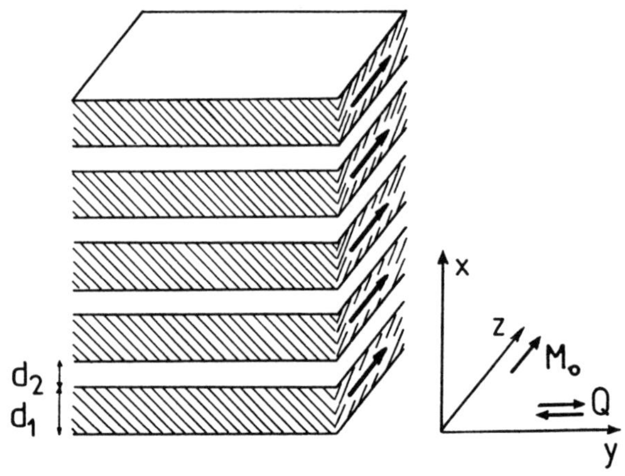

Fig. 3.6 Propagation geometry and the coordinate system used for description of magnetostatic modes.

Owing to the translational invariance in the film plane one can seek $\varphi(\mathbf{r}, t)$ in the form

$$\varphi(\mathbf{r}, t) = \Phi(x) \exp[i(\mathbf{Q} \cdot \mathbf{R} - \omega t)], \tag{3.35}$$

where **R** is the two-dimensional in-plane position vector. Inside the magnetic films the function $\Phi(x)$ obeys the following differential equation:

$$\frac{\partial^2\Phi(x)}{\partial x^2} - Q^2\Phi(x) = 0, \tag{3.36}$$

with $Q = \mid Q_y \mid$. The corresponding equation for non-magnetic films has the same form. To find $\Phi(x)$ in layered structures, one has to solve the corresponding equations for $\Phi(x)$ in both films in the elementary bilayer and then to taylor the solutions in adjacent films with the appropriate boundary conditions. This can be easily done within a transfer matrix formalism similar to that described in Sect.3.2.2, by writing $\Phi(x)$ inside the first film of the m-th elementary bilayer, $(m-1)L < x < (m-1)L + d_1$, in the form

$$\begin{aligned} \Phi_m(x) &= A_m^+ \exp[Q\,(x-(m-1)L)] \\ &\quad + A_m^- \exp[-Q\,(x-(m-1)L)], \end{aligned} \tag{3.37}$$

where L is the thickness of the elementary bilayer as a whole, $L = d_1 + d_2$. d_1 and d_2 are the thicknesses of the first (magnetic) and second (non-magnetic) layers in the elementary unit, whereas A_m^+ and A_m^- are constants. A similar expression holds also for the non-magnetic layers, but with new constants. The magnetostatic boundary conditions of continuity of the tangential component of the magnetic field and normal component of $\mathbf{b}(\mathbf{r},t) = \mathbf{h}(\mathbf{r},t) + 4\Pi\mathbf{m}(\mathbf{r},t)$ at each interface relate the constants in adjacent elementary bilayers and this relation defines the appropriate transfer matrix **T** in accordance with the relation (3.29). The matrix **T** is given explicitly by the formulae

$$\begin{aligned} T_{11(22)} &= \exp(\pm Qd_1) \\ &\times \left[\cosh(Qd_2) \pm \frac{1}{2}\left(\mu_v(\omega) + \frac{1}{\mu_\perp(\omega)}\right)\sinh(Qd_2)\right], \end{aligned} \tag{3.38a}$$

$$T_{12(21)} = \exp(\mp Q d_1) \sinh(Q d_2) \times \left(-\mu_\times(\omega) \mathrm{sign}(Q_y) \pm \frac{1 - \mu_\perp^2(\omega) - \mu_\times^2(\omega)}{2\mu_\perp(\omega)} \right), \quad (3.38b)$$

where the lower sign refers to the indices in parentheses, $\mathrm{sign}(Q_y) = 1$ for $Q_y > 0$ and $\mathrm{sign}(Q_y) = -1$ for $Q_y < 0$, whereas $\mu_v(\omega)$ is the magnetic permeability in Voigt configuration,

$$\mu_v(\omega) = \mu_\perp(\omega) - \frac{\mu_\times^2(\omega)}{\mu_\perp(\omega)}. \quad (3.39)$$

Having found the transfer matrix, one can derive the appropriate dispersion equations for the wave spectrum of infinite, semi-infinite or finite structures in terms of the matrix $\mathbf{T}$ and some additional functions which follow from the magnetostatic boundary conditions at the free surfaces [40]. For example, the spectrum of bulk superlattice modes is determined by the explicit dispersion equation

$$\cos(q_x L) = \cosh(Q d_1) \cosh(Q d_2) + \frac{1}{2} \left(\mu_v(\omega) + \frac{1}{\mu_\perp(\omega)} \right) \sinh(Q d_1) \sinh(Q d_2). \quad (3.40)$$

Similarly, the dispersion relation for surface modes in the semi-infinite stack takes the form

$$\mu_\perp(\omega) + \mu_\times(\omega) \mathrm{sign}(Q_y) + 1 = 0, \quad (3.41)$$

provided $d_1 > d_2$. If $d_1 < d_2$, no surface solution exists. In a similar way one can find the dispersion equation for finite structures.

Some important features of the spin-wave modes follow from the above general considerations. First of all, we emphasise that the

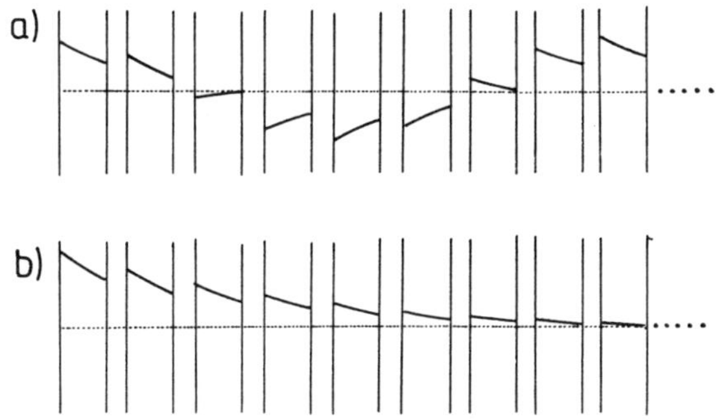

Fig. 3.7 Schematic illustration of the bulk (a) and surface (b) magnetostatic modes composed of the surface DE-modes in each individual ferromagnetic film.

superlattice modes discussed here (bulk, surface and those in finite structures) are composed of the surface modes in each individual magnetic film. In the case of bulk modes the envelope function varies oscillatory-like, whereas for surface modes it varies exponentially. This behaviour is shown schematically in Fig. 3.7. Some other features follow from the general dispersion equations (3.40) and (3.41). These are: (i) The surface modes exist only for $d_1 > d_2$, *i.e.*, if the magnetic films are thicker than the non-magnetic ones. If $d_1 < d_2$ no surface mode can propagate. (ii) Equation (3.41) coincides with the dispersion equation for surface modes in semi-infinite axial magnetic systems. Consequently, the surface modes in semi-infinite stacks of alternating magnetic and non-magnetic films have the same frequency as the surface modes in the corresponding semi-infinite magnetic material. (iii) If $\mu_\times(\omega) \neq 0$ then Eq. (3.41) is not invariant with respect to the reversal of the direction of wave propagation, $\mathbf{Q} \Rightarrow -\mathbf{Q}$, contrary to Eq. (3.40) which is symmetrical with respect to the transformation $\mathbf{Q} \Rightarrow -\mathbf{Q}$. Consequently, the spectrum of superlattice surface modes is non-reciprocal if $\mu_\times(\omega) \neq 0$, exactly as in semi-infinite homogeneous magnetic materials. On the other hand, the spectrum of bulk superlattice modes is symmetrical with respect to the reversal of the propagation direction.

3.3.2 Ferromagnetic/non-magnetic structures

The first layered structures investigated both experimentally [41,42] and theoretically [33,34] were composed of ferromagnetic and non-magnetic materials. If the external field is along the easy axis, the components of the frequency-dependent magnetic permeability tensor are equal;

$$\mu_{\perp}(\omega) = 1 + \frac{\Omega_0 \Omega_m}{\Omega_0^2 - \omega^2}, \tag{3.42a}$$

$$\mu_{\times}(\omega) = \frac{-\Omega_m \omega}{\Omega_0^2 - \omega^2}, \tag{3.42b}$$

where Ω_0 and Ω_m are defined as $\Omega_0 = \gamma(H_0 + H_0^{an})$ and $\Omega_m = 4\Pi\gamma M_0$, with $-\gamma$ being the magneto-mechanical factor (γ is positive by definition). The axial anisotropy is included by way of a static effective anisotropy field H_0^{an}. In many ferromagnets, however, the anisotropy field is small and can be neglected in usually applied fields, so we assume $\Omega_0 = \gamma H_0$.

Let us begin with the simplest structure consisting of two identical ferromagnetic films of thickness d_1, separated by a non-magnetic spacer of thickness d_2. This problem has first been analysed by Grünberg [43]. Consider a wave with the in-plane wavevector $\mathbf{Q}$ propagating in a single isolated film. The precessing magnetic dipoles then generate a macroscopic magnetic field with the frequency and wavevector of the spin wave. This field decays exponentially with the distance from the film. If there is another film at the distance d_2, it will couple to the field generated by the propagating mode. The coupling modifies the spectrum of excitations which now are collective states of both films. If d_2 is sufficiently large, for a given Q, the two films can be considered as decoupled. For small d_2 the modes interact, which lifts the mode degeneracy. This behaviour is shown in Fig. 3.8. For vanishing spacer thickness the upper mode becomes the DE-mode of a single film with doubled thickness, whereas the lower one merges into the bulk-type modes. The spin-wave spectrum is symmetrical with respect to reversal of the propagation direction. If the number of magnetic films in the stack increases, the spin-wave spectrum becomes more complicated. In the general case one finds as many branches of magnetostatic modes as many ferromagnetic films are contained in the structure [35]. Finally, for the infinite superlattice they form a band of bulk superlattice modes which are collective states of the whole stack. Such a band is shown in Fig. 3.9 (hatched region) for the parameters typical for YIG. Two waves propagating oppositely have the same frequency, but their wavefunctions are localized at opposite surfaces of the magnetic films. In semi-infinite stacks there is an additional surface mode for $d_1 > d_2$. This mode has the frequency $\omega_{DE} = \Omega_0 + \Omega_m/2$ and propagates only for $Q_y < 0$, which coincides with the DE-mode in the correponding semi-infinite ferromagnet. This mode is represented in Fig. 3.9 by the broken line.

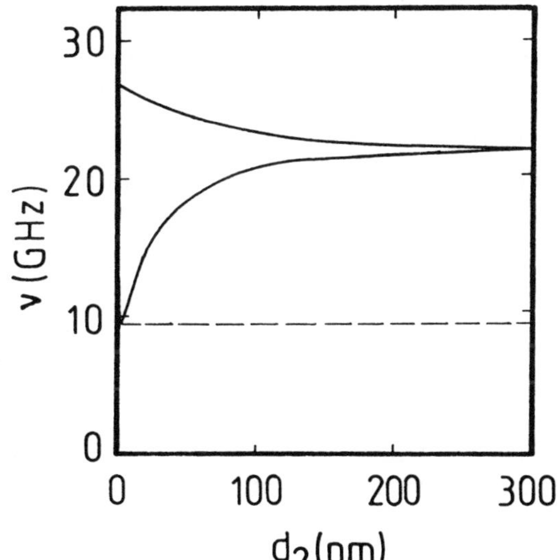

Fig. 3.8 DE-modes (full lines) of two identical films of thickness $d_1 = 15nm$ *versus* the spacer thickness d_2 for $H_0 = 0.5\,kG, 4\Pi M_0 = 22\,kG$, $g = 2$ and $Q = 1.82 \times 10^5\,cm^{-1}$. The broken line corresponds to volume modes in the ferromagnetic films.

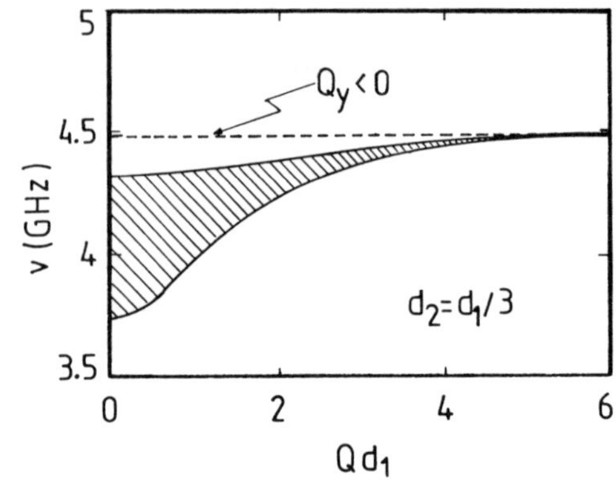

Fig. 3.9 Spectrum of bulk (hatched region) and surface (broken line) spin-wave modes in semi-infinite structures for $4\Pi M_0 = 1.78\ kG$, $H_0 = 445\ G$ and $g = 2$.

3.3.3 Antiferromagnetic/non-magnetic structures

The formulae discussed in Sect.3.3.1 apply as well to antiferromagnetic/non-magnetic structures with the easy-axis for sublattice magnetizations parallel to the film planes and for external fields weaker than the spin-flop transition field. The difference between ferromagnetic and antiferromagnetic systems resides in the magnetic permeability tensor, which for antiferromagnetic materials can be written in the form

$$\mu_\perp(\omega) = 1 + \frac{\Omega_a \Omega_m}{\Omega_1^2 - (\omega + \gamma H_0)^2} + \frac{\Omega_a \Omega_m}{\Omega_1^2 - (\omega - \gamma H_0)^2}, \tag{3.43a}$$

$$\mu_\times(\omega) = \frac{\Omega_a \Omega_m}{\Omega_1^2 - (\omega + \gamma H_0)^2} - \frac{\Omega_a \Omega_m}{\Omega_1^2 - (\omega - \gamma H_0)^2}, \tag{3.43b}$$

where Ω_1 is the antiferromagnetic resonance frequency in zero applied field,

$$\Omega_1 = \gamma \left[H_0^{an} \left(H_0^{an} + 2H_0^{exch} \right) \right]^{\frac{1}{2}}, \tag{3.44}$$

with H_0^{an} and H_0^{exch} denoting the static anisotropy and exchange field, respectively. In Eqs (3.43a) and (3.43b) Ω_m is defined similarly as for ferromagnets, but with M_0 denoting the sublattice magnetization. Finally, Ω_a is defined as $\Omega_a = \gamma H_0^{an}$.

The magnetostatic modes in a semi-infinite antiferromagnet and in a single antiferromagnetic film with in-plane static magnetization have been studied by Camley [44] and Stamps and Camley [45], respectively. In a semi-infinite antiferromagnet the surface modes exist for both $Q_y > 0$ and $Q_y < 0$, and the spectrum is reciprocal at $H_0 = 0$. The reciprocality is a simple consequence of the existence of two equivalent magnetic sublattices with opposite magnetizations and can also be proved on the basis of some general symmetry arguments. However, the reciprocality is destroyed for $H_0 > 0$. The surface modes with $Q_y > 0$ then shift down and the modes with $Q_y < 0$ shift up on the energy scale. In a single film there are two modes at arbitrary wavevector **Q**. If the film is thick enough (for a given Q), the corresponding wavefunctions do not overlap and the two modes are degenerate at $H_0 = 0$ [45]. The magnetic field lifts the degeneracy. For small thickness the modes couple to each other, so the degeneracy is lifted even for vanishing external magnetic field.

Consider now a stack of such antiferromagnetic films separated by a non-magnetic material. The single-film states delocalize due to the long-range nature of the dipolar forces and form bands of collective states of the system as a whole. The spectrum of bulk and surface modes in semi-infinite stacks are shown in Fig. 3.10(a) for $H_0 = 0$ and for $d_2 = d_1$, $d_2 > d_1$ and $d_2 < d_1$. The other parameters assumed there are typical for MnF_2. In each case there are two bands of bulk waves (the hatched regions), which develope from the two surface modes existing in a thin antiferromagnetic film. The bands are well separated for $d_2 \neq d_1$, but the gap vanishes at $d_2 = d_1$. If $d_2 < d_1$ then the surface mode occurs in the middle of the gap and exists for both $Q_y > 0$ and $Q_y < 0$. Its frequency is the same as for the corresponding semi-infinte antiferromagnet, *ie*,

$$\omega_{DE} = \gamma \left[H_0^{an}\left(H_0^{an} + 2H_0^{exch} + 4\Pi M_0\right)\right]^{\frac{1}{2}} . \qquad (3.45)$$

The case of $H_0 \neq 0$ is shown in Fig. 3.10(b). The band separation now increases with increasing H_0 and is non-zero also for $d_1 = d_2$. The magnetic field also destroys the reciprocality of the surface modes. The surface waves propagating with the wave vectors $Q_y < 0$ and $Q_y > 0$ now differ in frequency, similarly as in semi-infinite antiferromagnets.

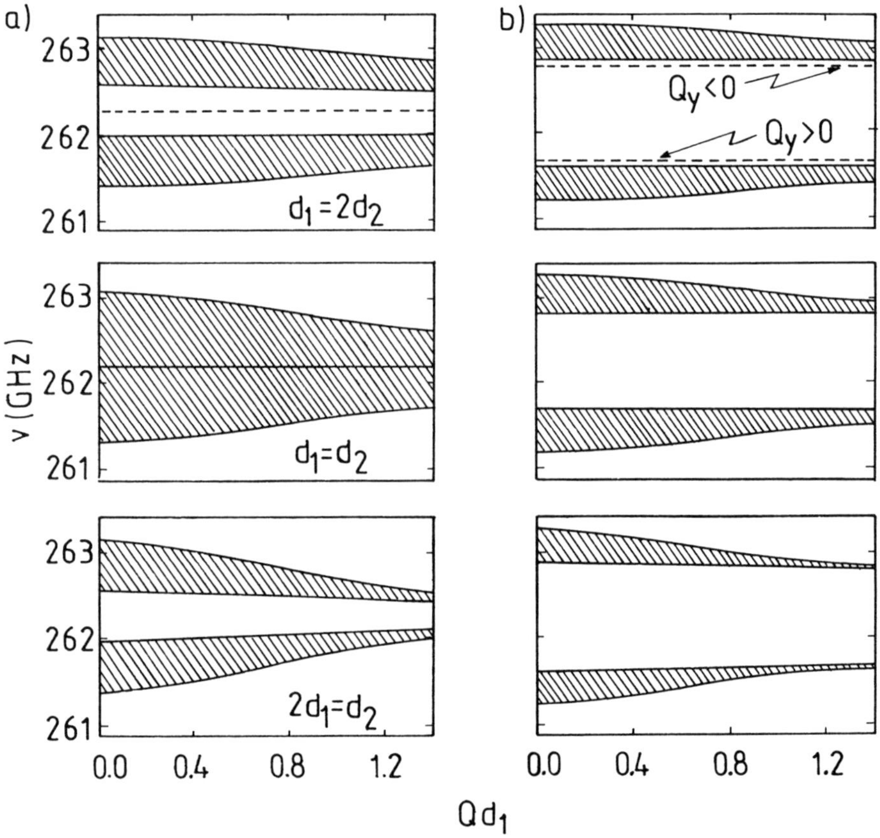

Fig. 3.10 (a) Bulk (hatched regions) and surface (broken lines) modes in semi-infinite antiferromagnetic/non-magnetic structures for $H_0^{an} = 7.87\ kG$, $4\Pi M_0 = 7.54\ kG$, $H_0^{exch} = 550\ kG$, $g = 2$, $H_0 = 0$ and d_2/d_1 as indicated. (b) The same as in (a) but for $H_0 = 0.2\ kG$.

3.3.4 Other structures

A very interesting example of layered structures, investigated both experimentally and theoretically, is the case of alternating ferromagnetic/non-magnetic structures with antiparallel alignment of the film magnetizations. Such an orientation can be obtained if the films couple antiferromagnetically across a thin inter-layer [46]. Antiparallel alignment of the magnetizations of two films can also be obtained in a magnetic field if the films have different coercive fields [47]. The simplest case of two ferromagnetic films with antiparallel alignment of the film magnetizations has been analysed by Grünberg [18]. Such a structure presents novel interesting features in comparison with the similar structure with parallel alignment of the magnetizations. The most interesting feature is the non-reciprocality of the spin-wave spectrum. The non-reciprocality is a simple consequence of some asymmetry in the wave propagation. The modes of a given wavevector are localized, say, at the internal surfaces, whereas the modes with opposite wavevectors propagate on external surfaces of the bilayer [48].

Consider now structures with both components ferromagnetic. The corresponding bilayer has been considered by Wolfram and DeWames [5]. Because of the lack of rotational symmetry, the appropriate spin-wave spectrum is non-reciprocal. Non-reciprocality also occurs for all multilayers containing a finite number of complete elementary bilayers. For a structure with the last unit incomplete, rotational symmetry is restored and, consequently, the spectrum is reciprocal. For infinite structures one finds two reciprocal bands, shown in Fig. 3.11 for parameters typical for Fe and Co, with the anisotropy fields neglected. The surface waves in the semi-infinite structures are shown in Fig. 3.11 by broken and dotted lines. In the general case there are two branches of such waves, which exist only at $Q_y < 0$. One of the two modes usually exists in a restricted range of wavevectors.

In all the cases discussed above the spectrum of the corresponding infinite structures was reciprocal. The reciprocality is a consequence of the fact that any two-component periodic structure has rotational

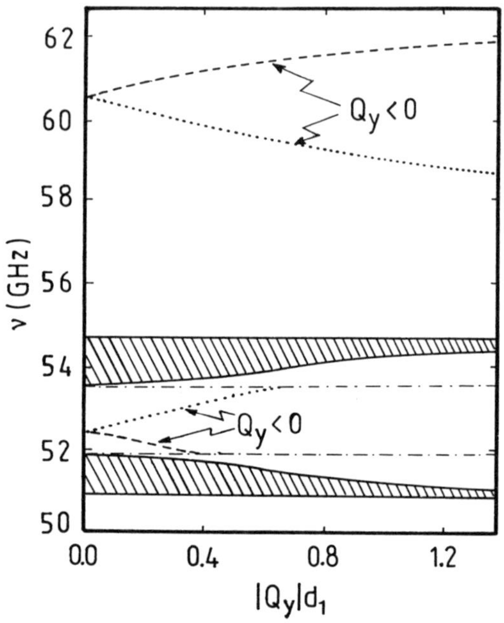

Fig. 3.11 Bulk (hatched regions) and surface (broken and dotted lines) modes of semi-infinite stacks composed of two different ferromagnetic materials for $d_1 = d_2$, $H_0 = 10\ kG$ and for parameters typical for Fe ($4\Pi M_0 = 21\ kG$ and $g = 2.15$) and Co ($4\Pi M_0 = 17.6\ kG$ and $g = 2.17$). The broken (dotted) curves correspond to Fe (Co) films at the top of the structure. Volume-type modes in the Fe- and Co-films are also indicated (— · — · — · —).

symmetry, although the corresponding finite structures do not necessarily have it. However, one can imagine structures for which not only surface but also bulk modes in infinite stacks are non-reciprocal. As one easily notes, rotational symmetry is absent, for example, in infinite structures with the elementary unit composed of three different ferromagnetic films [49]. The respective spectrum of bulk modes is shown in Fig. 3.12, where all three films were assumed to be ferromagnetic. The non-reciprocality is clearly evident. Similar behaviour can also be found if one of the three films is nonmagnetic.

The discussion of this Section was restricted to layered structures

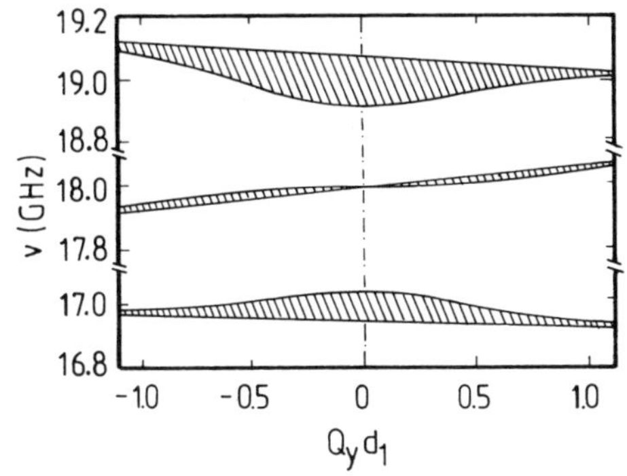

Fig. 3.12 Spin-wave spectrum of infinite structures composed of three different ferromagnetic materials for $d_1 = d_2 = d_3$, $H_0 = 5.3\,kG$, $4\Pi M_{01} = 3.2\,kG$, $4\Pi M_{02} = 2.4k\,G$, $4\Pi M_{03} = 1.6\ kG$ and $g_1 = g_2 = g_3 = 2$.

with the film magnetizations parallel to the film planes. One can also consider structures with magnetizations perpendicular to the film planes [37]. In such a configuration there is no analog of the surface DE-mode. Consequently, the exchange coupling plays a more important role in the wave spectrum.

3.4 DIPOLE-EXCHANGE MODES

As remarked in the Introduction, there is a region where the dipolar and exchange contributions to the spin-wave energy are comparable. This region is of particular interest because it is usually realized in experiments. Comparison of the theoretical and experimental results then provides important information about some parameters of the structure, such as for example the inter-layer exchange constant, or others. The exact solution of the problem could be obtained within the microscopic model with all terms of the Hamiltonian (1) considered on an equal footing. However, the problem can be simplified if one takes into account the fact that in this region the appropriate wavefunctions vary slowly on the scale of the lattice constant. It is then possible to go over from the discrete model to the macroscopic continuum description similarly as for the dipolar modes. However, the exchange coupling will be included into the considerations by way of some effective exchange field, whereas it was completely ignored for the dipolar modes. We will restrict our further discussion to layered structures composed of ferromagnetic materials.

3.4.1 Equation of motion for the magnetization

The classical equation of motion can be easily obtained from the quantum-mechanical equation by dealing with the spin variables as a continuous classical function, similarly as in the magnetostatic problem. This leads to the following equation for the time-derivative of the magnetization inside each ferromagnetic film [50]:

$$\frac{\partial \mathbf{M}(\mathbf{r},t)}{\partial t} = -\gamma \mathbf{M}(\mathbf{r},t) \times \mathbf{H}^{eff}(\mathbf{r},t), \tag{3.46}$$

where $\mathbf{H}^{eff}(\mathbf{r},t)$ is an effective field consisting of the internal magnetic field $\mathbf{H}^{i}(\mathbf{r},t)$, the anisotropy field $\mathbf{H}^{an}(\mathbf{r},t)$ and the effective exchange field $\mathbf{H}^{exch}(\mathbf{r},t)$:

$$\mathbf{H}^{eff}(\mathbf{r},t) = \mathbf{H}^{i}(\mathbf{r},t) + \mathbf{H}^{an}(\mathbf{r},t) + \mathbf{H}^{exch}(\mathbf{r},t). \tag{3.47}$$

In the general case, the internal field is a sum of the externally applied magnetic field, a constant demagnetization field and a dynamic part of the dipolar field, $\mathbf{h}(\mathbf{r},t)$. The demagnetization field vanishes for the static magnetization parallel to the film planes, so one can write

$$\mathbf{H}^{i}(\mathbf{r},t) = \mathbf{H}_0 + \mathbf{h}(\mathbf{r},t). \tag{3.48}$$

The anisotropy contribution to the effective field is given by

$$\mathbf{H}^{an}(\mathbf{r},t) = -(1/M_0)\nabla_{\mathbf{u}} E^{an}(\mathbf{u}), \tag{3.49}$$

where $E^{an}(\mathbf{u})$ is the anisotropy energy density, the exact form of which strongly depends on the structure and symmetry of the system, and $\mathbf{u}$ is a unit vector along the local magnetization, $\mathbf{u} = \mathbf{M}(\mathbf{r},t)/M_0$.

For brevity, the **r**- and t-dependence of **u** has been dropped. For example, the anisotropy energy in uniaxial ferromagnets has the form: $E^{an}(\mathbf{u}) = -K^u(\mathbf{e}\cdot\mathbf{u})^2$, where K^u is the uniaxial anisotropy constant and **e** is a unit vector along the symmetry axis. With this definition the case $K^u > 0$ corresponds to the case of an easy axis and $K^u < 0$ to that of an easy plane. For cubic ferromagnets the corresponding expression has the form $E^{an}(\mathbf{u}) = K(u_x^2u_y^2 + u_x^2u_z^2 + u_y^2u_z^2)$ in the coordinate system defined by the principal axes of the structure. Within the linear approximation one retains only terms linear in the dynamical part of the magnetization writing $\mathbf{H}^{an}(\mathbf{r},t) = \mathbf{H}_0^{an} + \mathbf{h}^{an}(\mathbf{r},t)$, where $\mathbf{H}_0^{an}$ is the static part of the anisotropy field and $\mathbf{h}^{an}(\mathbf{r},t)$ is its dynamical component, which is linear in $\mathbf{m}(\mathbf{r},t)$.

Finally, the effective exchange field for cubic materials is usually written in the form:

$$\mathbf{H}^{exch}(\mathbf{r},t) = (2A/M_0^2)\nabla^2\mathbf{m}(\mathbf{r},t), \tag{3.50}$$

where A is the exchange parameter. For materials with nn-exchange coupling this phenomenological parameter can be related to the microscopic exchange constant J (see Eq. (3.2)) by the formula

$$A = \frac{\sigma J S^2}{a}, \tag{3.51}$$

where $\sigma = 1/2$ for simple cubic (sc) lattice, $\sigma = 1$ for body-centered-cubic (bcc) lattice and $\sigma = 2$ for face-centered-cubic (fcc) lattice.

For a weakly excited system the equation of motion can be linearized with respect to the dynamical part of the magnetization $\mathbf{m}(\mathbf{r},t)$ and magnetic field $\mathbf{h}(\mathbf{r},t)$. This equation, together with the magnetostatic equations for $\mathbf{h}(\mathbf{r},t)$ form a closed system. Assuming the time-dependence in the form $\mathbf{m}(\mathbf{r},t) = \mathbf{m}(\mathbf{r})\exp(-i\omega t)$ (and a similar form for the scalar potential $\varphi(\mathbf{r},t)$), one obtains three independent equations for $m_x(\mathbf{r})$, $m_y(\mathbf{r})$ and $\varphi(\mathbf{r})$. Below, we write these equations for isotropic systems, where the magnetocrystalline anisotropy can be neglected:

$$i\tilde{\omega}m_x(\mathbf{r}) + \hat{L}m_y(\mathbf{r}) - \frac{M_0}{H_0}\frac{\partial\varphi(\mathbf{r})}{\partial y} = 0, \tag{3.52a}$$

$$i\tilde{\omega}m_y(\mathbf{r}) - \hat{L}m_x(\mathbf{r}) + \frac{M_0}{H_0}\frac{\partial\varphi(\mathbf{r})}{\partial x} = 0, \tag{3.52b}$$

$$\nabla^2\varphi(\mathbf{r}) - \frac{\partial m_x(\mathbf{r})}{\partial x} - \frac{\partial m_y(\mathbf{r})}{\partial y} = 0, \tag{3.52c}$$

where $\tilde{\omega} = \omega/\gamma H_0$ and

$$\hat{L} = \frac{2A}{M_0H_0}\nabla^2 - 1. \tag{3.53}$$

For $A = 0$ the problem reduces to the magnetostatic one. If $A \neq 0$ then Eqs (3.52) are differential equations of the second order, which include nonlocal contributions arising from the exchange interaction. In layered structures the solutions of these equations are of the form $\sim \exp(i\mathbf{Q}\cdot\mathbf{R} + ikx)$, where $\mathbf{Q}$ is the in-plane wavevector (real) and k is complex, in general. Since most relevant experiments are usually performed in the Voigt geometry, we assume that $\mathbf{Q}$ is perpendicular to $\mathbf{M}_0$ (see Fig. 3.6). For a given frequency one finds the following three different solutions for k^2:

$$k_{1(2)}^2 = \frac{M_0H_0}{2A}\left\{-1 - \frac{M_0}{2H_0} \pm \left[\left(\frac{M_0}{2H_0}\right)^2 + \tilde{\omega}\right]^{\frac{1}{2}}\right\} - Q^2, \tag{3.54a}$$

$$k_3^2 = -Q^2. \tag{3.54b}$$

Thus, the solutions of Eqs (3.52), corresponding to a given wavevector $\mathbf{Q}$, can be written in each layer as a linear combination of six independent terms [51-54],

$$m_x(\mathbf{r}) = \sum_{\nu=1}^{3}\left[C_\nu \exp(ik_\nu x) + D_\nu \exp(-ik_\nu x)\right]\exp(i\mathbf{Q}\cdot\mathbf{R}), \tag{3.55}$$

where C_ν and D_ν are constants. Similar expansions also hold for $m_y(\mathbf{r})$ and $\varphi(\mathbf{r})$ with new constants which, however, are not independent and can be related to C_ν and D_ν through simple relations [51,52].

3.4.2 Boundary conditions

The magnetostatic equation for the scalar potential and the linearized forms of the Landau-Lifshits equation of motion for the magnetization are all second-order differential equations. When considering a system with surfaces or interfaces one has to supplement these equations with appropriate boundary conditions which combine the scalar potential, magnetization and their first derivatives at the boundary. Such conditions follow from some basic physical laws. One kind of the boundary conditions are the magnetostatic interface conditions described previously. Other boundary conditions are independent of the dipolar field and follow from the torque balance at the surface or interface. For a ferromagnetic surface, they have been derived many years ago and are known as the Rado-Weertmann boundary conditions [55]. Taking surface pinning into account (surface anisotropy), one can write them in the form:

$$\mathbf{M} \times \left(\frac{1}{M_0} \nabla_{\mathbf{u}} E^{surf}(\mathbf{u}) - \frac{2A}{M_0^2} \frac{\partial \mathbf{M}}{\partial \mathbf{n}} \right) \bigg|_{surf} = 0, \qquad (3.56)$$

where $\partial/\partial\mathbf{n}$ is the derivative along the normal to the surface pointing from the surface into the magnetic system. $E^{surf}(\mathbf{u})$ is the surface anisotropy energy per unit area. This energy is usually assumed in the form of a uniaxial term with the symmetry axis normal to the surface. For clarity of notation we have dropped the $\mathbf{r}$- and t-dependence of $\mathbf{M}$ in Eq. (3.56). We will drop it also in the following boundary equations.

At the interface between two ferromagnetic materials there is an additional torque acting on the interface spins, which comes from the adjacent ferromagnet. The appropriate boundary conditions have been formulated for the first time by Hoffmann [56]. Recently, those conditions have been supplemented with an additional term which plays an essential role in the full coupling limit, when the inter-layer exchange coupling between two films of the same material is almost the same as inside the films [57,58]. Generally, we will write these conditions for two different materials in direct contact at $x = 0$. Let the half-space $x < 0$ ($x > 0$) be occupied by a material described by the parameters M_{01} (M_{02}) and A_1 (A_2). The exchange coupling between the materials is described by the phenomenological parameter A_{12}. For cubic systems with exchange coupling between nearest neighbours and for (100) interfaces, A_{12} is related to the microscopic exchange integral J_I by the expression

$$A_{12} = \frac{\nu J_I S_1 S_2}{a^2}, \tag{3.57}$$

where $\nu = 1/2$ for *sc*, $\nu = 2$ for *bcc* and $\nu = 4$ for *fcc* lattices. Under these assumptions one can write the general boundary conditions in the form

$$\mathbf{M}_1 \times \left(\frac{1}{M_{01}}\nabla_{\mathbf{u}_1} E_1^{int}(\mathbf{u}_1) + \frac{2A_1}{M_{01}^2}\frac{\partial \mathbf{M}_1}{\partial x}\right) - \frac{2A_{12}}{M_{01}M_{02}}\mathbf{M}_1 \times \mathbf{M}_2$$

$$- \frac{A_{12}d_a}{M_{01}M_{02}}\left(-\frac{\partial \mathbf{M}_1}{\partial x} \times \mathbf{M}_2 + \mathbf{M}_1 \times \frac{\partial \mathbf{M}_2}{\partial x}\right) = 0, \tag{3.58a}$$

$$\mathbf{M}_2 \times \left(\frac{1}{M_{02}}\nabla_{\mathbf{u}_2} E_2^{int}(\mathbf{u}_2) - \frac{2A_2}{M_{02}^2}\frac{\partial \mathbf{M}_2}{\partial x}\right) - \frac{2A_{12}}{M_{01}M_{02}}\mathbf{M}_2 \times \mathbf{M}_1$$

$$- \frac{A_{12}d_a}{M_{01}M_{02}}\left(\frac{\partial \mathbf{M}_2}{\partial x} \times \mathbf{M}_1 - \mathbf{M}_2 \times \frac{\partial \mathbf{M}_1}{\partial x}\right) = 0. \tag{3.58b}$$

where $\mathbf{M}_1$ and $\partial\mathbf{M}_1/\partial x$ is taken at $x = 0^-$ whereas $\mathbf{M}_2$ and $\partial\mathbf{M}_2/\partial x$ at $x = 0^+$, and d_a is the distance between two adjacent atomic planes, $d_a = a$ for *sc* and $d_a = a/2$ for *bcc* and *fcc* lattices.

If the materials are interspaced by a thin non-magnetic layer of thickness d, the boundary conditions can be written in the forms (3.58) albeit with $\mathbf{M}_1$ and $\partial\mathbf{M}_1/\partial x$ calculated at, say, $x = -d/2$, whereas $\mathbf{M}_2$ and $\partial\mathbf{M}_2/\partial x$ at $x = d/2$. However, the boundary conditions are not so simple when inter-layer exchange coupling is of long-range nature, as for example the coupling across a relatively thick metallic spacer. The boundary conditions (3.58) with A_{12} as a phenomenological parameter are correct for short-range coupling. If the exchange coupling is long-range, then one may write the effective boundary conditions in the form (3.58), but this is justified only for perturbations with oscillation period considerably longer than the typical range of exchange coupling.

Considering linear spin-wave modes we have to linearize the boundary conditions too. Assuming the interface anisotropy in the form of a uniaxial perpendicular anisotropy with the corresponding constants K_1^i and K_2^i for the interface atomic planes, one can write these equations as

$$\frac{A_1}{M_{01}}\frac{\partial m_{1x}}{\partial x}\bigg|_{x=0^-} + \frac{A_{12}-K_1^i}{M_{01}} m_{1x}(x=0^-) - \frac{A_{12}}{M_{02}} m_{2x}(x=0^+)$$
$$- \frac{d_a}{2}\left(\frac{A_{12}}{M_{01}}\frac{\partial m_{1x}}{\partial x}\bigg|_{x=0^-} + \frac{A_{12}}{M_{02}}\frac{\partial m_{2x}}{\partial x}\bigg|_{x=0^+}\right) = 0, \qquad (3.59a)$$

$$\frac{A_1}{M_{01}}\frac{\partial m_{1y}}{\partial x}\bigg|_{x=0^-} + \frac{A_{12}}{M_{01}} m_{1y}(x=0^-) - \frac{A_{12}}{M_{02}} m_{2y}(x=0^+)$$
$$- \frac{d_a}{2}\left(\frac{A_{12}}{M_{01}}\frac{\partial m_{1y}}{\partial x}\bigg|_{x=0^-} + \frac{A_{12}}{M_{02}}\frac{\partial m_{2y}}{\partial x}\bigg|_{x=0^+}\right) = 0, \qquad (3.59b)$$

$$\frac{A_2}{M_{02}}\frac{\partial m_{2x}}{\partial x}\bigg|_{x=0^+} - \frac{A_{12}-K_2^i}{M_{02}} m_{2x}(x=0^+) + \frac{A_{12}}{M_{01}} m_{1x}(x=0^-)$$
$$- \frac{d_a}{2}\left(\frac{A_{12}}{M_{01}}\frac{\partial m_{1x}}{\partial x}\bigg|_{x=0^-} + \frac{A_{12}}{M_{02}}\frac{\partial m_{2x}}{\partial x}\bigg|_{x=0^+}\right) = 0, \qquad (3.59c)$$

$$\frac{A_2}{M_{02}}\frac{\partial m_{2y}}{\partial x}\bigg|_{x=0^+} - \frac{A_{12}}{M_{02}}m_{2y}(x=0^+) + \frac{A_{12}}{M_{01}}m_{1y}(x=0^-)$$

$$-\frac{d_a}{2}\left(\frac{A_{12}}{M_{01}}\frac{\partial m_{1y}}{\partial x}\bigg|_{x=0^-} + \frac{A_{12}}{M_{02}}\frac{\partial m_{2y}}{\partial x}\bigg|_{x=0^+}\right) = 0. \qquad (3.59\text{d})$$

Similarly, one can write the Rado-Weertmann conditions in a linearized form which can also be obtained from the above formulae by assuming $A_{12} = 0$ and replacing the interface anisotropy constants by the appropriate surface ones.

The terms proportional to $A_{12}d_a$ in Eqs (3.58) and (3.59) are just the additional ones. With these terms neglected, Eqs (3.58) or Eqs (3.59) reduce to those of Hoffmann. However, the Hoffmann form leads to some difficulties in the full coupling limit ($d = 0$). Assume $A_1 = A_2 = A$, $M_{01} = M_{02} = M$, $K_1^i = K_2^i = 0$ and $A_{12}d_a = A$, which corresponds to the full coupling limit. On neglecting the additional terms, one finds that in the full coupling limit $\partial\mathbf{m}/\partial x$ is continuous at $x = 0$, but $\mathbf{m}$ is discontinuous at the interface and

$$\mathbf{m}(x=0^+) - \mathbf{m}(x=0^-) = d_a\frac{\partial\mathbf{m}}{\partial x}\bigg|_{x=0}. \qquad (3.60)$$

This jump, however, is rather small. It is equal to the change in $\mathbf{m}$ on the distance between the atomic planes, and this should be small in the region where the continuum approximation applies. For weakly coupled films the discontinuity of $\mathbf{m}$ at $x = 0$ can be quite large [52], much greater than the step introduced when neglecting the additional terms in Eqs (3.58) or Eqs (3.59). Thus, the difference between Eqs (3.57) or Eqs (3.58) and the Hoffmann form is small in the weak coupling limit.

One could also assume that at direct contact the two ferromagnetic materials are separated by d_a [51]. The Hoffmann form is then applicable in the full coupling limit, but this approach is not consistent with the notion of two continua at direct contact when, from the macroscopic point of view, the distance between them vanishes.

3.4.3 Dipole-exchange modes in ferromagnetic multilayers

Having found the general solution for the dynamical variables and the appropriate boundary conditions, we can calculate the eigenmodes of any layered ferromagnetic structure. Basically, one could develop a formalism similar to that for the magnetostatic and exchange modes. In the present case the relevant transfer matrix would be of the order of 6×6. This formalism, however, has not been worked out yet. Instead, a direct diagonalization of the appropriate eigenvalue matrix is usually performed in order to obtain the corresponding spin-wave spectrum.

In most experimentally investigated structures the ferromagnetic films were separated by a number of atomic planes of some nonmagnetic material. Exchange coupling between the films is very sensitive to the thickness of the spacer, contrary to dipolar coupling, which is almost insensitive in this range of thicknesses. This allows one to neglect, in many cases, the spacer thickness and to consider the structure as if the ferromagnetic films were in direct contact. The spacer is then included indirectly by way of the inter-layer exchange constant.

As the simplest example of layered structures we again consider two identical ferromagnetic films of thickness d, exchange-coupled at the interface. Fig. 3.13 shows the lowest four modes of the bilayer *versus* the inter-layer exchange parameter A_{12} for vanishing in-plane wavevector, *i.e.*, for the situation typical for ferromagnetic or spin-wave resonance. At $A_{12} = 0$ the two layers are exchange- and dipolar-decoupled. Consequently, the modes are degenerate at $A_{12} = 0$. At increasing A_{12} the degeneracy is lifted and the odd modes shift up on the energy scale and finally saturate in the full coupling limit, *i.e.*, when the inter-layer coupling is equal to that inside the films and the bilayer is equivalent to a single homogeneous film with doubled thickness. The even modes, on the other hand, are insensitive to the inter-layer coupling. This behaviour is similar to that found when considering the exchange modes in symmetrical bilayers and can be easily understood if one examines the corresponding mode profiles. These are shown in Fig. 3.13 for three different values of A_{12} and for

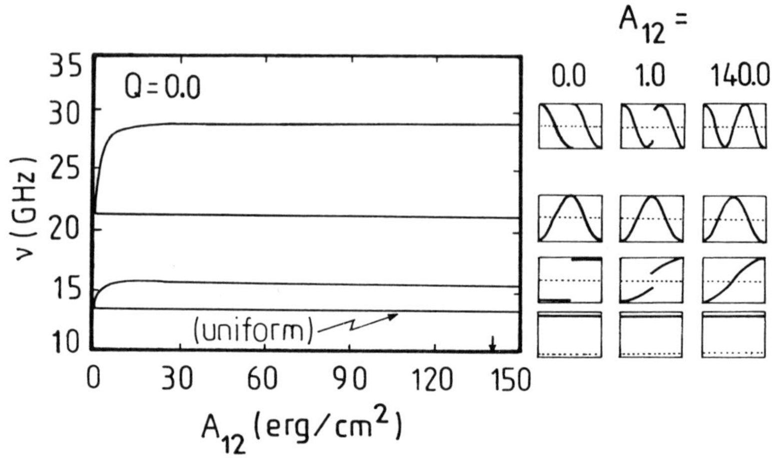

Fig. 3.13 Frequencies and mode profiles of the four lowest modes of a symmetrical bilayer for $Q = 0$, $H_0 = 1.0\ kG$, $4\Pi M = 21\ kG$, $A = 2.1 \times 10^{-6}\ erg/cm$, $g = 2.1$, $d = 44\,nm$ and for vanishing bulk, surface and interface anisotropy. The full-coupling limit value of A_{12}, A_{12}^c, is indicated by the arrow.

vanishing interface and surface anisotropy. The mode profiles of the even modes are continuous at the interface. Thus, such modes do not feel the inter-layer coupling. The odd modes, on the other hand, have discontinuous profiles at the interface and, consequently, are strongly affected by the inter-layer exchange coupling. For $A_{12} = A_{12}^c$ (A_{12}^c is the full coupling limit of A_{12}) the modes are equivalent to those of a single uniform film with doubled thickness. The lowest mode in Fig. 3.13 is the uniform mode whereas the others are usually referred to as exchange modes. If the films are sufficiently thin, then the single-film levels are well separated. For weak inter-layer coupling the lowest two levels can then be described within the subspace obtained from the uniform modes of both films and the influence of the higher energy levels can be ignored. Such an approximation is equivalent to the assumption that the magnetization of an individual film behaves like a single magnetic moment. Thus the problem reduces to that of a simple two-level system. Similar arguments apply also to structures

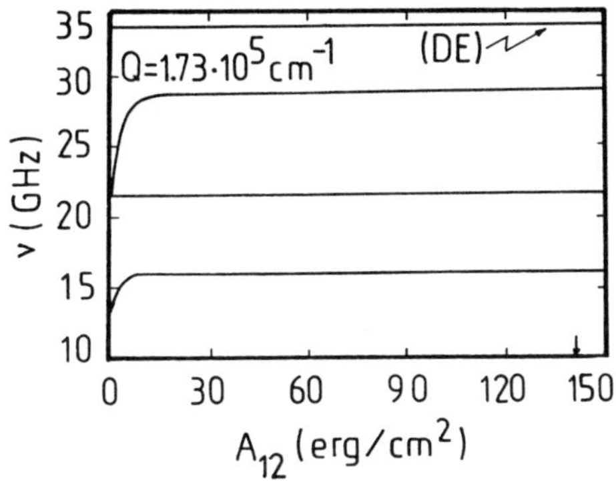

Fig. 3.14 The same as in Fig. 3.13 but for $Q = 1.73 \times 10^5\ cm^{-1}$.

with many equal ultra-thin ferromagnetic films. The lowest manifold of energy levels, which developes from the single-film uniform modes, can be described by ignoring higher levels. This, however, is justified only for relatively weak inter-layer coupling. An approximation of this kind is very often used in theoretical descriptions of certain experimental results [59-63]. For stronger inter-layer coupling and/or for film thicknesses at which the level separation is not great enough, one has to include the internal dynamics of the films, *i.e.*, one has to include the influence of higher modes [51,64-66].

For increasing Q the mode spectrum of a bilayer changes considerably [51,67,68]. The uniform mode of Fig. 3.13 goes over into the DE-mode of the structure as a whole. This is shown in Fig. 3.14 for the same parameters as in Fig. 3.13 with the exception of Q which now is equal to $1.73 \times 10^5\ cm^{-1}$ (typical for waves excited in Brillouin scattering experiments) instead of $Q = 0$. The uniform mode at $Q = 0$ (the lowest mode in Fig. 3.13) now changed into the DE-mode (the upper mode in Fig. 3.14). The DE-mode is almost independent of the inter-layer exchange coupling. The other modes (exchange modes), shift slightly upwards in comparison with those of Fig. 3.13.

For the symmetrical bilayers discussed above the wave spectrum is reciprocal. For similar bilayers but with antiparallel alignment

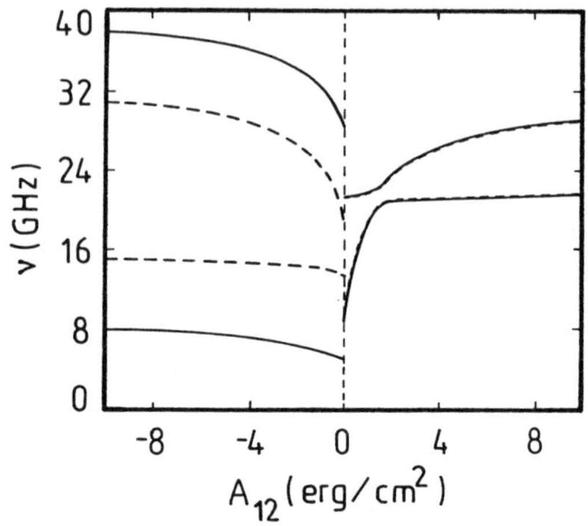

Fig. 3.15 Two lowest modes (the first exchange mode and the DE-mode) for a bilayer with parallel ($A_{12} > 0$) and antiparallel ($A_{12} < 0$) alignment of the film magnetizations for $Q_y = 1.73 \times 10^5\ cm^{-1}$ (full lines) and $Q_y = -1.73 \times 10^5\ cm^{-1}$ (broken lines). The other parameters assumed here are those which give the best fit to the experimental data obtained in $Fe/Cr/Fe$ structures [52].

of the film magnetizations, the corresponding spin-wave spectrum is non-reciprocal, similarly as for the magnetostatic modes discussed in Sect.3.3.4. In Fig. 3.15 we see the two lowest modes of an exchange coupled bilayer *versus* the inter-layer exchange constant for both $A_{12} > 0$ and $A_{12} < 0$ and for two opposite wavevectors, $Q_y = \pm 1.73 \times 10^5\ cm^{-1}$. The discontinuity of the wave spectrum at $A_{12} = 0$ results from the change of the ground state configuration from parallel at $A_{12} > 0$ to antiparallel at $A_{12} < 0$ (in vanishing external field, $H_0 = 0$). For $A_{12} > 0$ the corresponding wave spectrum is reciprocal, whereas for $A_{12} < 0$ it is non-reciprocal.

Dipole-exchange modes in superlattices and multilayers have been thoroughly analysed by Hillebrands [53,54] and Vayhinger and Kronmüller [67,68]. Their basic features are similar to those of spin-wave modes in bilayers. Considering for example the Q-dependence of the wave frequencies one finds a characteristic anticrossing (repulsion) behaviour in those regions where the dispersion curves of surface-like modes should intersect the dispersion curves of the exchange modes—similarly as in bilayers or single films.

In many practical realizations of exchange-coupled bilayers and multilayers the ferromagnetic films are separated by several atomic planes of a non-magnetic metal. The exchange coupling across the inter-layer is then significantly smaller than the corresponding bulk coupling and varies considerably with the spacer thickness. Comparison of the experimental data with those calculated numerically allows to determine the inter-layer exchange constant A_{12}. Such a comparison, however, is possible only for the weak coupling limit, where the exchange modes vary significantly with the inter-layer coupling. In the strong inter-layer coupling limit, the modes are almost independent of the exchange constant A_{12} and, consequently, one cannot determine A_{12} uniquely from such a comparison [51,52].

3.5 FINAL REMARKS

The problems discussed in the preceding Sections are not the only ones we face when considering excitations in magnetic layered structures. Our discussion was restricted to periodically layered structures, though this is not the only possibility. One can now produce layered systems with arbitrarily designed parameters, strictly controlled during the growth process. An important class of layered systems are quasi-periodic and randomly layered structures. Such systems offer a new possibility of the experimental verification of certain theoretical models described by one-dimensional Hamiltonians. It is well known that all states are localized in one-dimensional random systems. This is true also for randomly layered magnetic structures [69]. Quasi-periodic systems, like Fibonacci sequences, are intermediate between completely periodic and completely random systems. They have a particularly interesting excitation spectrum in the form of a Cantor set [70,71].

When considering periodically layered structures we restricted ourselves to the case when the material parameters changed step-wise or almost step-wise (with the exception of two interfacial atomic planes). Obviously, this is an idealization. In some structures, particularly in those with considerable interdiffusion, sinusoidal-like [72] or trapezoid-like variation of the appropriate properties and parameters is more realistic than step-like variation.

There are yet other problems and methods [74–78] which are not discussed here due to limited space. Additional ones will certainly arise as new structures are grown.

Acknowledgements. The author would like to thank Dr M. Zimpel, Dr K. Flatau and Dr L. Smardz for proofreading the manuscript. The work was supported by the Research Committee under Grant No 2 0294 91 01.

REFERENCES

1. D.L. Mills, in *Light Scattering in Solids V: Superlattices and Other Microstructures*, ed. M. Cardona and G. Güntherodt (Springer, Berlin, 1989), p. 13.
2. M.H. Grimsditch, ibid., p. 285.
3. P. Grünberg, ibid., p. 303.
4. M.G. Cottam and D.R. Tilley, *Introduction to surface and superlattice excitations* (Cambridge University Press, Cambridge, 1989).
5. T. Wolfram and R.E. DeWames, Prog. Surf. Science **2** (1972) 233.
6. H. Puszkarski, Prog. Surf. Science **9** (1979) 191.
7. C.E. Patton, Phys. Reports **103** (1984) 252.
8. P. Grünberg, Prog. Surf. Science **18** (1985) 1.
9. R.E. Camley, Surf. Science Reports **7** (1987) 103.
10. R. Świrkowicz, Physica B **159** (1989) 304; Physica B 160 (1990) 329.
11. D.L. Mills, in *Surface Excitations*, ed. V.M. Agranovich and R. Loudon (Elsevier, Amsterdam, 1984), p. 379.
12. J. Barnaś, Solid State Commun. **61** (1987) 405; J. Phys.: Cond. Matter **2** (1990) 7173.
13. N. Raj and D.R. Tilley, Phys. Rev. B **36** (1987) 7003.
14. J.F. Cochran and J.R. Dutcher, J. Appl. Phys. **63** (1988) 3814.
15. S.W. McKnight and C. Vittoria, Phys. Rev. B **36** (1987) 8574; C. Vittoria, Phys. Rev. B **37** (1988) 23387.
16. E.L. Albuquerque, P. Fulco, E.F. Sarmento and D.R. Tilley, Solid State Commun. **58** (1986) 41.

17. L. Dobrzynski, B. Djafari-Rouhani and H. Puszkarski, Phys. Rev. B **33** (1986) 3251.
18. J. Barnaś, J. Phys.C: Solid State Phys. **21** (1988) 1021.
19. J. Barnaś, to be published in Phys. Rev. B; J. Barnaś and L. Smardz, to be published in J. Magn. Magn. Materials.
20. D. Mercier, J.C.S. Lévy, M.L. Watson, J.S.S. Whiting and A. Chambers, Phys. Rev. B **43** (1991) 3311.
21. J.C.S. Lévy and H. Puszkarski, J. Phys.: Condens. Matter **3** (1991) 5247.
22. H. Puszkarski, Acta Physica Polonica A **74** (1988) 701.
23. H. Puszkarski and L. Dobrzynski, Phys. Rev. B **39** (1989) 1819.
24. F. Garcia-Moliner and V. Velasco, Phys. Scripta **34** (1986) 257.
25. L. Dobrzynski, Surf. Science Reports **6** (1986) 119.
26. R.E. Camley and D.R. Tilley, Phys. Rev. B **37** (1988) 3413.
27. J. Barnaś and P. Grünberg, J. Magn. Magn. Materials **98** (1991) 57.
28. J.G. LePage and R.E. Camley, Phys. Rev. B **40** (1989) 9113.
29. J.G. LePage and R.E. Camley, Phys. Rev. Letters **65** (1990) 1152.
30. L.L. Hinchey and D.L. Mills, Phys. Rev. B **33** (1986) 3329; Phys. Rev. B **34** (1986) 1689.
31. K.M. Leung, J. Appl. Phys. **61** (1987) 4070.
32. See also the first chapter of this book.
33. R.E. Camley, T.S. Rahman and D.L. Mills, Phys. Rev. B **27** (1983) 261.
34. P. Grünberg and K. Mika, Phys. Rev. B **27** (1983) 2955.
35. P.R. Emtage and M.R. Daniel, Phys. Rev. B **29** (1984) 212.
36. K. Mika and P. Grünberg, Phys. Rev.B **31** (1985) 4465.
37. R.E. Camley and M.G. Cottam, Phys. Rev. B **35** (1987) 189.
38. G. Rupp, W. Wettling and W. Jantz, Appl. Phys. A **42** (1987) 45.
39. M. Villeret, S. Rodriguez and E. Kartheuser, Phys. Rev. B **39** (1989) 2583.
40. J. Barnaś, J. Phys.C: Solid State Phys. **21** (1988) 4097.
41. M. Grimsditch, M.R. Khan, A. Kueny and I.K. Schuller, Phys. Rev. Letters **51** (1983) 498.
42. B. Hillebrands, P. Baumgart, R. Mock, G. Güntherodt, A. Boufelfel and C.M. Falco, Phys. Rev. B **34** (1986) 9000.
43. P. Grünberg, J. Appl. Phys. **51** (1980) 4338.
44. R.E. Camley, Phys. Rev. Letters **45** (1980) 283.
45. R.L. Stamps and R.E. Camley, J. Appl. Phys. **56** (1984) 3497.
46. P. Grünberg, R. Schreiber, Y. Pang, M.B. Brodsky and H. Sowers, Phys. Rev. Letters **57** (1986) 2442.
47. P.X. Zhang and W. Zinn, Phys. Rev. B **35** (1987) 5219.

48. P. Grünberg, J. Appl. Phys. **52** (1981) 6824.
49. J. Barnaś, Phys. Status Solidi (b) **145** (1988) K55.
50. A.I. Akhiezer, V.G. Bar'yakhtar and S.V. Peletminskii, *Spin Waves* (Nauka, Moscow, 1967).
51. M. Vohl, J. Barnaś and P. Grünberg, Phys. Rev. B **39** (1989) 12003.
52. J. Barnaś and P. Grünberg, J. Magn. Magn. Materials **82** (1989) 186.
53. B. Hillebrands, Phys. Rev. B **37** (1988) 9885.
54. B. Hillebrands, Phys. Rev. B **41** (1990) 530.
55. G.T. Rado and J.R. Weertman, J. Phys. Chem. Solids **11** (1959) 315.
56. F. Hoffmann, Phys. Status Solidi **41** (1970) 807.
57. J. Barnaś, J. Magn. Magn. Materials **102** (1991).
58. D.L. Mills, to be published in Phys. Rev. B.
59. J.F. Cochran, B. Heinrich and A.S. Arrott, Phys. Rev. B **34** (1986) 7788.
60. J.F. Cochran and J.R. Dutcher, J. Appl. Phys. **64** (1988) 6092.
61. B. Heinrich, S.T. Purcell, J.R. Dutcher, K.B. Urquhart, J.F. Cochran and A.S. Arrott, Phys. Rev. B **38** (1988) 12879.
62. A. Layadi and J.O. Artman, J. Magn. Magn. Materials **92** (1990) 143.
63. A.Z. Maksymowicz, J.S.S. Whiting, M.L. Watson and A. Chambers, J. Magn. Magn. Materials **94** (1991) 109.
64. G. Spronken, A. Friedmann and A. Yelon, Phys. Rev. B **15** (1977) 5141.
65. C. Vittoria, Phys. Rev. B **32** (1985) 1679.
66. R.P. van Stapele, F.J.A.M.Greidanus and J.W. Smits, J. Appl. Phys. **57** (1985) 1282.
67. K. Vayhinger and H. Kronmüller, J. Magn. Magn. Materials **62** (1986) 159.
68. K. Vayhinger and H. Kronmüller, J. Magn. Magn. Materials **72** (1988) 307.
69. J. Barnaś, Phys. Status Solidi (b) **151** (1989) K71.
70. S. Xiong, J. Phys. C: Solid State Phys. **20** (1987) L167.
71. M. Kolar and M.K. Ali, J. Phys.: Condens. Matter **1** (1989) 823.
72. J.A. Morkowski and A. Szajek, J. Magn. Magn. Materials **71** (1988) 299.
73. W. Maciejewski, Thin Solid Films **146** (1987) L37.
74. D. Schwenk, F. Fishman and F. Schwabl, Phys. Rev. B **38** (1988) 11618.
75. W. Schmidt, J. de Physique **49** (Coll. C8) (1989) 1689.
76. H.T. Diep, Phys. Letters A **138** (1989) 69.
77. N.S. Almeida and D.L. Mills, Phys. Rev. B **38** (1988) 6698.
78. G. Vecris and J.J. Quinn, Phys. Rev. B **43** (1991) 8303.

CHAPTER 4

SPIN-WAVE POLARITONS IN FILMS AND MULTILAYERS

D.R. TILLEY

Department of Physics, University of Essex,
Colchester CO4 3SQ, UK

Contents

4.1 INTRODUCTION

The gyromagnetic susceptibility tensor for a ferromagnet, Eq. (1.41), was derived in Sec. 1.2.3. It was subsequently applied, in Chapters 1 to 3, for a discussion of dipolar and dipole-exchange effects, since these can be described by the magnetostatic form of Maxwell's equations. As is seen from Table 1 in Chapter 1, the magnetostatic régime covers an intermediate range of wave-vector values which is accessible to several experimental probes and is therefore very important in practice. In this Chapter we turn to the region of small wave vector $q \sim \omega/c$ in which retardation effects are important and therefore Maxwell's equations in their full form must be used. The corresponding retarded, or polariton, modes have so far been little studied experimentally, although as will be discussed in Sec. 4, they should be accessible in some cases.

It is convenient to use the permeability tensor $\bar{\bar{\mu}}$ of Eq. (1.48) rather than the susceptibility. Except for the special case of an antiferromagnet without an applied static field, for which the tensor components are given by Eq. (1.56), $\bar{\bar{\mu}}$ has the distinctive property of off-diagonal elements $\pm i\chi_2$ which are pure imaginary in the absence of damping. Thus $\bar{\bar{\mu}}$ cannot be diagonalised by a rotation of axes through a real angle, although it is diagonalised by transformation to a rotating-wave representation, $h_{\pm} = h_x \pm ih_y$. This has the consequence that for propagation in the bulk along the direction of the static field, the propagating eigenmodes are the two signs of circular polarization [1]. For plane polarized radiation, the plane of polarization is rotated. This effect forms the basis for the ferrite devices like circulators and isolators which are important in microwave engineering [2-4].

For the ferromagnet, it is seen from Eq. (1.42) that both the diagonal and the off-diagonal elements of $\bar{\bar{\mu}}$ have poles at the ferromagnetic resonance frequency ω_H, Eq. (1.43), while for the two-sublattice antiferromagnet the pole is at the antiferromagnetic resonance frequency

$$\omega_{AF} = \left[\omega_A(\omega_A + 2\omega_E)\right]^{1/2} \tag{4.1}$$

as seen in Eq. (1.56). There is an important distinction in practice between these frequencies. For typical applied field values, ω_H is in the microwave frequency region, $\omega_H/2\pi c \leq 1$ cm^{-1}(=30 GHz); as will be discussed in detail in Chapter 6, this means that surface modes on ferromagnets may be studied by Brillouin scattering. For antiferromagnets, on the other hand, Eq. (4.1) shows that both anisotropy and exchange fields are involved in resonance; for most materials $\omega_A << \omega_E$, so that Eq. (4.1) simplifies to $\omega_{AF} = (2\omega_A\omega_E)^{1/2}$. Substitution of typical values shows that ω_{AF} is in the far-infrared part of the spectrum. For example, for MnF_2, which has very low anisotropy, $\omega_{AF}/2\pi c$ = 8 cm^{-1}, while for FeF_2, with much higher anisotropy, $\omega_{AF}/2\pi c$ = 48 cm^{-1}.

A number of generalisations of these expressions for $\bar{\bar{\mu}}$ may be found in the literature. For the two-sublattice antiferromagnet in an applied field H_o parallel to the ordering direction, a non-zero off-diagonal element μ_2, which is proportional to H_o, is induced [5,6]. In addition, the resonance frequency splits into two values $\omega_{AF} \pm \gamma H_o$, and poles at both these frequencies are found in both μ_1 and μ_2. The much more complicated case of a two-sublattice antiferromagnet with an applied field in a general direction was discussed by Almeida and Mills [7]; they derived the permeability tensor and applied it to a discussion of surface magnetostatic modes. Attention should also be drawn to the discussion of the nonlinear response of antiferromagnets by the same authors [8]. The permeability tensor for a four-sublattice structure like $KMnF_3$ was evaluated by Barnaś and Stasch [9].

The two-sublattice ferrimagnet has both a ferromagnetic resonance and an exchange resonance at a much higher frequency; $\bar{\bar{\mu}}$ retains the gyromagnetic form, with both μ_1 and μ_2 having poles at both resonant frequencies [1,10].

Because of a competition between ferromagnetic nearest-neighbour and antiferromagnetic next-nearest neighbour exchange, some rare-earth metals have a magnetic order in some temperature interval which is helical or conical [11,12]. For example, in Dy between 85K and 179K the spins are all ferromagnetically aligned in-plane within one close-packed plane, but the alignment direction rotates by an angle ϕ from one close-packed plane to the next; this is the helical state. In H_o below 20K the same description applies except that the alignment direction within each plane is tilted up towards the c-axis; this is the cone state. The permeability tensor for these states has been derived recently and applied to a study of the magnetostatic surface modes [13].

The expressions for $\bar{\bar{\mu}}$ given in Secs 1.2.3 and 1.2.5, and in many of the references, do not include the effects of spin-wave damping. Neglect of damping is usually satisfactory for discussion of dispersion relations $\omega(q)$ for magnetostatic or polariton modes. For some calculations, however, for example of far-infrared reflectivity, it is necessary to include damping. This can be done phenomenologically by inclusion of a dissipative term $-\alpha\mathbf{M}$ in torque equations like Eq. (1.36).

Application of a static field $\mathbf{H}_0$ or the existence of magnetic ordering modifies the electric permittivity as well as the magnetic permeability. In fact, as pointed out by Landau and Lifshitz [14], whilst it is meaningful to consider frequency dependence of the permeability in the microwave and far-infrared parts of the spectrum, for shorter wavelengths optical phenomena are determined solely by the permittivity. In general, in the presence of a magnetic field this becomes a tensor $\bar{\bar{\varepsilon}}$ with the same structure as the permeability $\bar{\bar{\mu}}$ in Eq. (1.48); this $\bar{\bar{\varepsilon}}$ is called gyrotropic. A very general discussion of the way in which the form of $\bar{\bar{\varepsilon}}$ and its symmetry properties are governed by the breaking of time-reversal invariance by the applied field is given in ref 14. It is the gyrotropic $\bar{\bar{\varepsilon}}$ which determines the surface magneto-optic Kerr effect (SMOKE) which has become important in investigations of thin magnetic films and which underlies magneto-optic recording techniques [15]. In addition, many of the properties to be discussed here, for example nonreciprocal propagation, $\omega(-q) \neq \omega(q)$, are parallelled in the gyrotropic dielectric. However, these matters are beyond the scope of this chapter.

In Sec 1.2.4 there is a brief discussion of bulk polariton modes in a gyromagnetic medium. Further discussion may be found in ref 1; here we pass straight on to the topic of surface and interface polaritons.

4.2 SURFACE MAGNETIC POLARITONS

The phrase "surface polariton" describes an electromagnetic mode localised at the plane interface between two media. Its properties are given by solutions of Maxwell's equations which have plane-wave dependence $\exp(iqy - i\omega t)$ in the propagation direction y within the plane and decay exponentially with distance from the interface [16]. For magnetic systems the choice of relative directions of magnetic ordering direction , propagation direction and normal to the interface leads to a large number of possibilities. By far the most commonly discussed arrangement is the Voigt geometry, in which the static field $\mathbf{H}_o$ is parallel to the interface and the propagation direction is transverse to $\mathbf{H}_o$; most of what follows is concerned with the Voigt geometry.

We consider the interface between vacuum ($x > 0$) and a magnetic medium ($x < 0$) with permeability given by Eq (1.48); the normal to the plane is taken as the x axis, and the polariton propagates in the y direction. We seek solutions of Maxwell's equations and the boundary conditions of the form

$$\mathbf{h} = \mathbf{h}_1 \exp(iqy)\exp(-\kappa_1 x)\exp(-i\omega t) \qquad x > 0 \tag{4.2}$$

$$\mathbf{h} = \mathbf{h}_2 \exp(iqy)\exp(\kappa_2 x)\exp(-i\omega t) \qquad x < 0 \tag{4.3}$$

in which exponential decay with distance from the interface is assumed. It is readily shown from the requirement that these should satisfy Maxwell's equations that $h_z = 0$ and that E_z is the only non-vanishing component of $\mathbf{E}$. That is, the surface polariton is TE or s-polarised. It also follows from Maxwell's equations that

$$\kappa_1^2 = q^2 - \omega^2/c^2 \tag{4.4}$$

$$\kappa_2^2 = q^2 - \varepsilon\mu_v\omega^2/c^2 \tag{4.5}$$

where ε is the dielectric constant of the medium and

$$\mu_v = (\mu_1^2 - \mu_2^2) \,/\, \mu_1 \tag{4.6}$$

is called the Voigt permeability. In comparison with Eq. (1.48) we put $\mu_1 = 1 + \chi_1$ and $\mu_2 = \chi_2$.

The divergence equation $\nabla \,.\, (\bar{\bar{\mu}}\, \mathbf{h})$ gives the ratio h_{ix}/h_{iy} in each medium and E_{iz}/h_{iy} is found, for example, from the $\nabla \times \mathbf{h}$ equation. Thus one field amplitude h_{iy} determines the others. The boundary conditions of continuous h_y and E_z determine independent expressions for the ratio h_{1y}/h_{2y}, and equating these ratios we find the dispersion relation

$$\kappa_2 = -\kappa_1\mu_v + q\mu_2/\mu_1 \tag{4.7}$$

This may be solved numerically together with Eqs. (4.4) and (4.5) to give the dispersion relation $\omega(q)$.

The above derivation is due to Hartstein et al [17], who went on to discuss the dispersion curves for a ferromagnet. They took

$$\mu_1 = \mu_\infty[1 + \omega_M\omega_H/(\omega_H^2 - \omega^2)] \tag{4.8}$$

$$\mu_2 = \mu_\infty\omega_M\omega/(\omega_H^2 - \omega^2) \tag{4.9}$$

$$\mu_3 = \mu_\infty \tag{4.10}$$

that is, the form given by Eqs. (1.41), (1.42) and (1.48) multiplied by a "background permeability" μ_∞ which would be due to higher frequency resonances, like the exchange resonance in a ferrimagnet. Their dispersion curve is shown in Fig. 4.1, and merits some discussion. First, the propagation is non-reciprocal, since $+q$ and $-q$ are not equivalent. This is a consequence of Eq. (4.7), in which q appears as a linear factor. The $+q$ mode starts at the point (q_v,ω_v), where

$$\omega_v = [\omega_H(\omega_H + \omega_M)]^{1/2} \tag{4.11}$$

$$q_v = (\omega_v/c)[(\omega_H + \omega_M)/\omega_M]^{1/2} \tag{4.12}$$

It is asympototic ($q \to \infty$) to the Damon-Eshbach surface magnetostatic mode frequency ω_s, given by

$$\omega_s = [\omega_H + \mu_\infty(\omega_H + \omega_M)]/(\mu_\infty + 1) \tag{4.13}$$

The limit $q \to \infty$, more strictly $q >> \omega/c$, always gives the magnetostatic frequency, since it can be found by putting $c = 0$ (magnetostatic approximation) at the outset.

The $-q$ mode appears only with $\mu_\infty > 1$; it starts on the vacuum photon line $\omega = c|q|$ at frequency

$$\omega_- = (\mu_\infty - 1)^{-1/2}\left[\omega_H^2 + \mu_\infty\omega_H(\omega_H + \omega_M)\right]^{1/2} \tag{4.14}$$

and merges with the upper bulk continuum at a finite value of q and ω. This means that it has no magnetostatic counterpart.

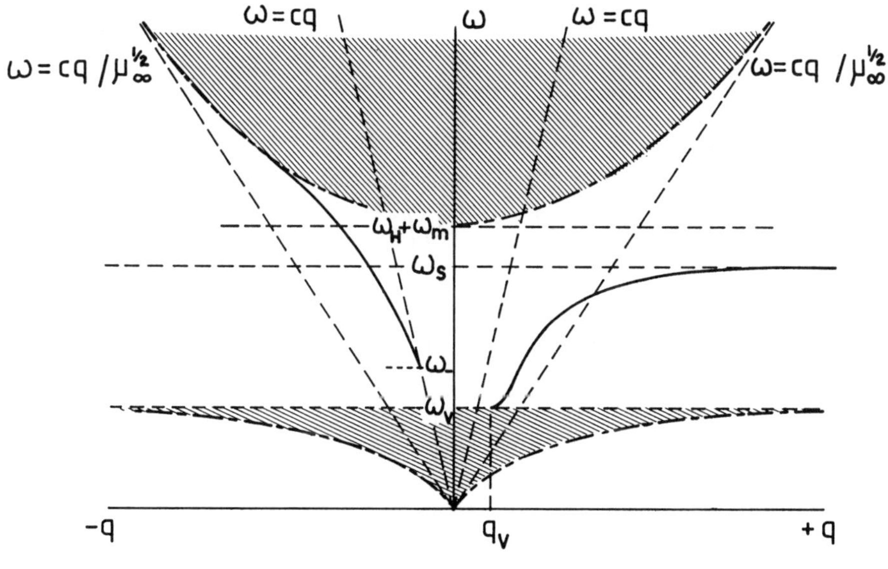

Fig.4.1 Surface-polariton dispersion curves (full lines) for the +q and −q modes on a ferromagnet in the Voigt geometry. The bulk continuum regions (solution of Eq.(1.50)) are shown shaded. After ref.17.

There has not been a complete discussion of surface polaritons with propagation in a general direction in the surface. Hartstein et al [17] give a brief qualitative discussion, but only for $\varepsilon = 1$. They state that propagation occurs only in a range of angles $\phi < \phi_c$ about the Voigt direction, where ϕ_c is given by the same expression as holds in the magnetostatic limit:

$$\sin\phi_c = \left[\omega_H/(\omega_H + \omega_M)\right]^{1/2} \tag{4.15}$$

Some curves to illustrate this are shown subsequently in Fig. 4.7. Fukui et al [18], however, include the dielectric constant ε of the ferromagnet, and show that for the Faraday configuration (**q** along $\mathbf{H}_o$) there is a propagating TM surface mode, with dispersion relation

$$q^2 = \frac{\omega^2}{c^2}\,\frac{\varepsilon(\varepsilon - \mu_1)}{\varepsilon - 1} \tag{4.16}$$

The corresponding dispersion curve for YIG, with $\varepsilon = 5.529$, is shown in Fig. 4.2.

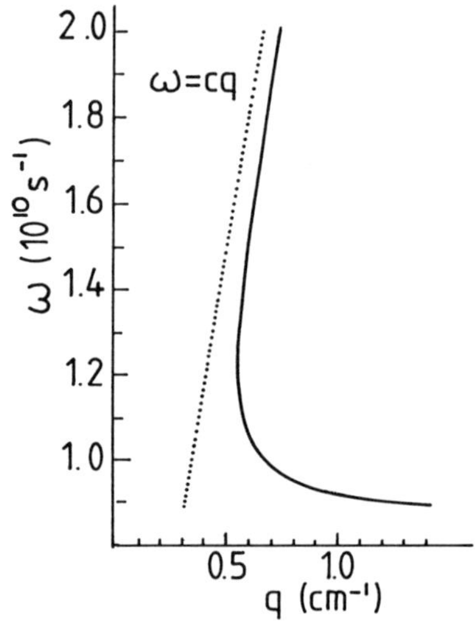

Fig.4.2 Dispersion curve for the TM surface polariton in the Faraday geometry on YIG. After ref.18.

The formal results of Eqs. (4.4), (4.5) and (4.7) continue to apply to antiferromagnets, but as was appreciated by Camley and Mills [19] and independently by Cao and Caillé [20] the implications are somewhat different. In the absence of an applied field the permeability tensor is diagonal, as already noted. In consequence, the dispersion equation can be written in the explicit form [5,19]

$$q^2 = (\omega^2/c^2)\mu_1(\varepsilon-\mu_1)/(1-\mu_1^2) \tag{4.17}$$

An example of the dispersion curve is shown in Fig. 4.3. Since there is no off-diagonal permeability component μ_2 (and therefore the term linear in q in Eq. (4.7) vanishes) the propagation is reciprocal, $\omega(-q) = \omega(q)$. The surface polariton in this case does have a magnetostatic counterpart.

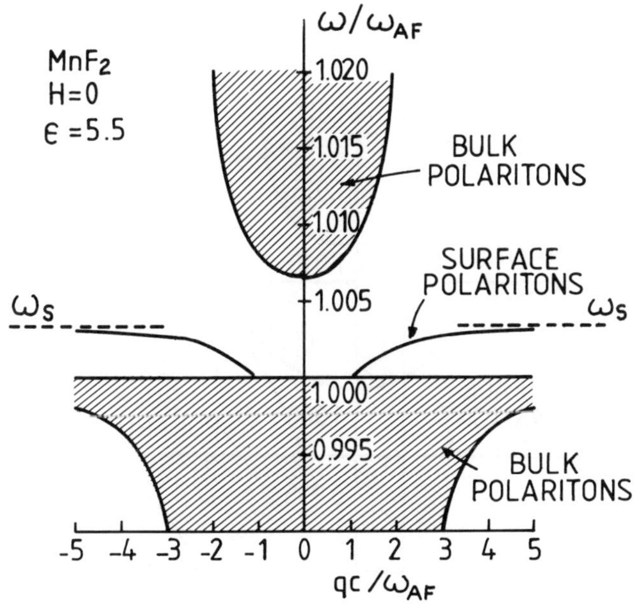

Fig.4.3 Surface-polariton dispersion curves and bulk continua on an antiferromagnet in the Voigt geometry with no applied field. The axes are scaled by the resonance frequency w_{AF} and $\varepsilon = 5.5$, $H_A = 3.8$ kG, $H_E = 550$kG, M = 0.6kG corresponding to MnF_2. After ref.19.

The striking effect of even a modest applied field is illustrated by Fig. 4.4. Compared with Fig. 4.3, the lower bulk continuum is compressed and the upper bulk continuum splits into two. Surface-mode propagation in the lower "window" is now non-reciprocal, both modes having magnetostatic counterparts. In the upper "window", which appears due to splitting of the bulk continuum, a surface polariton with no magnetostatic counterpart appears.

It is worth commenting on the very fine frequency scale in both Figs. 4.3 and 4.4, covering a range of only about 3% of ω_{AF} in all. This results from the rather weak coupling to the electromagnetic field which is implicit in Eq. (1.56), and it has important implications for experiment, as will be seen in Sec. 4.5. The use of dimensionless units in Figs. 4.1, 4.3 and 4.4 conceals the very different frequency scales which were already pointed out in Sec. 4.1: for

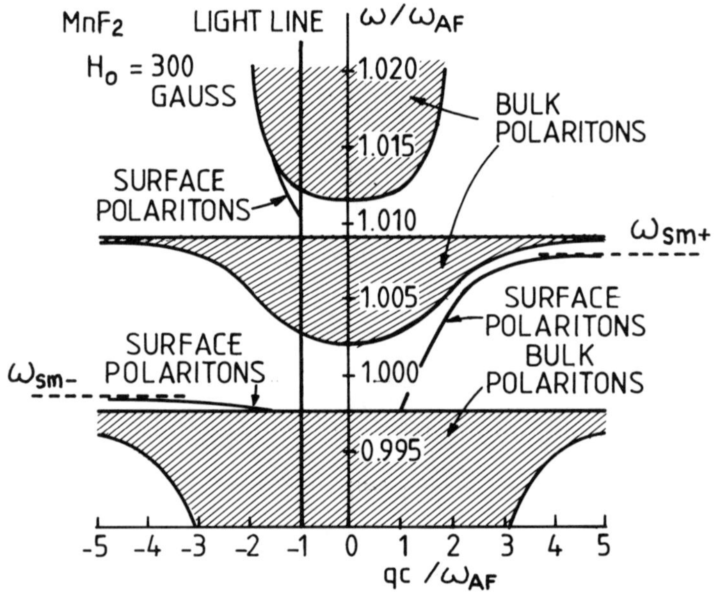

Fig.4.4 As Fig.4.2, but with an applied field H_0 = 300 Gauss in the ordering direction. After ref.19.

typical ferromagnets the frequencies in Fig. 4.1 are in the microwave region, while for the particular case of MnF_2 $\omega_{AF}/2\pi c$ is 8 cm^{-1}.

Various extensions of these basic results may be noted. It is possible that in the vicinity of the surface of a ferromagnet the magnetization is reduced below its bulk value; the influence of such a magnetization profile on the polariton dispersion curve is discussed by Thibaudeau and Caillé [21]. Camley and Mills [19] consider, in addition to the insulating antiferromagnet, a metallic antiferromagnet characterised by a negative dielectric constant. Stamps and Camley [22] include the effects of damping on antiferromagnetic surface polariton dispersion curves and in addition calculate the related Green functions. Barnaś [23] considers the interface between an antiferromagnet (without applied field) and a polar dielectric in which the dielectric function has a resonance, typically in the far infrared region. As well as substantial modifications to the Voigt-geometry mode, he finds that a TH mode propagates in the Faraday geometry, **q** parallel to **M**; its dispersion depends on both the magnetic permeability and the dielectric function of the bounding medium. Yaniv [24] gives a discussion of the polaritons propagating along the interface between two ferromagnets.

4.3 FILM POLARITONS

We now extend the calculations of the previous section to the case of a film. In this case, it will be seen that the nature of the modes depends upon the nature of the wave-vector component q_x normal to the interfaces within the film. For imaginary q_x one has surface-type modes, essentially a perturbation of the modes discussed in the previous section, while real q_x corresponds to guided modes. Properties of the surface-type modes were first investigated by Gerson and Nadan [25] and independently somewhat later by Karsono and Tilley [26]. Discussion of the guided modes was originated by Marchand and Caillé [27].

We still consider the Voigt geometry; the film is assumed to occupy the region $-L < x < 0$, with identical non-magnetic media in $x < -L$ and $x > 0$. As before, we seek solutions of Maxwell's equations that decay exponentially with distance form the film in either bounding medium. It follows from Maxwell's equations, as previously, that the only non-vanishing field components are h_x, h_y and E_z. Expressions for **h** are taken in the form

$$\mathbf{h} = \mathbf{h}_1 \exp(iqy)\exp(-i\omega t)\exp(-\kappa_1 x) \qquad x > 0 \qquad (4.18)$$

$$\mathbf{h} = \exp(iqy)\exp(-i\omega t)[\mathbf{A}\exp(iq_x x) + \mathbf{B}\exp(-iq_x x)] \qquad -L < x < 0 \qquad (4.19)$$

$$\mathbf{h} = \mathbf{h}_3 \exp(iqy)\exp(-i\omega t)\exp(\kappa_1 x) \qquad x < -L \qquad (4.20)$$

Here κ_1 is given by Eq. (4.4), and q_x by the negative of Eq. (4.5):

$$q_x^2 = -\kappa_2^2 = \varepsilon\mu_v\omega^2/c^2 - q^2 \qquad (4.21)$$

An important difference between the film and the semi-infinite medium is that q_x can now be either pure real or pure imaginary, whereas for the semi-infinite medium it can only be imaginary, corresponding to real κ_2. For real q_x, it is seen from Eq. (4.19) that the fields are oscillatory within the film; as in the analogous case of a dielectric film [16] the mode is then called a guided wave. For imaginary q_x, the fields are a combination of decaying exponentials at each surface, so that in effect one is dealing with a surface polariton at one surface whose properties are modified by the presence of the other surface. This again has an analogue in dielectric films. Eq. (4.21) shows that the inequalities $\varepsilon\mu_v\omega^2/c^2 > q^2$ and $\varepsilon\mu_v\omega^2/c^2 < q^2$ define "windows" in the ($q\omega$) plane for guided and surface-polariton type modes.

Eqs. (4.18) to (4.20) are written on the assumption that the media in $x > 0$ and $x < -L$ are both the same (and with Eq. (4.4) that they are both vacuum, so that the film is free standing). The generalisation to the case when the media are different, as would occur for a film on a substrate, for example, is not difficult.

The derivation of the dispersion relation proceeds in a similar way to that for the semi-infinite medium. All the field amplitudes may be determined from h_{1y}, A_y, B_y and h_{3y}. The four boundary conditions at the two interfaces give four homogeneous equations between these four amplitudes, and the solvability condition gives the dispersion equation, which may be written in the form

$$\left(\kappa_1^2\,\mu_v^2 + \kappa_2^2 - q^2\mu_2^2/\mu_1^2\right)\tanh(\kappa_2 L) + 2\kappa_1\kappa_2\mu_v = 0 \tag{4.22}$$

It is seen that as $L \to \infty$ this reduces to eq. (4.7), or more accurately to the square of Eq. (4.7). In addition, it is the generalisation of the Damon-Eshbach result for the magnetostatic modes of a film to include retardation.

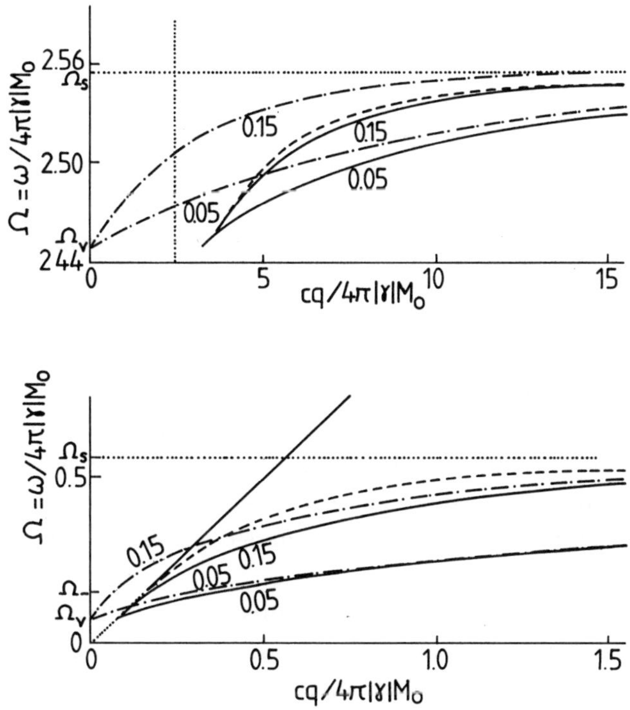

Fig.4.5 Dispersion curves for surface-polariton type modes of a free-standing ferromagnetic film. Background permeability $\mu_\infty = 1.25$. The axes are scaled in terms of the frequency $\omega_M = 4\pi\gamma M_o$, Eq.(1.43). Top: $\omega_H/\omega_M = H_o/4\pi M_o = 2$; Bottom: $\omega_H/\omega_M = 0.005$. The curves are marked by values of the dimensionless film thickness $4\pi\gamma M_o L/C$. Full lines are the retarded film modes of Eq.(4.22) and broken lines are the +q mode of Fig.4.1 (semi-infinite medium). Also shown, as chain-dotted lines, are the corresponding magnetostatic modes. The vacuum light line $\omega = cq$ is shown as a dotted line. After ref.26.

Two examples of surface-polariton type dispersion curves (κ_2 real) for a ferromagnetic film are shown in Fig. 4.5. The dispersion curves occupy the same region $\omega_v < \omega < \omega_s$ and $q > \omega/c$ as the +q mode of Fig. 4.1, and naturally approach that curve as L increases. For large q, the curves approach the magnetostatic limit, although it is seen that on the scale of q values chosen for illustration the corrections due to retardation are quite significant. For small q, the magnetostatic-mode curves have the unphysical property that they cross the vacuum light line, whereas the retarded-mode curves stay to the right of the light line.

The non-reciprocal character of the modes on the semi-infinite medium, shown in Fig. 4.1, has already been discussed. For the film, the dispersion curves themselves are reciprocal, $\omega(-q) = \omega(q)$, as is seen by the appearance of q^2 rather than q in Eq. (4.21). However, the field localisation is non-reciprocal, in that the field profile for the +q mode is biassed towards one surface and the field profile for the −q mode is biassed towards the other surface. This is discussed in more detail elsewhere [1,5].

The guided-mode solutions of Eq. (4.22) occupy the windows defined by $\varepsilon\mu_v\omega^2/c^2 > q^2$ and also $q^2 > \omega^2/c^2$; the latter is required for exponential decay of the fields in the bounding media. These inequalities define two windows, as is also found for a dispersive dielectric when the dielectric function has a single pole [16]. Dispersion curves for the guided waves in these windows are shown in Fig. 4.6.

A substantial extension of these results is due to Lima and Oliveira [28,29], who consider a general direction of propagation and in addition derive Green functions rather than merely dispersion equations. Fig. 4.7 shows dispersion curves for different directions of propagation on a thick film ($L \to \infty$) with the same parameters as those used for Fig. 4.5. It is seen that as **q** moves away from the Voigt direction, the frequency interval occupied by the surface polariton shrinks from both ends. This is similar to what happens with the Damon-Eshbach mode.

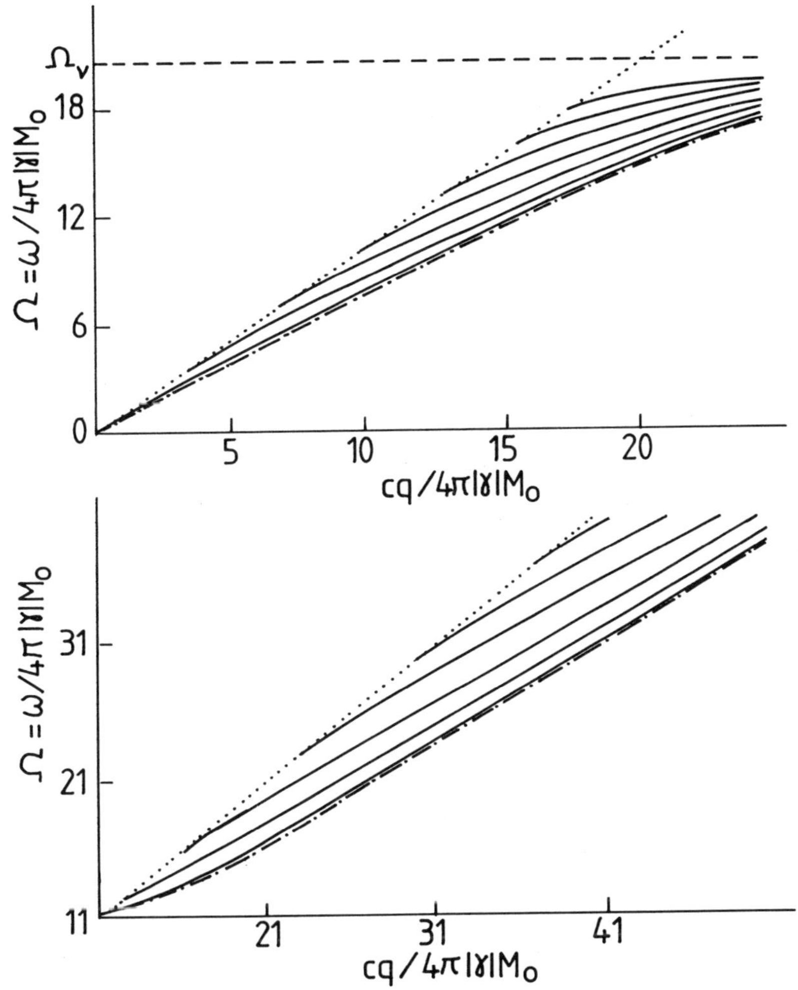

Fig.4.6 Guided-mode dispersion curves (full lines) for a free-standing ferromagnetic film. The vacuum light line $q = \omega/c$ is shown dotted and the bulk-polariton boundaries $q^2 = \varepsilon\mu_v\omega^2/c^2$ are shown chain-dotted. Top: Lower frequency window, $\mu_\infty = 1.75$, $\omega_H/\omega_M = 2$, $4\pi\gamma M_0 L/c = 1$. Bottom: Upper frequency window, $\mu_\infty = 1.75$, $\omega_H/\omega_M = 10$, $4\pi\gamma M_0 L/c = 0.5$. After ref.27.

The formal results for the surface-type modes of an antiferromagnetic film are given in a brief paper by Oliveira [30], while a discussion of guided modes is given by Cao and Caillé [31]; the former allows for a general direction of propagation, while the latter is for the Voigt direction only. Other calculations that may be mentioned are the detailed study by Thibaudeau and Caillé [32] of a dielectric film bounded by a ferromagnet and an antiferromagnet, and the study by Li et al [33] of a ferromagnetic film on a metal substrate; the latter was previously discussed by Gerson and Nadan [25].

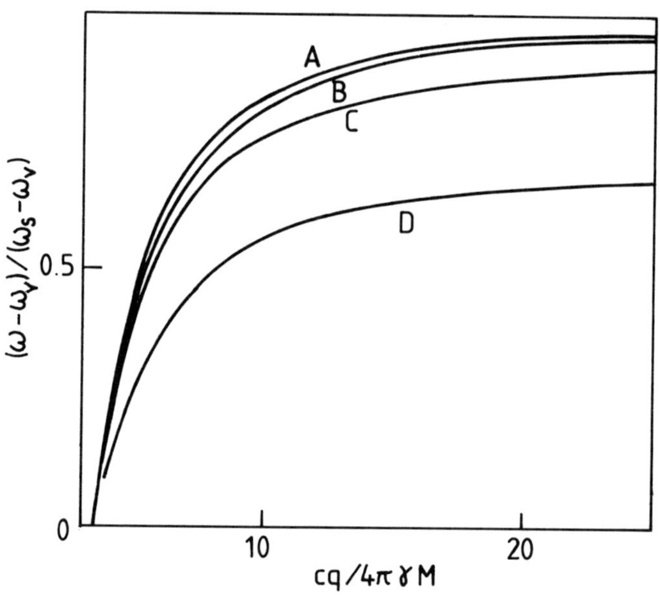

Fig.4.7 Dispersion curves for surface-type modes on a ferromagnetic film of thickness L for various angles θ between $\mathbf{q}$ and $\mathbf{H}_0$. Material parameters as in Fig.4.7. A and B, $\theta = 90^o$ (Voigt geometry); A,L $\rightarrow \infty$, B, $4\pi\gamma M_0 L/c = 0.15$. (These two curves appear in Fig.4.5). C, L $\rightarrow\infty$, $\theta = 80^o$; D, L $\rightarrow \infty$, $\theta = 70^o$. After ref.28.

4.4 SUPERLATTICE POLARITONS

Study of the retarded modes of magnetic superlattices began [34,35] a little later than investigations of magnetostatic [36,37] and exchange-coupled [38] modes. All the work that has been reported to date starts from a "bulk-slab" model in which each layer of the superlattice is presumed to have the dielectric permittivity and magnetic permeability of the corresponding bulk material. This should be adequate when each layer is many monolayers thick, but may break down for thin layers or when a magnetic ordering specific to the superlattice is found. For example, in the Fe/Gd system, the exchange interaction at the interface is antiferromagnetic, and consequently in some regions of the H-T phase diagram a "twisted phase" develops in which the spin alignments near an interface turn away from the applied field direction [39]. Another example is the Dy-Gd system [40] in which for certain H and T values alternate Dy spirals have opposite helicity. In such cases it will be necessary in effect to calculate a permeability for the whole superlattice. An analogous situation arises in short-period semiconductor superlattices, in which the dielectric response in the reststrahl frequency region is governed by the confined optic phonons of the superlattice structure itself [41]. Here we restrict attention to the bulk-slab model.

The calculation of the dispersion equation for superlattice excitations is conveniently expressed in terms of a transfer matrix [42,43]. As an example we consider s-polarized polaritons in the superlattice depicted in Fig. 4.8.

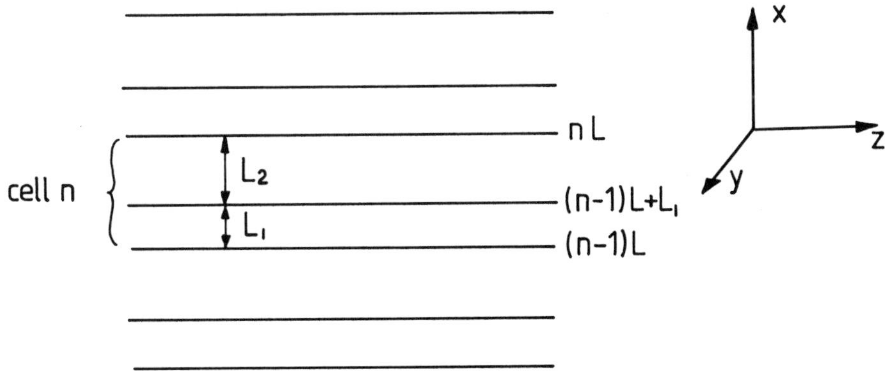

Fig.4.8 Notation for superlattice calculation.

Layers of thickness L_1 and L_2 alternate; both are taken to be magnetic media with ordering along z. We restrict attention to modes propagating in the x-y plane (Voigt geometry), so that **E** has the single component E_z, which in medium α of cell n may be written in the form

$$E_{nz}^{(\alpha)} = \left[a_n^{(\alpha)} \exp\left(\kappa_\alpha (x-nL)\right) + b_n^{(\alpha)} \exp\left(-\kappa_\alpha (x-nL)\right)\right] \exp\left[i(qy-\omega t)\right] \quad (4.23)$$

By phase matching at the boundaries, the y component of wave vector takes the common value q throughout the superlattice; κ_1 and κ_2 are given by Eq.(4.21) with the appropriate values of μ_v. Expressions for the magnetic-field components $B_{nx}^{(\alpha)}$ and $H_{ny}^{(\alpha)}$ are derived from Eq. (4.23) by means of Maxwell's equations; using the fact that these components are continuous at the interfaces one can relate the amplitudes $a_{n+1}^{(1)}$ and $b_{n+1}^{(1)}$ to those in the previous unit cell $a_n^{(1)}$ and $b_n^{(1)}$:

$$\begin{pmatrix} a_{n+1}^{(1)} \\ b_{n+1}^{(1)} \end{pmatrix} = T \begin{pmatrix} a_n^{(1)} \\ b_n^{(1)} \end{pmatrix} \quad (4.24)$$

The explicit expressions for the elements of the transfer matrix T are given elsewhere [35,43,44]. The standard derivation [16,43] shows that the Bloch wave vector component Q in the x direction is given by

$$\cos(QL) = \frac{1}{2}\,\mathrm{Tr}T \quad (4.25)$$

and substitution for T gives the explicit form

$$\cos(QL) = \cosh(\kappa_1 L_1)\cosh(\kappa_2 L_2) + \left(\Delta_1^{(1)}\Delta_1^{(2)} + \frac{1}{2}\Delta_2^{(1)}\Delta_3^{(2)} + \frac{1}{2}\Delta_3^{(1)}\Delta_2^{(2)}\right)\sinh(\kappa_1 L_1)\sinh(\kappa_2 L_2) \quad (4.26)$$

where

$$\Delta_1^{(\alpha)} = q\,\mu_2^{(\alpha)} / \kappa_\alpha \mu_1^{(\alpha)} \tag{4.27}$$

$$\Delta_2^{(\alpha)} = \left(\kappa_\alpha^2 \mu_1^{(\alpha)2} - \mu_2^{(\alpha)2} q^2\right) / \,i\,q\,\kappa_\alpha \mu_1^{(\alpha)} \left(\mu_1^{(\alpha)2} - \mu_2^{(\alpha)2}\right) \tag{4.28}$$

$$\Delta_3^{(\alpha)} = -q\left[\mu_1^{(\alpha)2} - \mu_2^{(\alpha)2}\right] / \,i\kappa_\alpha \mu_1^{(\alpha)} \tag{4.29}$$

Among other special cases, Eq.(4.26) applies to magnetic/non-magnetic superlattices, which had been considered previously for both ferromagnets [34] and antiferromagnets [45].

With Q real, Eq.(4.26) describes bulk-type polariton modes of the whole superlattice, with wave vector (Q,q) in the x-y plane. Examples of dispersion curves for a ferromagnet/non-magnet superlattice with $L_1 = L_2$ are shown in Fig.4.9. Although the curves are generally similar, an additional branch appears for propagation normal to the layers.

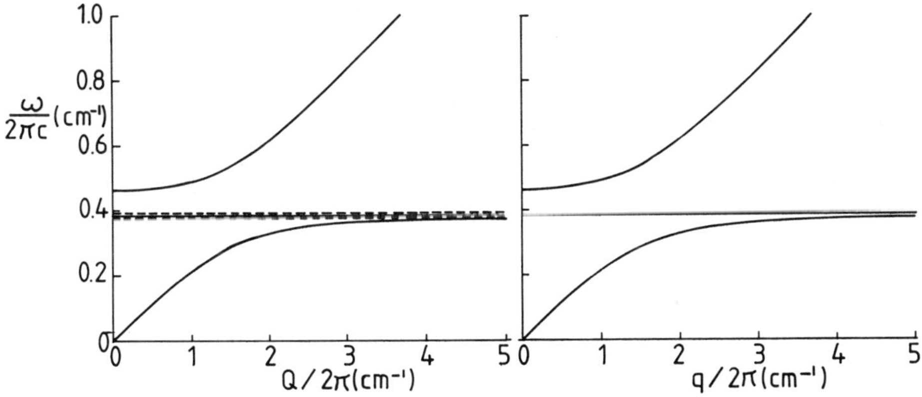

Fig.4.9 Long-wavelength bulk dispersion curves for an infinite YIG/YAG superlattice with $L_1 = L_2$. Left: propagation normal to layers; Right: propagation parallel to layers. After ref.35.

It is not part of Bloch's theorem that Q in Eq.(4.25) should be real. In particular, a pure imaginary $Q = i\alpha$ corresponds to exponential decay of the field amplitudes in successive unit cells:

$$\begin{pmatrix} a_{n+1}^{(1)} \\ b_{n+1}^{(1)} \end{pmatrix} = \exp(-\alpha L) \begin{pmatrix} a_{n}^{(1)} \\ b_{n}^{(1)} \end{pmatrix} \tag{4.30}$$

This means that an imaginary Q can correspond to a surface mode on a semi-infinite superlattice. The derivation of the dispersion relation for such a surface mode may be outlined as follows[43,44]. Eq.(4.25) determines α as a function of q and ω, $\alpha = \alpha(q,\omega)$. The electromagnetic boundary condition (H_y continuous) at the free surface gives a further relation between α, q and ω, which enables the surface-mode dispersion equation $\omega = \omega(q)$ to be determined. Barnaś[44] gives several examples, of which one is shown in Fig.4.10. The frequency scale, given in kG, corresponds to the resonance frequency of 2 cm^{-1} quoted previously. Surface modes are also discussed by

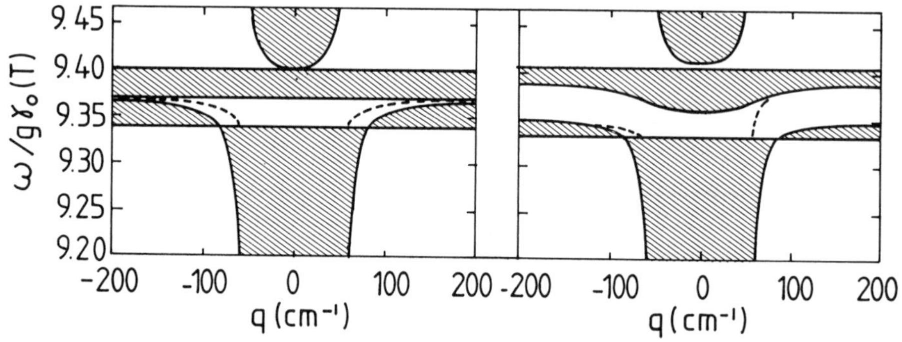

Fig.4.10 Spectrum of bulk polaritons (hatched regions) and surface polaritons (broken curves) in an antiferromagnetic MnF_2/non-magnet superlattice with $L_1 = L_2 = 5 \times 10^{-4}$cm. Left: $H_o = 0$, Right: $H_o = 200$ Gauss. After ref.44.

How and Vittoria[46,47], both for ferromagnet/non-magnet superlattices, which are a special case of what has been discussed here, and for ferromagnet/non-magnet superlattices in which the directions of magnetization alternate. Surface modes for antiferromagnet/non-magnet superlattices are considered by Zhu and Cao[45].

So far we have been concerned with propagation transverse to the ordering direction, assumed to lie within the layers. Expressions in terms of a 4 x 4 transfer matrix for both bulk and surface modes in a general propagation direction are given by Barnaś[48]. A 4 x 4 transfer matrix also occurs for a superlattice in a metallic waveguide, discussed by Zhu and Cao[49,50]. The case of a structure without periodicity, for which there is therefore no Bloch theorem, is discussed by Vecris and Quinn[51], who give a formal prescription for calculating the dispersion curves for a finite assembly of layers.

In the far infrared frequency range in which magnetic polaritons occur, the wavelength is very long compared with a superlattice period. Therefore in Eq.(4.26) the quantities QL, $\kappa_1 L_1$ and $\kappa_2 L_2$ may all be expected to be small, so that the equation can be expanded to the lowest significant order in these quantities. When this is done for the somewhat simpler case of dielectric superlattices, it is found [52] that the superlattice can be described as a uniaxial "effective medium" in which the dielectric-tensor components are given as spatial averages of the dielectric constants of the constituent layers. The same description can be arrived at from the observation that the tangential component of **E** and the normal component of **D** are continuous at the interfaces[53,54]. In the same way an effective-medium description of a magnetic superlattice can be obtained [35,55]. The full expressions are quite complicated; the method may be illustrated for the antiferromagnet/antiferromagnet superlattice with no magnetic field applied, which is easier since $\bar{\bar{\mu}}$ is diagonal. In the axes of Fig.4.8, if the ordering is in the z direction and propagation in the x-y plane, then the rf field components H_y and B_x are continuous at the interfaces. Since the wavelength is long compared with the period, these components have the same values in adjacent layers, say

$$H_y = \overline{H}_y \tag{4.31}$$

$$B_x = \overline{B}_x \tag{4.32}$$

In medium α, the other components are given by

$$B_y^\alpha = \mu_\alpha \overline{H}_y \tag{4.33}$$

$$H_x^\alpha = \mu_\alpha^{-1} \overline{B}_x \tag{4.34}$$

Defining the averages as $\overline{B}_y = \left(L_1 B_y^1 + L_2 B_y^2\right)/L$ etc., we find

$$\overline{B}_y = \mu_{yy} \overline{H}_y \tag{4.35}$$

$$\overline{H}_x = \mu_{xx}^{-1} \overline{B}_x \tag{4.36}$$

where

$$\mu_{yy} = (L_1\mu_1 + L_2\mu_2)/L \tag{4.37}$$

$$\mu_{xx}^{-1} = (L_1\mu_1^{-1} + L_2\mu_2^{-1})/L \tag{4.38}$$

These give the effective-medium description of the superlattice. In keeping with the macroscopic symmetry, μ_{yy} is distinct from μ_{xx}, and it is seen that μ_{yy} has poles where μ_1 or μ_2 have poles, while μ_{xx} has zeros where μ_1 or μ_2 have zeros.

The implications of Eqs.(4.37) and (4.38) are discussed by Almeida and Tilley[56]. They consider in particular an antiferromagnet/non-magnet superlattice. The usual type of calculation[16] shows that surface polaritons on this superlattice have the dispersion equation

$$q^2 = \frac{\omega^2}{c^2} \frac{\bar{\mu}_{xx}\left(\varepsilon_m \bar{\mu}_{yy} - \bar{\varepsilon}\right)}{\bar{\mu}_{xx}\bar{\mu}_{yy} - 1} \tag{4.39}$$

where ε_m is the dielectric constant of the bounding medium and $\bar{\varepsilon} = (L_1\varepsilon_1 + L_2\varepsilon_2)/L$ is the yy element of the effective-medium dielectric tensor.

Some dispersion curves corresponding to this for the particular case of FeF_2/ZnF_2 are shown in Fig.4.11. It is seen that the dispersion curves occupy a narrow frequency interval near ω_{AF}, and this interval decreases as the volume fraction $f_1 = L_1/L$ of magnetic material decreases. For $f_1 > 0.5$ the mode persists to $q \to \infty$, so that it has a magnetostatic counterpart, but for $f_1 < 0.5$ although there is still a surface polariton it terminates at a finite q, with no magnetostatic limit. This is reminiscent of the earlier result [36,37,57,58] that in a ferromagnet/non-magnet superlattice a magnetostatic mode occurs only for $f_1 > 0.5$.

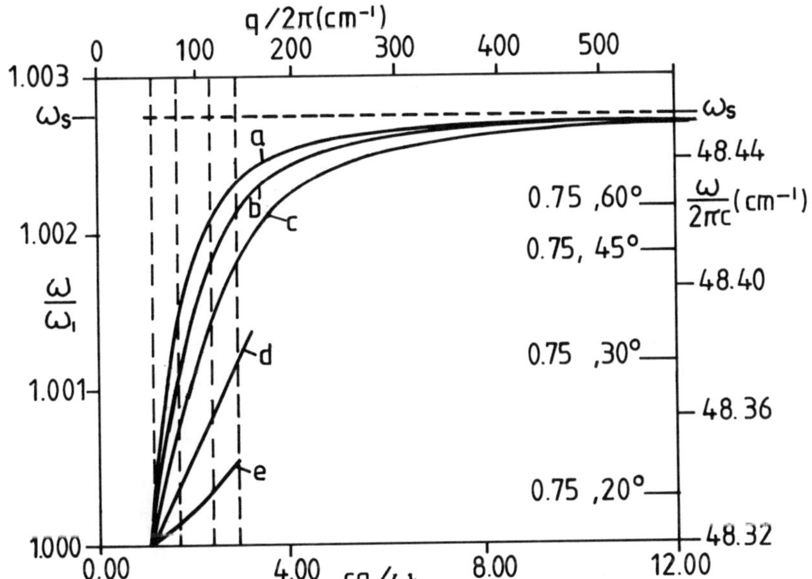

Fig.4.11 Surface-polariton dispersion curves for FeF_2/ZnF_2 superlattices interfaced with vacuum ($\varepsilon_m = 1$). The values of L_1/L are: (a) 1.0 (FeF_2), (b) 0.75, (c) 0.5, (d) 0.25, (e) 0.1. Also shown for later reference are ATR scan lines $q = (n_p \sin\theta)\, \omega/c$ with $n_p = 3.4$ (Si) and $\theta = 20^o, 30^o, 45^o, 60^o$. Points of intersection of these scan lines with the dispersion curve for $L_1/L = 0.75$ are marked on the right-hand frequency axis for later comparison with Fig.4.13. After ref.56.

4.5 EXPERIMENTAL SITUATION

As has been seen in the previous sections, there is a considerable theoretical literature on surface magnetic polaritons. However, no direct experimental study of these modes has as yet been reported. This is in striking contrast to magnetostatic surface modes, for which Brillouin scattering in particular has proved to be a very effective probe.

The work that is most closely related is the study of nonreciprocal reflection from an antiferromagnetic surface by Remer et al[59]. They measured oblique (45°)-incidence reflectivity from the surface of MnF_2 with an applied field parallel to the surface normal to the plane of incidence (Voigt geometry). The radiation source had a fixed frequency of 264GHz, and spectra were measured as reflectivity versus magnetic field. As seen in Fig.4.12 the key point is that the reflectivity changes when the direction of the magnetic field is reversed. This non-reciprocity is closely related to the nonreciprocal propagation of the surface polaritons, as seen in Fig.4.4 for example; a detailed discussion is given by Remer et al[59].

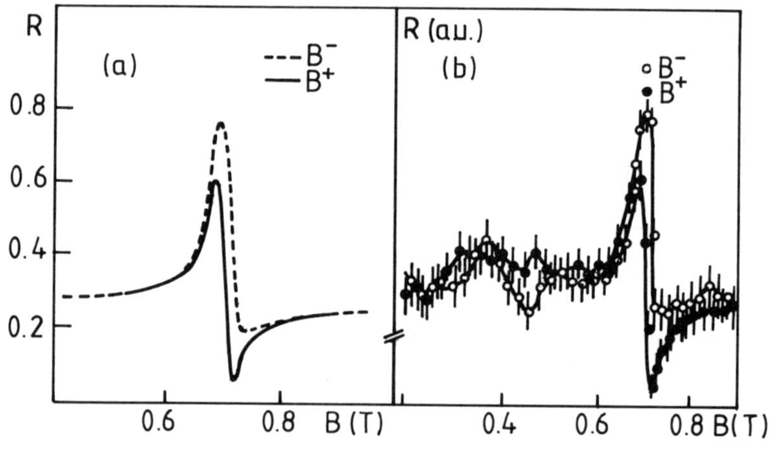

Fig.4.12 (a) Calculated and (b) measured reflectivity at 45° incidence and frequency 264 GHz off MnF_2 in the Voigt geometry. After ref.59.

A number of theoretical papers discuss possible experiments. Lima and Oliveira[28] discuss inelastic light scattering in general, and Cao and Caillé[31] discuss specifically near-forward Raman scattering as a proble of magnetic guided polaritons in a film. However, it must be recognised that the original experiments on Raman scattering from polaritons in semiconductor films [60,61] were something of a tour de force and the technique has not been applied subsequently. Application to magnetic modes would certainly be a severe challenge.

The standard techniques of grating coupling and attenuated total reflection (ATR) have been proposed by a number of authors. In ATR, shown schematically in Fig.4.13, radiation is incident at angle θ in a transparent prism, typically Si in the far infrared. It is arranged that $\theta > \theta_c$, where $\theta_c = \sin^{-1}(1/n_p)$ is the critical angle for total internal reflection. The in-plane wavevector component in the prism, and therefore throughout the set up, is

$$q = n_p(\omega/c)\sin\theta \tag{4.40}$$

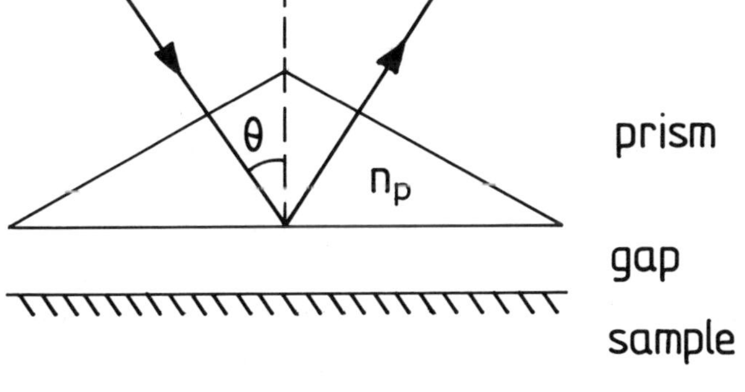

Fig.4.13 To illustrate the principle of attenuated total reflection. Radiation is incident at angle $\theta > \sin^{-1}(1/n_p)$ in a prism of refractive index n_p, which is separated by a gap from the surface of the sample.

and since $n_p \sin\theta > 1$, $q > \omega/c$. Thus if the angle is fixed and the frequency is scanned, the $(q\omega)$ plane is traversed along the scan line given by Eq.(4.40); several examples of such scan lines are shown in Fig.4.11. Since the scan line lies to the right of the light line $q = \omega/c$ it may cross surface-polariton dispersion curves. A sharp dip in reflectivity is expected at the crossings, since the incident radiation can couple into the surface modes rather than be reflected. In grating coupling, the period D of a grating on the surface of the sample adds Bragg vectors $2\pi n/D$ to q.

Fukui et al[18] discuss the use of ATR for surface modes on ferromagnets, specifically YIG. The difficulty is that typical ferromagnetic resonance frequencies are quite low, for example less than 1 cm^{-1} in YIG. Consequently samples would have to be very large, and in addition the frequencies are below the range of conventional far-infrared spectrometers.

This leaves ATR study of antiferromagnetic surfaces as the most promising proposal, as was first pointed out by Camley and Mills[19] and Cao and Caillé[20]. As an aid to discussion, Fig.4.14 shows calculated ATR curves for a FeF_2/ZnF_2 superlattice in the effective-medium approximation, corresponding to dispersion curve (b) and the scan lines of Fig.4.11. The curves are drawn for a frequency scan at fixed angle, which is automatically produced by a Fourier-transform instrusment [62]. As indicated by the arrows on the frequency axis, the dips do indeed occur at the intersection of the scan lines with the dispersion curve. An important feature of Fig.4.14 is the very fine scale of the frequency axis, which reflects the scale on Figs.4.3 and 4.4. In the Michelson interferometer normally used for Fourier-transform spectroscopy, the frequency resolution is proportional to the inverse of the traverse of the moving mirror; a resolution of order 0.01 cm^{-1} as shown in Fig.4.l4 implies a traverse of order 1 m. An alternative strategy, following Fig.4.12, would be to employ a fixed-frequency source and scan the magnetic field, but the implications of this have not yet been discussed.

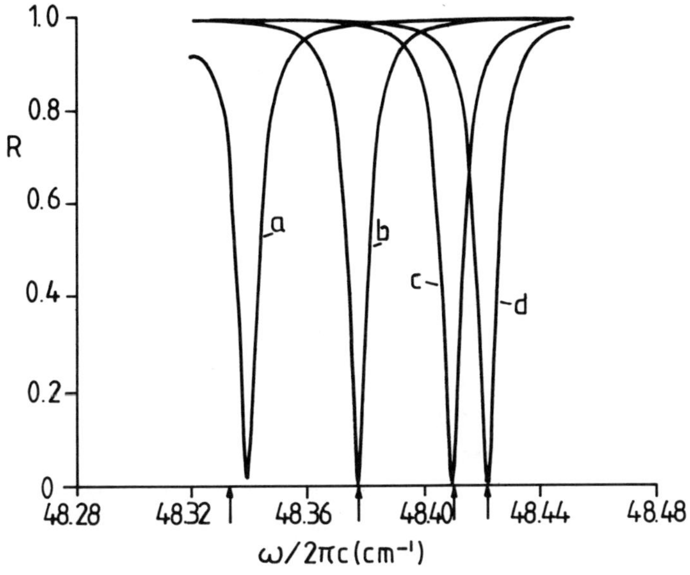

Fig.4.14 Theoretical ATR spectra for a 75% FeF_2/25°ZnF_2 superlattice without applied magnetic field. A Si prism (n_p = 3.4, θ_c = 17.2°) is assumed; angles of incidence (θ) and gap values (d in cm) for the different curves are:

Curve	a	b	c	d
θ	20	30	45	60
d	0.04	0.03	0.02	0.015

REFERENCES

1. E.F.Sarmento and D.R.Tilley, Chapter 16 of *Electromagnetic Surface Modes*, ed. A.D.Boardman (Wiley, New York 1982).
2. O.P.Gandhi, *Microwave Engineering and Applications* (Pergamon, New York 1981).
3. S.Y.Liao, *Microwave Devices and Circuits* (Prentice-Hall, New Jersey, 3rd edn.1990).
4. P.Hlawiczka, *Gyrotropic Waveguides* (Academic Press, London 1981).
5. R.E.Camley, *Surface Science Reports*, **7** (1987) 103.
6. D.L.Mills, in *Surface Excitations*, ed. V.M.Agranovich and R.Loudon (North-Holland, Amsterdam, 1984).
7. N.S.Almeida and D.L.Mills, *Phys.Rev. B* **37** (1988) 3400.
8. N.S.Almeida and D.L.Mills, *Phys.Rev.B* **36** (1987) 2015.
9. J.Barnaś and A.Stasch, *Solid State Commun.* **48** (1983) 975.
10. F.A.Oliveira, A.F.Khater, E.F.Sarmento and D.R.Tilley, *J.Phys.C: Solid State Physics* **12** (1979) 4021.
11. B.R.Cooper, in *Solid State Physics*, ed. F.Seitz, D.Turnbull and H.Ehrenreich (Academic Press, New York) **21** (1968) 393.
12. R.J.Elliott, in *Magnetism*, ed. G.T.Rado and H.Suhl (Academic Press, New York) **IIA** (1965) 385.
13. N.S.Almeida and D.R.Tilley, *Phys.Rev. B* **43** (1991) 11145.
14. L.D.Landau and E.M.Lifshitz, *Electrodynamics of Continuous Media* (Pergamon, Oxford 1960).
15. J.Zak, E.R.Moog, C.Liu and S.D.Bader, *J.Magn.Magnetic Mater.* **89** (1990) 107.
16. M.G.Cottam and D.R.Tilley, *Introduction to Surface and Superlattice Excitations* (Cambridge University Press, Cambridge 1989).
17. A.Hartstein, E.Burstein, A.A.Maradudin, R.Brewster and R.F.Wallis, *J.Phys.C* **6** (1973) 1266.
18. M.Fukui, H.Dohi, J.Matsuura and O.Tada, *J.Phys.C* **17** (1984) 1783.
19. R.E.Camley and D.L.Mills, *Phys.Rev. B* **26** (1982) 1280.
20. S.Cao and A.Caillé, *Solid State Commun.* **42** (1982) 233.
21. C.Thibaudeau and A.Caillé, *Solid State Commun.* **54** (1985) 537.
22. R.L.Stamps and R.E.Camley, *Phys.Rev.B* **40** (1989) 596.

23. J.Barnaś, *Acta Physica Polonica* **A59** (1981) 449.
24. A.Yaniv, *J.Phys.C* **14** (1981) 2549.
25. T.J.Gerson and J.S.Nadan, *IEEE Trans.Microwave Theory and Techniques* **MTT-22** (1974) 757.
26. A.D.Karsono and D.R.Tilley, *J.Phys.C* **11** (1978) 3487.
27. M.Marchand and A.Caillé, *Solid State Commun.* **34** (1980) 827.
28. N.P.Lima and F.A.Oliveira, *Solid State Commun.* **47** (1983) 921.
29. N.P.Lima and F.A.Oliveira, *J.Phys.C* **19** (1986) 5381.
30. F.A.Oliveira, *Phys.Lett.* **95A** (1983) 245.
31. S.Cao and A.Caillé, *Solid State Commun.* **43** (1982) 411.
32. C.Thibaudeau and A.Caillé, *Phys.Rev.B* **32** (1985) 5907.
33. Li Yibing, Li Shaoping and He Huahui, *J.Phys.C* **21** (1988) 2369.
34. J.Barnaś, *Solid State Commun.* **61** (1987) 405.
35. N.Raj and D.R.Tilley, *Phys.Rev.B* **36** (1987) 7003.
36. P.Grünberg and K.Mika, *Phys.Rev.B* **27** (1983) 2955.
37. R.E.Camley, T.S.Rahman and D.L.Mills, *Phys.Rev.B* **27** (1983) 261.
38. E.L.Albuquerque, P.Fulco, E.F.Sarmento and D.R.Tilley, *Solid State Commun.* **58** (1986) 41.
39. R.E.Camley and D.R.Tilley, *Phys.Rev.B* **37** (1988) 3413.
40. R.E.Camley, J.Kwo, M.Hong and C.L.Chien, *Phys.Rev.Letts.* **64** (1990) 2703.
41. T.Dumelow, A.A.Hamilton, T.J.Parker, D.R.Tilley, C.T.B.Foxon, D.Hilton and K.J. Moore, *Int.J.Infrared and Millimeter Waves* **11** (1990) 901.
42. J. Barnaś, *J.Phys.C* **21** (1988) 1021.
43. N.Raj and D.R.Tilley, *The Electrodynamics of Superlattices,* Chapter 7 of *The Dielectric Function of Condensed Systems,* ed. L.V.Keldysh, D.A.Kirzhnitz and A.A.Maradudin (Elsevier, Amsterdam 1989).
44. J. Barnaś, *J.Phys: Condens.Matter* **2** (1990) 7173.
45. N.Zhu and S.Cao, *Phys.Lett.A* **124** (1987) 515.
46. H.How and C.Vittoria, *Phys.Rev.B* **39** (1989) 6823.
47. H.How and C.Vittoria, *Phys.Rev.B* **39** (1989) 6831.
48. J. Barnaś, *Phys.Stat.Sol.(b)* **165** (1991) 529.
49. N.Zhu and S.Cao, *Physics Letters A* **124** (1987) 515.

50. N.Zhu and S.Cao, *J.Phys: Condens.Matter* **1** (1989) 7977.
51. G.Vecris and J.J.Quinn, *Phys.Rev.B* **43** (1991) 8303.
52. N.Raj and D.R.Tilley, *Solid State Commun.* **55** (1985) 373.
53. V.M.Agranovich and V.E.Kravstov. *Solid State Commun.* **55** (1985) 85.
54. W.M.Liu, G.Eliasson and J.J.Quinn, *Solid State Commun.* **55** (1985) 533.
55. N.S.Almeida and D.L.Mills, *Phys.Rev.B* **38** (1988) 6698.
56. N.S.Almeida and D.R.Tilley, *Solid State Commun.* **73** (1990) 23.
57. M.Grimsditch, M.R.Khan, A.Kueny and I.K.Schuller, *Phys.Rev.Lett.* **51** (1983) 498.
58. A.Kueny, M.R.Khan, I.K.Schuller and M.Grimsditch, *Phys.Rev.B* **29** (1984) 2879.
59. L.Remer, B.Lüthi, H.Sauer, R.Geick and R.E.Camley, *Phys.Rev.Lett.* **56** (1986) 2752.
60. J.Y.Prieur and S.Ushioda, *Phys.Rev.Lett.* **34** (1975) 1012.
61. S.Ushioda and R.Loudon, in *Surface Polaritons,* ed. V.M.Agranovich and D.L.Mills (North-Holland, Amsterdam 1982).
62. T.Dumelow, A.A.Hamilton, K.A.Maslin, T.J.Parker, D.R.Tilley, R.B.Beall, C.T.B.Foxon, J.J.Harris, D.Hilton and K.J. Moore. *Proc.Int.Workshop on Fourier Transform Infrared Spectroscopy,* ed. E.F.Vansant (Antwerp 1990) 141.

CHAPTER 5

GREEN'S FUNCTIONS IN MAGNETIC MULTILAYERED STRUCTURES

R.E. Camley and R.L. Stamps

Department of Physics, University of Colorado,
Colorado Springs, Colorado 80933-7150, USA

Contents

5.1. INTRODUCTION

Previous chapters have discussed the spin waves which exist in layered structures. A natural question is how these spin waves can be observed. Physically, one way is to apply some external driving field to the system and to observe the response. Typically the response is largest when the characteristics of the driving field (eg. the frequency or wavevector) matches that of an allowed spin wave. The mathematical description of the response to a driving field is most conveniently carried out through the use of a Green's function, and thus this chapter is devoted to the methods of calculating Green's functions for magnetic layered structures. In addition to describing responses to external probes, the Green's functions also provide a wealth of information about the excitations themselves.

Green's functions (or response functions) have become an important instrument in much of theoretical physics. The Green's function method represents a powerful tool which can be employed to obtain information on the excitations of the system, the thermal fluctuations in the system and, most directly, the response of the system to externally applied driving fields and currents. [1]

In order to understand the dynamic behavior of a particular system, one typically seeks to describe and characterize the allowed normal modes of the system. All of this information is contained in a Green's function. The presence of a surface [2] can strongly influence the dynamics of a system and one of the main goals of this chapter is to use a Green's function method to examine the dynamic excitations of magnetic systems whose geometry includes one or more surfaces or interfaces. We will illustrate with several examples just how the geometry of a particular system can be "built into" the Green's function with appropriately weighted phase factors.

To appreciate the power of the Green's function method, it is helpful to characterize the information that can be obtained by solving the equations of motion as is typically done to find a dispersion relation and contrast this with the information obtained from the Green's function methods. In the dispersion relation method one starts with a set of homogeneous equations of motion; no driving terms are included. In a planar geometry, one searches for plane wave solutions. These are oscillatory in time with frequency ω and are characterized by a two-dimensional wavevector $\mathbf{Q}$ which is parallel to the surface. Substitution of this form of solution into the equations of motion yields the dispersion relation relating $\mathbf{Q}$ and ω and the eigenfunction solution as well, i.e. the profile of the wave for those particular values of $\mathbf{Q}$ and ω. Boundary conditions may place further constraints on the dispersion relation.

In a response function development we start with inhomogeneous equations of motion that contain a point source at position $\mathbf{r}' = (x',y',z')$. We find the response function or Green's function $G(\mathbf{r}, \mathbf{r}')$ which gives the response of the medium, including boundary effects, at the position $\mathbf{r}$ due to the source located at $\mathbf{r}'$. At the simplest level this Green's function can then be used to find the response to **any** driving field by simply integrating over all source points. However, through the use of linear response theory [2-3], the Green's function also provides a great deal of additional information. It can be used to find the thermal averaged value of fluctuations in the material or to find certain correlation functions. The Green's functions are thus ideally suited for light scattering problems [4] where the amount of scattered light depends on the thermally generated spin waves in the material.

The Green's functions can be found by a number of different methods and we will illustrate some of these methods in this chapter. We note that one method for constructing the Green's function is to use an eigenfunction expansion using the eigenfunctions found in a dispersion relation calculation. So the Green's functions and dispersion relations are clearly intimately connected.

In contrast to the dispersion relation results, however, the Green's functions can provide information about the excitations in a material for **any** value of **Q** and ω and not only for those values of **Q** and ω which satisfy the dispersion relation. It is thus possible to use Green's functions to examine the effects of dissipation and other loss mechanisms on the propagation of excitations in a system. This means that a Green's function approach is especially useful when attempting to define attenuation lengths and lifetimes for excitations, as well as studying the scattering and interaction of excitations in a system due to some perturbation or inhomogeneity [1].

In summary we see that the Green's function method is a powerful tool which complements the information found from a simple dispersion relation calculation. In the next section we shall illustrate several methods which can be used to find Green's functions in magnetic systems for semi-infinite materials and multilayers. In the following sections we will discuss some applications of Green's functions to the nonlinear combination of spin-waves and to the problem of Brillouin light scattering from spin waves in ferromagnets and in ferromagnetic multilayers. Finally, we will illustrate the eigenfunction expansion method of constructing Green's functions for a ferromagnetic monolayer including exchange, dipolar and anisotropy energies. We will use the Green's functions to calculate the behavior of the magnetization as a function of temperature and to calculate the Brillouin light scattering spectrum.

5.2. SIMPLE EXAMPLES OF GREEN'S FUNCTIONS FOR SEMI-INFINITE GEOMETRIES

In this section we illustrate some basic techniques for calculating the response functions for both a semi-infinite homogeneous material and a semi-infinite superlattice. In order to keep these examples as simple as possible, we will consider only the long-wavelength dipolar waves in the magnetostatic limit in this section. The Green's functions appropriate to a semi-infinite geometry or thin film have been calculated for spins coupled through exchange interactions [5,6], dipolar interactions [7], and both dipole and exchange interactions [8]. Surface response functions for magnetic polaritons have also been discussed [9]. Many of the results for ferromagnetic and antiferromagnetic surface response functions are contained in the excellent review article by Cottam and Maradudin [2].

5.2.1 The Homogeneous Magnetic Material

The semi-infinite geometry is illustrated below.

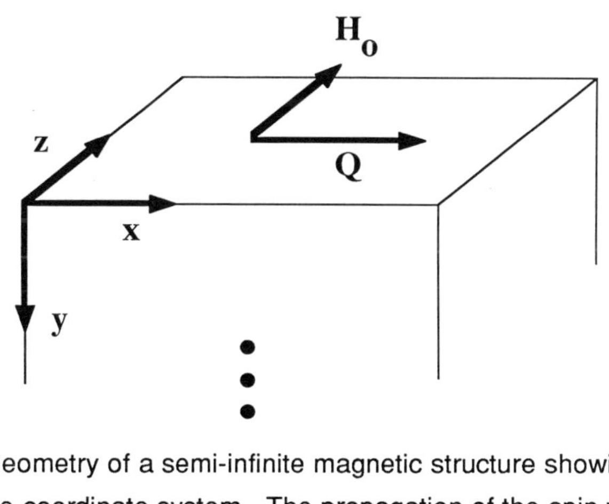

Fig. 5.1. Geometry of a semi-infinite magnetic structure showing the applied field and the coordinate system. The propagation of the spin wave is perpendicular to the applied field.

The surface is the xz plane with the z axis being the magnetization axis and the direction along which an external field H_o is applied. In this case the torque equations governing the spin motion can be used to obtain constituitive relations between B and H as has been discussed in previous chapters. This relationship is given by [10]

$$\begin{pmatrix} B_x \\ B_y \\ B_z \end{pmatrix} = \begin{pmatrix} \mu_1 & i\mu_2 & 0 \\ -i\mu_2 & \mu_1 & 0 \\ 0 & 0 & 1 \end{pmatrix} \begin{pmatrix} H_x \\ H_y \\ H_z \end{pmatrix} \tag{5.1}$$

where

$$\mu_1 = 1 + \frac{4\pi\gamma^2 H_o M}{\gamma^2 H_o^2 - \omega^2} \tag{5.2}$$

$$\mu_2 = \frac{4\pi\gamma\omega M}{\gamma^2 H_o^2 - \omega^2} \tag{5.3}$$

for a simple ferromagnet. Since similar expressions for μ hold for ferrimagnets [11] and antiferromagnets [12], the form of the results obtained here is quite general. In cases where anisotropy fields lower the symmetry the tensor μ will have a different form, but the method discussed here will still apply.

For the special case when no dynamical variables depend on z (and thus propagation is along the x axis) the static form of Maxwell's equations reduce to a rather simple equation [13]

$$\mu_1\left(\frac{\partial^2}{\partial x^2} + \frac{\partial^2}{\partial y^2}\right)\phi(x,y) = 0 \tag{5.4}$$

The magnetic scalar potential $\phi(x,y)$ is related to **B** in the usual way

$$\mathbf{B} = -\nabla\phi \tag{5.5}$$

The associated Green's function equation is

$$\mu_1\nabla^2 G(\mathbf{r},\mathbf{r}') = \delta(\mathbf{r}-\mathbf{r}') \tag{5.6}$$

where **r'** represents the position of the source term and **r** represents the observation point. (In the special case considered here **r** = (x,y,0) and **r'** = (x',y',0) since no z dependence is included.) Once we have found the Green's function, we can, as usual, solve the inhomogeneous version of Eq. (5.4)

$$\mu_1\left(\frac{\partial^2}{\partial x^2} + \frac{\partial^2}{\partial y^2}\right)\phi(x,y) = F(x,y) \tag{5.7}$$

where F(x,y) represents the "driving terms". These may be due, for example, to a spatially dependent external field with time dependence $e^{i\omega t}$. The solution for ϕ(x,y) in this case is then given by

$$\phi(x,y) = \int G(x,x',y,y';\omega)F(x',y')dx'dy' \tag{5.8}$$

That this is the solution to the inhomogeneous equation can be easily verified by using Eqs. (5.6)-(5.7)

We can obtain an explicit relationship between the driving fields F(x,y) appearing in Eq. (5.7) and any externally applied fields in the following way. We let the total H field be given by

$$\mathbf{h}_{\text{total}} = \mathbf{h}_{\text{ext}} + \mathbf{h}_{\text{d}} \tag{5.9}$$

where $\mathbf{h}_d$ is the field set up by the precessing dipoles. The inhomogeneous equations of motion for ϕ in the presence of an external driving field $\mathbf{h}_{ext}$ are obtained from the magnetostatic form of Maxwell's equations. We have

$$\nabla \cdot \mathbf{b} = \nabla \cdot (\mathbf{h}_{total} + 4\pi\mathbf{m}) = 0 \tag{5.10}$$

$$\nabla \times \mathbf{h}_{total} = 0 \tag{5.11}$$

Since $\mathbf{h}_{ext}$ is produced by some external current source, the divergence and curl of $\mathbf{h}_{ext}$ both vanish. As a result the curl of $\mathbf{h}_d$ also vanishes and we can still write the dipolar field as $\mathbf{h}_d = -\nabla\phi$. The magnetostatic Maxwell Equations then reduce to the single statement

$$\nabla \cdot (-\nabla\phi + 4\pi\mathbf{m}) = 0 \tag{5.12}$$

However $\mathbf{m}$ is related to $\mathbf{h}_{total}$ through the susceptibility:

$$\mathbf{m} = \overleftrightarrow{\chi}\mathbf{h}_{total} \tag{5.13}$$

The susceptibility for a given magnetic material is a known tensor and is related to μ in the usual way

$$\overleftrightarrow{\mu} = \overleftrightarrow{I} + 4\pi\overleftrightarrow{\chi} \tag{5.14}$$

where $\mathbf{I}$ is the identity matrix. With these substitutions, we obtain

$$\nabla \cdot (\overleftrightarrow{\mu}(-\nabla\phi) = 4\pi\overleftrightarrow{\chi}\mathbf{h}_{ext} \tag{5.15}$$

Within a homogeneous material Eq. (5.15) is identical to our general inhomogeneous Eq. (5.7) thus allowing us to identify F(x,y) with the right hand side of Eq. (5.15) .

In a finite system, such as a semi-infinite ferromagnetic, μ_1 will be a function of position. This means that the Green's function will satisfy a different differential equation in each different region of space. Furthermore, there will also be boundary conditions connecting the Green's functions in different regions.

We will solve the for the Green's function in the presence of boundaries by:

1) Finding a particular solution to the inhomogenous equation in the material
2) Finding solutions to the homogeneous equation both inside and outside the material
3) Matching the complete solutions (inside it is the sum of the particular and homogeneous solutions) at the boundaries.

First we solve the inhomogeneous equation. Since we have translational invariance along the surface, we start by introducing the Fourier transform of the Green's function parallel to the surface. Again for simplicity we consider a special case where the propagation wavevector parallel to the surface, **Q**, is directed only perpendicular to the field and thus is along the x-axis.

$$G(\mathbf{r},\mathbf{r}') = \frac{1}{2\pi}\int_{-\infty}^{\infty} e^{iQ(x-x')}G(Q,y,y')dQ \tag{5.16}$$

At this point we will also specify that the propagation is always along the +x axis so that Q is a positive number. Propagation in the opposite direction can be considered simply by reversing the direction of the applied magnetic field. In our problem such a reversal is manifested simply as a change in sign of μ_2.

Similarly we introduce the Fourier expansion for the delta function

$$\delta(\mathbf{r}-\mathbf{r}') = \delta(y-y')\frac{1}{2\pi}\int_{-\infty}^{\infty} e^{iQ(x-x')}dQ \tag{5.17}$$

We substitute the expressions for the Green's function and delta function into Eq. (5.6) and obtain

$$\mu_1\left(\frac{d^2}{dy^2} - Q^2\right)G_P(y,y') = \delta(y-y') \tag{5.18}$$

where the subscript P indicates that this is the particular solution. We solve this last equation under the assumption of an infinite, homogeneous medium. Thus we expect the Green's function to be a function of the distance |y-y'| only. Then we may also introduce the Fourier expansion in the y direction

$$G_P(y,y') = G_P(y-y') = \frac{1}{2\pi}\int_{-\infty}^{+\infty}dq_y \; e^{iq_y(y-y')}G_P(q_y) \tag{5.19}$$

Substituting this into Eq. (5.18) we obtain $G_P(q_y)$ as

$$G_P(q) = -\frac{1}{\mu_1}\left(\frac{1}{q_y^2 + Q^2}\right) \tag{5.20}$$

We may now substitute $G_P(q)$ into Eq. (5.19) and obtain G_P (y-y'). The integration necessary is done by contour integration and with the condition that the argument should go to zero for large values of y-y'. We find

$$G_P(y-y') = -\frac{1}{\mu_1}\left(\frac{e^{-Q|y-y'|}}{2Q}\right) \tag{5.21}$$

We must now add the solutions to the homogeneous equation to the particular solution found above. It is easy to see that these solutions are

Inside: $G(y) = A(y') e^{-Qy}$ (5.22)

Outside: $G(y) = B(y')e^{+Qy}$ (5.23)

The signs in the exponential terms are chosen so that the potential decays to zero infinitely far from the interface.

The total solution for the Green's function is thus given by

$$\text{Inside:} \quad -\frac{1}{\mu_1}\left(\frac{e^{-Q|y-y'|}}{2Q}\right) + A(y')e^{-Qy} \tag{5.24}$$

$$\text{Outside:} \quad B(y')e^{+Qy} \tag{5.25}$$

The unknown coefficients A(y') and B(y') are determined through the use of the boundary conditions. These conditions come from the usual electromagnetic boundary conditions that the normal component of **b** and the tangential components of **h** are continuous. These conditions require that the magnetic scalar potential obeys

$$\phi^{in} = \phi^{out} \tag{5.26}$$

$$\left(-i\mu_2\frac{\partial}{\partial x} + \mu_1\frac{\partial}{\partial y}\right)\phi^{in} = \frac{\partial}{\partial y}\phi^{out} \tag{5.27}$$

The boundary conditions for the Green's function may be obtained directly from these equations using the relationship between ϕ and G given by Eq. (5.8). Away from the source point, the Green's function simply represents the magnetic scalar potential at the position y. We take the source point y' to be in the material and away from the surface. Thus in the boundary conditions we take y' > y in our Green's function since the boundary is at y = 0. The boundary condition equations reduce to

$$-\frac{1}{\mu_1}\left(\frac{e^{-Qy'}}{2Q}\right)+ A(y') = B(y') \tag{5.28}$$

and

$$\mu_2 Q\left[-\frac{1}{\mu_1}\left(\frac{e^{-Qy'}}{2Q}\right) + A(y')\right] + \mu_1\left[-\frac{1}{\mu_1}\left(\frac{e^{-Qy'}}{2Q}\right)Q - A(y')Q\right] = B(y')Q \tag{5.29}$$

From the last two equations we can easily obtain A(y'):

$$A(y') = -\frac{1}{\mu_1}\left(\frac{e^{-Qy'}}{2Q}\right)\left[\frac{\mu_1+\mu_2-1}{1+\mu_1-\mu_2}\right] \tag{5.30}$$

Finally the Green's function for the case where the source and observation points are in the material is given by

$$G(y,y') = -\frac{1}{\mu_1}\left(\frac{e^{-Q|y-y'|}}{2Q}\right) - \frac{1}{\mu_1}\left[\frac{\mu_1+\mu_2-1}{1+\mu_1-\mu_2}\right]\left(\frac{e^{-Q(y+y')}}{2Q}\right) \tag{5.31}$$

This equation contains features characteristic of Green's functions in finite geometries. The first term on the right hand side represents the propagation of a wave directly from the source point y' to the observation point y and thus contains the distance |y-y'|. This term represents the **bulk** contribution to the Green's function and has a pole at the frequency defined by the bulk dispersion relation. Here when $1/\mu_1 = 0$ we have a frequency given by

$$\omega = |\gamma|\sqrt{H_o(H_o + 4\pi M)} = |\gamma|\sqrt{H_oB} \tag{5.32}$$

which is the frequency for a bulk magnetostatic mode propagating perpendicularly to the applied field in a ferromagnet. The second term on the right hand side represents a wave propagating from the source point y' up to the surface and then back to the observation point y. As a result this terms contains the distance y+y'. This term therefore includes the contribution of the **surface** to the Green's function. We note that the second term also has a pole at the frequency appropriate to a surface wave. The condition that the denominator in the second term is zero, ie $1+\mu_1 - \mu_2 = 0$ is equivalent to

$$\omega = |\gamma|(H_o + 2\pi M) \tag{5.33}$$

which gives the dispersion relation for a surface wave propagating perpendicular to the applied field. When the external field is reversed and thus μ_2 --> $-\mu_2$ there is no pole for the surface term. This demonstrates the nonreciprocal nature [14] of the magnetostatic surface wave as discussed in previous chapters.

The Green's function obtained above is in a mixed representation [2] in that it has both position dependence (for the y coordinate perpendicular to the surface) and wavevector dependence (for the wavevector **Q** parallel to the surface). One can obtain expressions for Green's functions which are either completely positionally dependent or wavevector dependent, and the particular representation should be chosen so as to simplify the problem at hand. For our purposes, the mixed representation will be the most convenient.

We can obtain some feeling for the relative contribution of the bulk and surface excitations to the spin fluctuations at any distance d away from the surface by using a form of the fluctuation dissipation theorem [8]. The fluctuation dissipation theorem relates the Green's functions found above, which describe the driven linear response of the system, to the probability of exciting the normal modes of the system through small random thermal fluctuations **in** the system. In general, the probability of finding a thermally induced excitation with wavevector Q and energy hω a distance d down from the surface is related to the imaginary part of the **complex** Green's function G(y=y'=d,Q, ω +i/τ). The small imaginary frequency that has been added to the frequency in the argument of the Green's function is a device whose sign insures that the signal reaches the observation point **after** leaving the source point. In principle one takes the limit as the imaginary frequency tends toward zero, but in practice a nonzero imaginary frequency is often retained and used to represent true dissipation processes in the material.

In Fig. 5.2 we plot the imaginary part of the Green's function as a function of frequency and depth into the material. We see in this figure that the imaginary part of G shows significant structure at two frequencies, those appropriate for the bulk and surface spin waves.

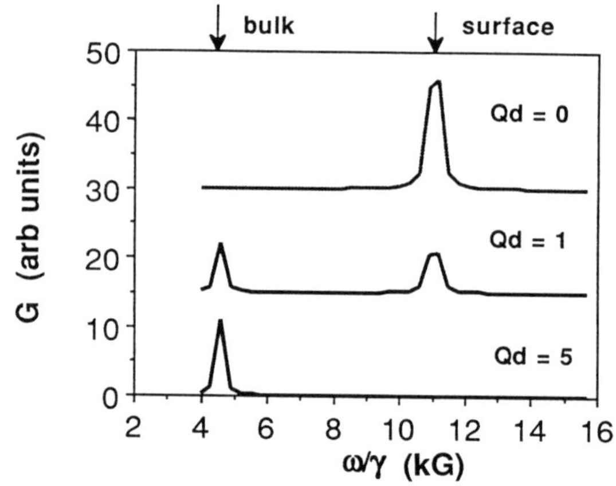

Fig. 5.2. The imaginary part of the Green's function as a function of frequency for different values of Qd. Qd=0 represents the fluctuation spectrum at the surface while Qd=5 represents the fluctuations a few wavelengths deep into the material. The parameters for the calculation are M_o = 1.6 kG, H = 1 kG and the damping parameter $1/\tau\gamma = .1$ kG.

For this propagation direction there is only one bulk spin wave peak since all the bulk spin waves are degenerate in frequency; we shall see in the superlattice structure that this degeneracy is broken. The peaks have finite width due to the inclusion of damping. Note that the size of surface wave peak decreases as d (distance into the material) is increased. Also the bulk peak at d=0 is reduced since the surface wave "steals" some spectral density away from the bulk modes [8].

5.2.2 The Superlattice Structure

Green's functions for a number of different magnetic superlattice systems have appeared in the literature. The problems studied include ferromagnetic [15] and antiferromagnetic [16] systems with either pure dipolar or pure exchange coupling. Green's functions for magnetic polaritons in superlattices have also been studied [17]. A number of general formulations for superlattice response functions also exist [18,19].

The superlattice geometry is shown in Fig. 5.3. We use here a different method from that explored in the previous section which is more convenient for a general geometry. The reason that this is necessary is that even in the case of the infinitely extended superlattice there is no infinitesimal translational invariance in the direction perpendicular to the layers. This means that the Green's function can not be a simple function of |y-y'| since, for example, for the same distance |y-y'| y could be in one material and y' in the other or both y and y' could be in the same material.

The basic idea is that away from the source point the solutions for the Green's function must simply be those for the homogeneous differential equation for the magnetic scalar potential. These solutions should already satisfy the boundary conditions at each interface and at the surface of the semi-infinite superlattice.

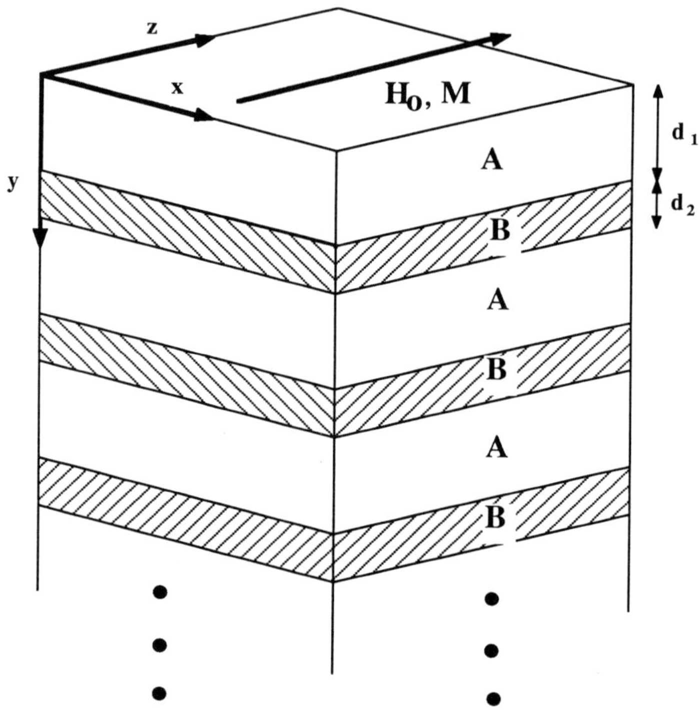

Fig. 5.3. The semi-infinite superlattice geometry.

We thus introduce two functions $\phi^>(y)$ and $\phi^<(y)$ which satisfy the homogeneous equations of motion and the boundary conditions for the magnetic scalar potential in the regions y>y' and y<y' respectively. Thus $\phi^>(y)$ and $\phi^<(y)$ must satisfy

$$\mu_1\left(\frac{d^2}{dy^2} - Q^2\right)\phi^{>,<}(y) = 0 \tag{5.34}$$

Explicit forms for $\phi^>(y)$ and $\phi^<(y)$ will be found later on. We expect our general solution for the Green's function to have the form

$$G(y,y') = A(y')\phi^{>}(y)\Theta(y-y') + B(y')\phi^{<}(y)\Theta(y'-y) \tag{5.35}$$

Here the step function, $\Theta(y - y')$, is unity if the argument is positive and zero otherwise. The prefactors are independent of y but will depend on y', the location of the singularity, as we will see later. To fix the prefactors in from of the solutions for the two regions y>y' and y<y' we will match the solutions across the singularity provided by the delta function at y = y'. Thus the particular values of A and B must be chosen so that the Green's function obeys Eq. (5.6). In fact, it is well known that the equation

$$\left[-Q^2 + \frac{\partial^2}{\partial y^2}\right]\Phi(y,y') = \delta(y-y') \tag{5.36}$$

has as a solution

$$\Phi(y,y') = \frac{1}{W}[\ \phi^{>}(y)\phi^{<}(y')\Theta(y-y') + \phi^{<}(y)\phi^{>}(y')\Theta(y'-y)\ ] \tag{5.37}$$

where W is the Wronskian and is given by

$$W = \left(\frac{\partial}{\partial y}\phi^{>}(y)\phi^{<}(y') - \frac{\partial}{\partial y}\phi^{<}(y)\phi^{>}(y')\right)_{y=y'} \tag{5.38}$$

The Green's function which satisfies Eq. (5.6) is then given quite simply by

$$G(y,y') = \frac{1}{\mu_1 W}[\ \phi^{>}(y)\phi^{<}(y')\Theta(y-y')$$
$$+\ \phi^{<}(y)\phi^{>}(y')\Theta(y'-y)\ ] \tag{5.39}$$

The Green's function given above is essentially a formal solution and independent of geometry. If we wanted to consider a semi-infinite ferromagnet, then we must only substitute the forms for the scalar potential $\phi^{>}(y)$ and $\phi^{<}(y)$ which are the solutions for the semi-infinite ferromagnet. One may show that the Green's function defined in Eq. (5.39) does reproduce the correct Green's functions for the infinite ferromagnet, the semi-infinite ferromagnet and the finite thickness film.

Before deriving the superlattice Green's functions it is useful to explicitly obtain some results for a simpler case. This will be helpful in understanding the calculational steps in the more involved superlattice problem. We thus re-examine the semi-infinite ferromagnet Green's functions, but now obtained in a different fashion. We start by writing expressions for $\phi^{>}(y)$ and $\phi^{<}(y)$. For $y>y'$ we will have a wave which "propagates" toward $+\infty$. For values of $y < y'$ there is a wave which propagates upward toward the surface and one which is reflected back toward the source point. Outside the ferromagnet where y is negative we have an exponentially decaying wave. Thus we take

$$\phi^{>}(y) = Ae^{-Qy} \tag{5.40}$$

$$\phi^{<}(y) = Be^{+Qy} + Ce^{-Qy} \tag{5.41}$$

$$\phi^{out}(y) = De^{+Qy} \tag{5.42}$$

From the boundary conditions at the surface, Eqs. (5.26) - (5.27), we may eliminate D and find C in terms of B:

$$C = -\left[\frac{1-\mu_1-\mu_2}{1+\mu_1-\mu_2}\right]B \tag{5.43}$$

There remain two unknowns A and B. However it is not necessary to determine these parameters since they appear in the Wronskian in such a way as to make the Green's function independent of A and B. After a little bit of algebra the Wronskian is found to be

$$W = 2QAB \tag{5.44}$$

Substitution of the expressions for $\phi^{>}(y)$, $\phi^{<}(y)$ and W into Eq (5.39) for the Green's function yields (after a little algebra) exactly the same expression, Eq. (5.31), for Green's function for the semi-infinite geometry that we found earlier.

We now return to the calculation for the superlattice. We thus need to find explicit forms for $\phi^{>}(y)$ and $\phi^{<}(y)$ which are appropriate for the superlattice. For the layered structure considered here, $\phi^{>}(y)$ will represent wave in the region y>y' propagating toward $+\infty$. Similarly, $\phi^{<}(y)$ is composed of two waves. One propagates (upwards) away from the source term at y' and towards the surface. The second wave is the reflected wave which propagates (down) away from the surface.

Since the method for finding such solutions has been presented in earlier chapters, we only present a brief outline here. As in the previous problem the scalar potential outside the material exponentially decays with distance away from the surface and is given by Eq. (5.42). The solution in the magnetic films is given by

$$\phi^{>}(y) = e^{iQ_{\perp}nL}\left(v_{+}^{>}e^{Q(y-nL-d_1)} + v_{-}^{>}e^{-Q(y-nL)}\right) \tag{5.45}$$

$$\phi^{<}(y) = e^{-iQ_\perp nL}(v_+^I e^{Q(y-nL-d_1)} + v_-^I e^{-Q(y-nL)})$$

$$e^{+iQ_\perp nL}(v_+^R e^{Q(y-nL-d_1)} + v_-^R e^{-Q(y-nL)}) \tag{5.46}$$

In these expressions $Q_\perp$ is the Bloch wavevector of the superlattice excitation, n indicates the unit cell in which y is located, d_1 and d_2 are the thicknesses of the films in the superlattice and $L = d_1 + d_2$ is the length of the unit cell. The Bloch wavevector $Q_\perp$ is obtained by solving the equation

$$\cos(Q_\perp L) = \cosh(Qd_1)\cosh(Qd_2) + \left(\frac{1+\Lambda^+\Lambda^-}{\Lambda^+ + \Lambda^-}\right)\sinh(Qd_1)\sinh(Qd_2) \tag{5.47}$$

The coefficients v^I and v^R govern the contribution of the waves incident on the surface and reflected from the surface respectively. The coefficients $v^>$ measure the contribution of the wave which propagates from y' down to $+\infty$. The v_+ and v_- coefficients for the different superscripts are related to each other through the boundary conditions at the interfaces within the superlattice, ie at $y = nL$ and $y = nL + d_1$. It can be shown that

$$v_+^{R,>} = A^R v_-^{R,>} \tag{5.48}$$

and

$$v_+^I = A^I v_-^I \tag{5.49}$$

where

$$A^R = \left[\frac{e^{Qd_1} - e^{iQ_\perp L}e^{-Qd_2}}{1 - e^{-iQ_\perp L}e^{Qd_1}e^{-Qd_2}}\right]\frac{1+\Lambda^+}{\Lambda^- - 1} \tag{5.50}$$

and

$$A^{I} = \left[\frac{e^{Qd_1} - e^{iQ_{\perp}L}e^{-Qd_2}}{1 - e^{+iQ_{\perp}L}e^{Qd_1}e^{-Qd_2}}\right]\frac{1+\Lambda^{+}}{\Lambda^{-}-1} \tag{5.51}$$

Finally, we note that Λ^{+} and Λ^{-} are given by

$$\Lambda^{\pm} = \mu_1 \mp \mu_2 \tag{5.52}$$

We pause here to review our progress so far. At this point we have the solutions to the homogeneous equations of motion and thus have almost all the pieces we need to construct the Green's function. We can imagine that we have specified the wavevector parallel to the surface, Q, and the frequency, ω, as input parameters. From Q and ω we can find the Bloch wavevector $Q_{\perp}$ which is consistent with these values. Then we can work back through Eqs (5.45) - (5.54) and put everything in terms of three arbitrary constants, v^{R}_{-}, v^{I}_{-}, and $v^{>}_{-}$. As in our simpler example above we only need to find a relationship between two of these three variables to completely determine the Green's function. Again, as in the simpler example, this is done through the use of the boundary conditions at the surface of the semi-infinite superlattice. We find

$$v_{-}^{R} = Av_{-}^{I} \tag{5.53}$$

where

$$A = \frac{\left(\Lambda^{-}-1\right)A^{I}e^{-Qd_1} - \left(1+\Lambda^{+}\right)}{\left(1-\Lambda^{-}\right)A^{R}e^{-Qd_1} + \left(1+\Lambda^{+}\right)} \tag{5.54}$$

With these results, explicit expressions for the Green's functions may be obtained for the semi-infinite layered material. These expressions are somewhat lengthy but are easily obtained in a straight-forward way. As an example we present the results for the important case where both y and y' are in the same layer n:

$$G(y,y')=\frac{1}{\mu_1 W}\{[F_1(F_3+F_5)e^{Q(y+y')}+F_1(F_4+F_6)e^{Q(y-y')}$$
$$+F_2(F_3+F_5)e^{-Q(y-y')}+F_2(F_4+F_6)e^{-Q(y+y')}]\Theta(y-y')$$

$$[F_1(F_3+F_5)e^{Q(y'+y)}+F_1(F_4+F_6)e^{Q(y'-y)}$$
$$+F_2(F_3+F_5)e^{-Q(y'-y)}+F_2(F_4+F_6)e^{-Q(y'+y)}]\Theta(y'-y)\} \qquad (5.55)$$

Here the Wronskian, W, is given by

$$W=2QV_-^I V_-^>(A^R-A^I)e^{-Qd_1} \qquad (5.56)$$

As expected, the Wronskian is independent of position. The various F terms are given by

$$F_1=V_-^> A^R e^{-Q(nL+d_1)} \qquad (5.57)$$

$$F_2=V_-^> e^{+QnL} \qquad (5.58)$$

$$F_3=V_-^I A^I e^{-Q(nL+d_1)} \qquad (5.59)$$

$$F_4=V_-^I\ e^{QnL} \qquad (5.60)$$

$$F_5=V_-^I\ A^R A\, e^{-Q(nL+d_1)} e^{2iQ_\perp nL} \qquad (5.61)$$

$$F_6=V_-^I\ A e^{QnL}\, e^{2iQ_\perp nL} \qquad (5.62)$$

We see above that the two arbitrary amplitudes V^I and $V^>$ are contained in both the F terms and in the Wronskian and thus cancel out. The Green's function written out above displays the explicit dependence on the position coordinates y and y'. This dependence is much more complicated than that seen in the simple semi-infinite geometry, reflecting the more complicated structure of the superlattice. As we will see, the explicit dependence on y and y' is necessary for the calculation of light scattering due to spin waves.

5.2.3 A Green's Function Tensor and Spin Correlation Functions

We now discuss how spin correlation functions can be obtained from the Green's functions calculated here. As remarked in the introduction, an application of a fluctuation dissipation theorem relates a Green's function to a temperature dependent correlation function. For a spin system, the useful correlation functions will measure correlations between spin or magnetization variables, and is related to a matrix of magnetic Green's functions. To derive the appropriate magnetic Green's function matrix from the scalar potential Green's functions calculated above, we define the magnetization by

$$\mathbf{m}(\mathbf{r}) = \int d^3\mathbf{r}'\, \overleftrightarrow{\mathbf{G}}_m(\mathbf{r},\mathbf{r}')\cdot \mathbf{h}_{ext}(\mathbf{r}') \tag{5.63}$$

Here $\mathbf{h}_{ext}$ is some external driving field and $\mathbf{G}_m$ is the Green's function tensor we are seeking. (Essentially $\mathbf{G}_m$ represents the generalized susceptibility.) On the other hand, the magnetization may also be calculated directly through the use of the relation $\mathbf{m} = \chi\, \mathbf{h}_{total}$. The total field is the sum of the external driving field, $\mathbf{h}_{ext}$, and the dipolar field (generated by the external field) which in the magnetostatic limit is $\mathbf{h}_d = -\nabla\phi$. Thus

$$\mathbf{m}(\mathbf{r}) = -\overleftrightarrow{\chi}\cdot\nabla\phi + \overleftrightarrow{\chi}\cdot \mathbf{h}_{ext}(\mathbf{r}) \tag{5.64}$$

The scalar potential ϕ is the response to the external driving field

$\mathbf{h}_{ext}$ and thus can be obtained from the scalar Green's function G derived earlier. In fact,

$$\phi = \int d^3\mathbf{r}' G(\mathbf{r},\mathbf{r}') 4\pi \nabla' \cdot (\chi \cdot \mathbf{h}_{ext}(\mathbf{r}')) \tag{5.65}$$

Using Eqs. (5.63) - (5.65) we can obtain the components of $\mathbf{G_m}$ in terms of the scalar potential Green's function, G:

$$[G_m(\mathbf{r},\mathbf{r}')]_{\alpha\beta} = 4\pi \sum_{\gamma,\lambda} \chi_{\alpha\gamma} \chi_{\lambda\beta} \nabla_\gamma \nabla'_\lambda G(\mathbf{r},\mathbf{r}') + \chi_{\alpha\beta} \delta(\mathbf{r}-\mathbf{r}') \tag{5.66}$$

This Green's function matrix can then be used in a fluctuation dissipation theorem to calculate the temperature dependent magnetic correlation functions in the following manner.

We consider spin-spin correlation functions of the form

$$S_{ij}(\mathbf{r},\mathbf{r}';t-t') = \langle S_i(\mathbf{r},t) S_j(\mathbf{r}',t')\rangle \tag{5.67}$$

Since the system has translational invariance parallel to the surface we may write $S_{ij}(\mathbf{r},\mathbf{r}';t-t')$ in terms of its Fourier transform

$$S_{ij}(\mathbf{r},\mathbf{r}';t-t') = \frac{1}{(2\pi)^3} \int d\mathbf{Q} d\omega\, S_{ij}(\mathbf{Q},\omega,y,y') e^{i\mathbf{Q}\cdot(\mathbf{R}-\mathbf{R}')} e^{-i\omega(t-t')} \tag{5.68}$$

The correlation function $S_{ij}(\mathbf{Q},\omega,y,y')$ is constructed from the appropriate Green's functions using linear response theory. For completeness, we simply quote the result here

$$S_{ij}(\mathbf{Q},\omega,y,y') = i[1+n(\omega)]\left\{(G_m(\mathbf{Q},\omega+i\eta,y,y')_{ij} - (G_m(\mathbf{Q},\omega-i\eta,y,y')_{ij}\right\} \tag{5.69}$$

In the above $n(\omega)$ is the is the Bose-Einstein distribution function and $\mathbf{G}_m$ is the Green's function tensor defined in this section. A study of the correlation functions $S_{ii}(\mathbf{Q},\omega,y=y')$, with i = x or y, gives a picture of the eigenfunction associated with the excitation. In particular the imaginary part of $S_{ii}(\mathbf{Q},\omega,y=y')$ measures the square of the amplitude of the Fourier component with wavevector **Q** and frequency ω of the thermal fluctuations in $S_i(\mathbf{r},t)$ a distance y away from the surface. Examples of the use of the correlation functions to study particular excitations are given in references [6, 8, 20]

5.3. EXAMPLES OF APPLICATIONS OF GREEN'S FUNCTIONS

In this section we briefly illustrate the application of Green's functions in two different problems: The nonlinear mixing of bulk and surface magnetostatic spin waves [21]; and calculations of Brillouin scattering intensities for light scattering from spin wave excitations in semi-infinite ferromagnets [4, 8, 22] and superlattices[15]. In both cases we try to concentrate on the fundamental ideas, rather than give rigorous and complete derivations.

5.3.1. Nonlinear mixing of bulk and surface spin waves

As has been noted in previous chapters, the equations of motion governing the spin system, Bloch's torque equations, are inherently nonlinear. Here we will first discuss an interesting effect where, due to the nonlinearity, one can mix two waves to produce a resonantly enhance third wave. The geometry considered is illustrated below in Fig. 5.4.

First we simply outline the conditions necessary for resonant enhancement of the generated wave. Two waves, one with wavevector **Q** and angular frequency $\omega(\mathbf{Q})$ and a second wave with wavevector **Q'** and frequency $\omega(\mathbf{Q'})$, can interact nonlinearly to produce a wave with wavector $\mathbf{Q} + \mathbf{Q'}$ and frequency $\omega(\mathbf{Q}) + \omega(\mathbf{Q'})$. This wave will be resonantly enhanced if

$$\omega(\mathbf{Q}) + \omega(\mathbf{Q'}) = \omega(\mathbf{Q} + \mathbf{Q'}) \tag{5.70}$$

In other words, there is a resonant enhancement if the resulting final frequency and wavevector lie on the dispersion relation for the linear medium.

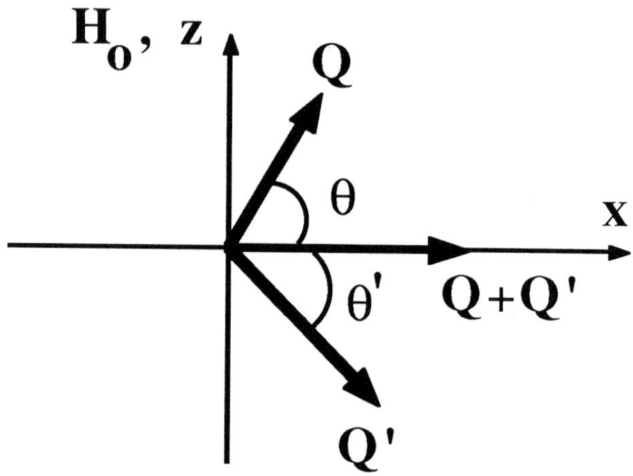

Fig. 5.4. The two incident waves with wavevector **Q** and **Q'** propagate at angles θ and θ' with respect to the x axis. The generated wave with wavevector **Q** + **Q'** is always directed along the x axis. We are looking at the surface plane from above.

Because of an unusual property of magnetostatic spin waves on a semi-infinite structure -- that the frequency depends only on the direction of propagation and not on the magnitude of the propagation wavevector -- the resonance condition may be satisfied in an unusual way. We can imagine fixing the direction of one of the input waves and of the nonlinearly generated wave. We then can vary the **direction** of the second input wave until the resonance condition is fulfilled. Note that the direction of the generated wave may be fixed because the magnitudes of the two incident wavevectors may be arbitrarily adjusted until the vector sum lies in the desired direction.

Explicit relationships for the resonance conditions can be obtained in terms of the angles of propagation θ and θ' shown in Fig. 5.4. We first consider two bulk waves propagating in the xz plane in an infinite medium. The general equation for the frequency of a bulk spin wave is given by

$$\omega_B(\theta)=\gamma H_o(H_o+4\pi M_s\cos^2\theta)^{1/2} \tag{5.71}$$

For propagation along the x axis, the frequency is thus $\gamma(H_oB)^{1/2}$ where the field B is given by $B = H_o + 4\pi M_s$. For convenience, we set the geometry of the two incident waves so that the output is always directed along x. The condition relating θ and θ' is then found from Eqs. (5.70) and (5.71). We find

$$\omega_B(\theta)+\omega_B(\theta')=\gamma(H_oB)^{1/2} \tag{5.72}$$

The condition for the resonant enhancement of a generated surface wave may be determined in a similar manner. For the semi-infinite geometry, the frequency of the surface mode in a simple ferromagnet is given by

$$\omega_S(\theta)=\frac{1}{2}\gamma\left(\frac{H_o}{\cos\theta}+B\cos\theta\right) \tag{5.73}$$

As has been noted, the surface waves are restricted in propagation direction so that only values where $\theta < |\theta_c|$ are allowed. The critical angle θ_c is given by the equation $\cos(\theta_c) = (H_o/B)^{1/2}$. The resonance condition for the two surface waves is then given by

$$\omega_S(\theta)+\omega_S(\theta')=\frac{1}{2}\gamma(H_o+B) \tag{5.74}$$

While these arguments describe the conditions necessary to produce a resonantly enhanced wave, the strength of the enhancement is still unknown. We can calculate the strength of this resonant enhancement using Green's functions to solve the nonlinear equations of motion. The nonlinear equations of motion are given to second order in the fields by

$$\begin{bmatrix} \omega_H & \partial/\partial t & \omega_M(\partial/\partial x) \\ -\partial/\partial t & \omega_H & \omega_M(\partial/\partial y) \\ 4\pi(\partial/\partial x) & 4\pi(\partial/\partial y) & -\nabla^2 \end{bmatrix} \begin{bmatrix} M_x \\ M_y \\ \phi \end{bmatrix} = \begin{bmatrix} \gamma M_x(\partial/\partial z)\phi \\ \gamma M_y(\partial/\partial z)\phi \\ (2\pi/M_s)(\partial/\partial z)(M_x^2 + M_x^2) \end{bmatrix} \tag{5.75}$$

Here m_x and m_y are the time-dependent, spatially fluctuating, transverse components of magnetization. ϕ is the magnetic scalar potential, and $\omega_H = \gamma H_o$ while $\omega_M = \gamma M_s$. The upper two equations are found from Bloch's equations and the lower is found from the magnetostatic Maxwell equations. Note that the nonlinear terms appear on the right hand side as "driving terms."

The form of the above equations suggest the use of a Green's function **tensor** which satisfies the following equations:

$$\begin{bmatrix} \omega_H & \partial/\partial t & \omega_M(\partial/\partial x) \\ -\partial/\partial t & \omega_H & \omega_M(\partial/\partial y) \\ 4\pi(\partial/\partial x) & 4\pi(\partial/\partial y) & -\nabla^2 \end{bmatrix} \begin{bmatrix} g_{xx} & g_{xy} & g_{xz} \\ g_{yx} & g_{yy} & g_{yz} \\ g_{zx} & g_{zy} & g_{zz} \end{bmatrix} = \begin{bmatrix} \delta(\mathbf{r}-\mathbf{r}') & 0 & 0 \\ 0 & \delta(\mathbf{r}-\mathbf{r}') & 0 \\ 0 & 0 & \delta(\mathbf{r}-\mathbf{r}') \end{bmatrix} \tag{5.76}$$

We can solve for the Green function tensor in a similar manner to the first example in section 5.2.1. We shall not give the details here, but refer the reader to reference [2] where explicit expressions for the Green's function tensor are given.

Having obtained the Green's function tensor we can solve Eq. (5.75) in an iterative manner. First the solutions to the linear equations are found, using the appropriate boundary conditions. These zero-order solutions will be the usual bulk or surface spin waves. Then one substitutes these solutions into the right-hand-side (the nonlinear terms) and solves for the new waves generated by the nonlinear mixing. The solution for the nonlinearly generated wave can be written

$$u_\alpha^{(g)}(\mathbf{r},t) = \sum_\beta \int d\mathbf{r}\, G_{\alpha\beta}(\mathbf{r},\mathbf{r}';t-t')F_\beta(u^{(d)}(\mathbf{r}',t')) \tag{5.77}$$

Here we have switched to a slightly more abstract notation in order to simplify the writing of this equation. $u_1 = M_x$, $u_2 = M_y$ and $u_3 = \phi$. The superscript (g) indicates the generated wave while the superscript (d) is for the driving waves. The vector **F** represents the nonlinear driving terms given on the right hand side of Eq. (5.75).

In order to obtain a feeling for the efficiency of the generation of the new wave, we can calculate the time average of the square of one of the components of the magnetization for both the generated wave and the sum of the input waves. If we take the component of magnetization along x_1 then these values will be denoted by $<(m_1{}^2)_g>$ for the generated wave and $<(m_1{}^2)_i>$ for the incident wave. The ratio $<(m_1{}^2)_g>/<(m_1{}^2)_i>$ then clearly shows resonant behavior.

In our example we use the geometry of Fig. 5.4 for both bulk and surface waves. For the ferromagnet we take yittrium iron garnet with $M_s = 140$ G and apply a field of $H_o = 50$ G. We fix $\theta = 50^o$ and plot the ratio $<(m_1{}^2)_g>/<(m_1{}^2)_i>$ as a function of the angle θ' for the second incident wave. We have also added a phenomenological linewidth to the problem by adding to $\omega(\mathbf{Q}) + \omega(\mathbf{Q}')$ a term $i\Delta\omega$ in the Green's functions. $\Delta\omega$ is related to the linewidth ΔH (the full width at half maximum in ferromagnetic resonance) by $\Delta\omega = \gamma\Delta H/2$. For the curves here we take a linewidth of 1 G. The results of this calculation are presented in Fig. 5.5.

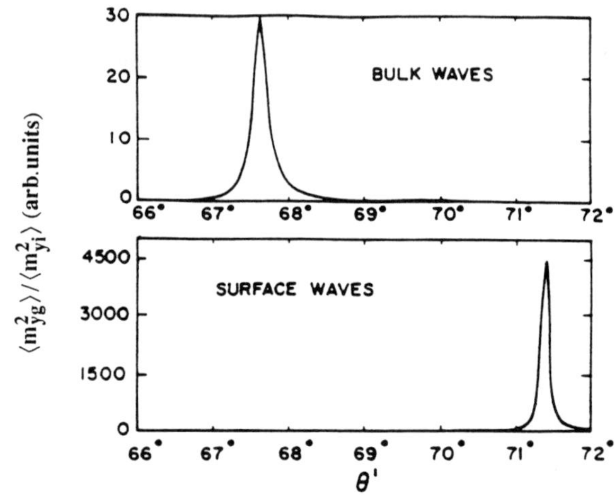

Fig. 5.5. The strength of the nonlinearly generated wave as a function of the angle θ'. The angle for the other incident wave is fixed at 50°. Note that the resonance angles for bulk and surface waves are quite different [21].

We note that the angles of the clear resonances seen in Fig. 5.5 are in agreement with the angles determined by the resonance conditions determined earlier for bulk and surface spin waves.

5.3.2 Light Scattering from Spin Waves

In our second example of the use of Green's functions we will consider Brillouin scattering from spin-wave excitations in magnetic systems. Essentially this is also a nonlinear process where the incident light sees deviations in the dielectric constant due to the thermally generated spin waves. These deviations cause the scattering of the incoming light.

The relationship between the change in the dielectric constant and the spin wave is given by [23]

$$\delta\varepsilon_{\mu\upsilon}(\mathbf{r},t)=\sum_{\lambda}K_{\mu\upsilon\lambda}S_{\lambda}(\mathbf{r},t)+\sum_{\lambda,\delta}G_{\mu\upsilon\lambda\delta}S_{\lambda}(\mathbf{r},t)S_{\delta}(\mathbf{r},t) \tag{5.78}$$

Here $\delta\varepsilon_{\mu\nu}(\mathbf{r},t)$ is the change in the dielectric tensor at time t and position **r** due to the spin wave **S**(**r**,t). We note that although the second term on the right-hand-side of Eq. (5.78) looks like a contribution from a two spin wave process, it also includes first order terms if S_λ or S_δ is equal to S_z and we replace the z component of **S** by the saturation value. While second order processes are dominate in some systems (antiferromagnets for example), for light scattering from ferromagnets the most important processes involve deviations in the dielectric tensor which are **linear** in the spin components.

We can obtain a feeling for the calculation of the scattered light by considering how this deviation in the dielectric tensor changes the wave equation. For our problem Maxwell's Equations are given by

$$\nabla\cdot\mathbf{D}=0 \tag{5.79}$$

$$\nabla\cdot\mathbf{B}=0 \tag{5.80}$$

$$\nabla\times\mathbf{E}=-\frac{1}{c}\frac{\partial\mathbf{B}}{\partial t} \tag{5.81}$$

$$\nabla\times\mathbf{H}=+\frac{1}{c}\frac{\partial\mathbf{D}}{\partial t} \tag{5.82}$$

The constituitive relations are

$$\mathbf{B}=\mu\mathbf{H} \tag{5.83}$$

and

$$\mathbf{D}=(\varepsilon+\delta\varepsilon)\mathbf{E} \tag{5.84}$$

By taking the curl of Eq. (5.81) in the usual way and using Eqs. (5.82)-(5.84) and the constituitive relations, we obtain a **driven wave equation** given by

$$\nabla^2\mathbf{E}+\frac{\omega^2\varepsilon\mu}{c^2}\mathbf{E}=-\frac{i\omega\mu}{c^2}\frac{\partial}{\partial t}(\delta\varepsilon\cdot\mathbf{E})-\frac{1}{\varepsilon}\nabla[\nabla\cdot(\delta\varepsilon\cdot\mathbf{E})] \tag{5.85}$$

The driving terms are proportional to the deviation in the dielectric tensor which carries the information about the spin waves.

We next write the electric field as the sum of the incident wave and the scattered wave; i.e. $\mathbf{E} = \mathbf{E}^{(i)} + \mathbf{E}^{(s)}$. Then treating $\mathbf{E}^{(s)}$ as a small perturbation we obtain a driven wave equation for the scattered field in terms of the incident fields:

$$\nabla^2\mathbf{E}^{(s)}+\frac{\omega^2\varepsilon\mu}{c^2}\mathbf{E}^{(s)}=-\frac{i\omega\mu}{c^2}\frac{\partial}{\partial t}\left(\delta\varepsilon\cdot\mathbf{E}^{(i)}\right)-\frac{1}{\varepsilon}\nabla\left[\nabla\cdot(\delta\varepsilon\cdot\mathbf{E}^{(i)})\right] \tag{5.86}$$

As we have seen earlier a convenient method to solve inhomogeneous equations is through the use of a Green's function, and the solution for the scattered wave can be written as

$$\mathbf{E}^{(s)}(\mathbf{r},t)=\left(\frac{\omega_o}{c}\right)^2\sum_{\beta\gamma}\int\frac{d\mathbf{r}'dt'}{4\pi}\mathbf{D}(\mathbf{r},\mathbf{r}';t-t')\delta\ddot{\varepsilon}(\mathbf{r}',t')\cdot\mathbf{E}^{(i)}(\mathbf{r}',t') \tag{5.87}$$

Although this is a lengthy equation, its structure can be easily appreciated. The right hand side contains the Green' function for Maxwell Equations $\mathbf{D}(\mathbf{r},\mathbf{r}';t-t')$ which gives the response at position **r** of the material due to a source located at **r'**. The driving terms $\delta\varepsilon\,\mathbf{E}$ are the same as those appearing in the driven wave equation obtained earlier.

In a Brillouin scattering experiment we measure the **average intensity** of the scattered light due to thermally generated spin waves. We obtain the average intensity from our expression for the scattered electric field by multiplying the scattered field by its complex conjugate and performing a thermal average. The resulting expression contains a number of spin-spin correlation functions of the form

$$S_{ij}(\mathbf{r},\mathbf{r}';t-t') = \langle S_i(\mathbf{r},t)S_j(\mathbf{r}',t')\rangle \tag{5.88}$$

with these terms coming from the expansion of the change in the dielectric tensor in terms of the dynamic spin variables. As we have seen in Section 5.2.3, it is helpful to write the correlation function in terms of its Fourier transform. When this is done, the intensity of the scattered light can then be expressed in terms of $S_{ij}(\mathbf{Q},\omega,y,y')$ as

$$\begin{aligned} <|E|^2> \; = \int dy \int dy' \; [&r_{xx}S_{xx}(\mathbf{Q},\omega,y,y') + r_{yy}S_{yy}(\mathbf{Q},\omega,y,y') \\ &+ r_{yx}S_{yx}(\mathbf{Q},\omega,y,y') + r_{xy}S_{xy}(\mathbf{Q},\omega,y,y')] \end{aligned} \tag{5.89}$$

where the r_{ij} are functions that depend on the reflectivity and transmisivity of the material, the angle of incidence, the scattering angle and the polarization of the incident wave and scattered waves. The important point is that the scattered intensity is directly related to the magnetic correlation function and this correlation function is constructed from the appropriate Green's functions using linear response theory as we have seen in Section 5.2.3.

We present an example of the light scattering spectrum for a semi-infinite ferromagnet in Fig. 5.6 and and example for a magnetic multilayer in Fig. 5.70. In both figures, the wavevector for the incident light lies in the xy plane so that the spin wave excitations which produce the backscattering travel in the x direction due to momentum conservation parallel to the surface. An external magnetic field is applied in the z direction.

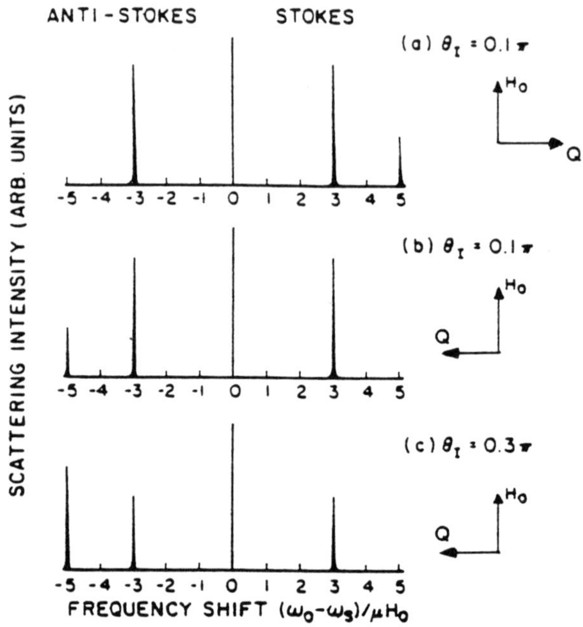

Fig. 5.6. Theoretical light scattering spectrum for semi-infinite EuO. Note that as the wavevector is reversed the high frequency peak changes from the Stokes to the anti-Stokes side. θ_I is the angle of incidence for the incoming light. The applied field is 3 kG and $4\pi M_S = 24$ kG [8].

The intensity for backscattered light is shown as a function of the frequency shift away from the frequency of the illuminating beam. A negative shift in frequency corresponds to the creation of a spin wave (called a Stokes process) which is equivalent to a spin wave travelling in the x direction in this example. A positive shift in frequency corresponds to the destruction of a spin wave (called an anti-Stokes process), which is equivalent to a spin wave travelling in the -x direction.

The spectrum for scattering from a semi-infinite ferromagnet shown in Fig. 5.6 shows the nonreciprocal features of the surface wave very clearly. The lower frequency peak appears on both the Stoke's and Anti-Stoke's sides and corresponds to the narrow band of bulk excitations. The other, high frequency peak which appears only on the Anti-Stokes side is due to the surface wave. A reversal of the applied field direction would move the peak to the Stoke's side, which is an extremely useful experimental "test" for a ferromagnetic surface wave.

In Fig 5.7, the spectrum for Brillouin light scattering from a ferromagnetic superlattice is shown. The superlattice is also assumed to be semi-infinite and built by alternating layers of ferromagnetic Ni and nonmagnetic Mo. The scattering geometry is the same as in Fig. 5.6, but the spectrum is much more involved. Instead of a narrow, degenerate band of bulk modes on both sides, there now appears a broad band of modes whose intensity decreases with frequency. These modes are in fact the collective bulk bands of the superlattice structure which are formed from the coupled surface modes of the individual magnetic layers. The nonreciprocal feature which appears only on one side of the spectrum at Ω_S is the collective surface mode of the superlattice, again formed from the coupled surface modes of the individual magnetic layers. We see in this figure that this mode does indeed switch from one side of the spectrum to the other when the magnetic field is reversed. It is also interesting to note that not only does this mode exist for a restricted set of propagation directions, it also only exists when the thickness of the nonmagnetic layers is less than the thickness of the magnetic layers (when no anisotropies are present in the magnetic layers).

In conclusion we note that the theory is in good agreement with experimental results. These experiments include work on thick films of insulators [24], metals [25], thin metallic films [26] and on metallic superlattices [27-29]. A review of experimental work on Brilluoin light scattering is given in chapter 6 of this book. We also point out that methods other than Green's functions have been used to calculate light scattering spectra [30].

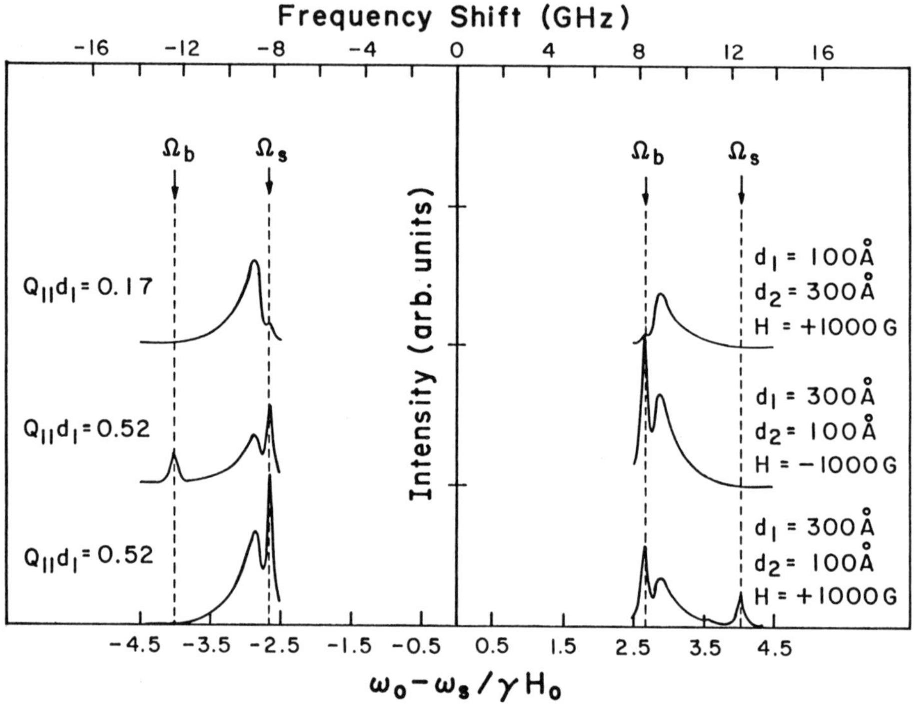

Fig. 5.7. Theoretical light scattering spectrum from a Ni/Mo superlattice. $\omega_0 - \omega_S$ is the shift in frequency between the incident light and the scattered light. Ω_b is the lowest frequency of the superlattice bulk modes and Ω_S is the frequency for the superlattice surface modes. Note that when $d_1 > d_2$ a surface wave exists while for $d_1 < d_2$ there is no surface wave [15].

5.3.3 Green's Functions and Monolayers

In the previous examples, the Green's functions were found by either solving a set of inhomogeneous differential equations or by constructing them out of the solutions to the homogeneous differential equations. While these are natural methods of dealing with systems which can be modeled by boundary value problems where the important variables are treated as continuous functions of position, one often encounters systems whose variables are more appropriately described by discrete functions of position. In order to find the Green's functions for such systems, it is useful to use another technique in which the Green's functions are constructed from the eigenfunctions and eigenvalues found by solving for the dynamic excitations of the system.

Our ultimate goal is to calculate the temperature dependent saturation magnetization and the Brillouin light scattering spectrum for a planar ferromagnet including the effects of a strong perpendicular anisotropy. To illustrate the eigenfunction expansion method, we will construct the magnetic correlation functions as follows. We will:

1. Write the equations of motion for the time and spatially varying magnetization and solve for the normal modes of the planar array of spins;
2. Form the response functions from the eigenvectors and eigenvalues corresponding to the normal modes and construct the appropriate magnetic Green's functions;
3. Use the fluctuation dissapation theorem to relate the Green's functions to magnetic correlation functions.

The energies we will consider include a Heisenberg exchange interaction:

$$H_{ex} = \sum_{<i,j>} J_{ij}\, \mathbf{S}_i \cdot \mathbf{S}_j \tag{5.90}$$

where the brackets indicate that the sum is restricted to nearest neighbors only. Note that the subscripts **i** and **j** are vectors denoting a site in the planar lattice. A uniaxial anisotropy energy is specified in the direction normal to the film plane (along the y axis, as before):

$$H_{ani} = \frac{1}{g\mu_b} K \sum_{\mathbf{i}} (\hat{\mathbf{y}} \cdot \mathbf{S_i})^2 \tag{5.91}$$

where K measures the strength of the anisotropy energy. A Zeeman term for an external static magnetic field directed along the z axis is also included:

$$H_{Ze} = g\mu_b H_o \sum_{\mathbf{i}} \hat{\mathbf{z}} \cdot \mathbf{S_i} \tag{5.92}$$

In the long wavelength region, the equations of motion for the Hamiltonian formed from the sum of these three energies have the form of Bloch's equations of motion:

$$\frac{\partial}{\partial t} \mathbf{S_i} = \mathbf{S_i} \times \mathbf{H_i} \tag{5.93}$$

The fields $\mathbf{H_i}$ are effective fields acting on the spin at site **i** determined by commuting $\mathbf{S_i}$ with the total Hamiltonian. The torque equations, Eqs. (5.93), provide a valid description of the spin dynamics for long wavelength excitations which, as we will see below, dominate the thermal properties of the saturation magnetization.

An interesting feature of a two dimensional Heisenberg ferromagnet is that ferromagnetic order is unstable in the absence of anisotropies or dipolar interactions [31]. In fact, long range ferromagnetic order has been observed experimentally in a number of systems [32-35]. Although anisotropies are present in most experimental systems of interest, dipolar coupling between the spins can stabilize ferromagnetic order in two dimensions even in the absence of magnetic anisotropies [35-38]. We thus also include dipolar demagnetizing fields into the equations of motion also using a Green's function method.

The method we present is based on a technique applied by Kalinikos, et. al., [39] to the problem of dipole-exchange modes in ferromagnetic systems. The basic idea of this method is to start with a Green's function appropriate to an infinite empty space. In a vacuum the static Maxwell's equations are given by $\nabla \times \mathbf{h}_d = 0$ and $\nabla \cdot \mathbf{h}_d = 0$ As usual we can introduce a scalar potential ϕ defined by $\mathbf{h}_d = -\nabla\phi$, and then obtain the equation $\nabla^2\phi = 0$. We then introduce a Green's function which solves $\nabla^2 g_\phi = \delta(\mathbf{r}-\mathbf{r}')$. For a planar geometry it is most convenient to work in a mixed representation and we take for the Green's function the form given in Eq. (5.21) with $\mu_1 = 1$.

The magnetic material, in this case a thin film, is introduced as a driving field in the region of the material. The advantage of this method is that we do not have to explicitly concern ourselves with boundary conditions since the film is defined simply as a collection of sources. First we use $\nabla \cdot \mathbf{b} = 0$ to obtain an equation of motion for the magnetostatic potential in the material and obtain

$$\nabla^2\phi = 4\pi\nabla \cdot \mathbf{m} \tag{5.94}$$

The right hand side of Eq. (5.94) provides the driving terms in the material. This equation can now be solved by the Green's function introduced above:

$$\phi(\mathbf{r}) = \int d^3\mathbf{r}'\, g_\phi(\mathbf{r},\mathbf{r}')\nabla' \cdot 4\pi\mathbf{m}(\mathbf{r}') \tag{5.95}$$

We then calculate $\mathbf{h}_d$ as

$$\mathbf{h}_d(\mathbf{r}) = -4\pi\int d^3\mathbf{r}'\, \nabla g_\phi(\mathbf{r},\mathbf{r}')\nabla' \cdot \mathbf{m}(\mathbf{r}') \tag{5.96}$$

Once we know g_ϕ, we can write the dipolar fields $\mathbf{h}_d$ as an integral equation in terms of **m**. To proceed further, we use the translational invariance of the system in the xz plane and expand all functions of x and z in Fourier series. The result is

$$\mathbf{h}_d(\mathbf{r}) = -2\pi \int d^2\mathbf{Q}\, e^{i\mathbf{Q}\cdot\mathbf{R}} \int dy' e^{-Q|y-y'|} \begin{bmatrix} q_x^2/Q & iq_x \mathrm{sgn}(y-y') \\ iq_x \mathrm{sgn}(y-y') & -Q \end{bmatrix} \mathbf{m}(y')$$

$$-4\pi \int d^2\mathbf{Q}\, e^{i\mathbf{Q}\cdot\mathbf{R}} \begin{bmatrix} 0 & 0 \\ 0 & 1 \end{bmatrix} \mathbf{m}(y) \tag{5.97}$$

The relationship between $\mathbf{h}_d$ and $\mathbf{m}$ given by Eq. (5.97) can be further simplified by noting that for a monolayer in the long wavelength regime, we are really only concerned with fields that are averaged over atomic thicknesses. With this in mind, we assume that $\mathbf{m}(y')$ is constant across the monolayer and perform the integration over y'. Finally we also average $\mathbf{h}_d(y)$ over the thickness "a" of the planar array of atoms. The result is

$$\mathbf{h}_d = -4\pi \int d^2\mathbf{Q}\, e^{i\mathbf{Q}\cdot\mathbf{R}} \begin{bmatrix} \frac{Qa}{2}\left(\frac{q_x^2}{Q^2}\right) & 0 \\ 0 & 1-\frac{Qa}{2} \end{bmatrix} \mathbf{m} \tag{5.98}$$

A failure of this continuum approach is that it does not properly describe the effects of a discrete lattice on the demagnetizing factor for the planar array of spins. If one evaluates the exact dipole sums for a square array of spins, then the 1 appearing in lower right hand corner of the matrix in Eq. (5.98) is replaced by (0.726) [38].

The result of using $\mathbf{H}_i+\mathbf{h}_d$ in Eq. (5.93) can be written in a simple matrix form:

$$\frac{\omega}{\gamma}\mathbf{m} = \begin{bmatrix} 0 & -iH_1 \\ iH_2 & 0 \end{bmatrix}\mathbf{m} \tag{5.99}$$

where

$$H_1 = H_o + \varepsilon_{\mathbf{Q}} + 4\pi M_s - 2\pi M_s Qa \tag{5.100}$$

$$H_2 = H_o + \varepsilon_{\mathbf{Q}} + 2\pi M_s (Qa)\frac{q_x^2}{Q^2} - \frac{K}{M_s} \tag{5.101}$$

and the exchange contribution is contained in $\varepsilon_{\mathbf{Q}}$:

$$\varepsilon_{\mathbf{Q}} = \frac{1}{g\mu_b}\sum_{<\mathbf{i},\mathbf{j}>} J_{\mathbf{i},\mathbf{j}}\left(e^{i\mathbf{Q}\cdot(\mathbf{R}_i - \mathbf{R}_j)} - 1\right) \tag{5.102}$$

The eigenvalues of Eq. (5.99) will be designated by Ω. The solutions to this equation are complicated by the presence of the uniaxial anisotropy and dipolar terms which make the matrix non-Hermitian. This means that there are *two* eigenvectors [40] for each eigenvalue Ω: one obtained by solving Eqn. (5.97) and one obtained by solving

$$\frac{\omega}{\gamma}\mathbf{m} = \mathbf{m}\begin{bmatrix} 0 & -iH_1 \\ iH_2 & 0 \end{bmatrix} \tag{5.103}$$

We designate the eigenvector solution of Eqn. (5.99) as the right eigenvector, $\mathbf{m}_r$, and the eigenvector solution of Eqn. (5.103) as the left eigenvector, $\mathbf{m}_l$.

The response functions are defined as before as solutions to the equations of motion driven by a point source. The Fourier transformed equation reads:

$$\frac{1}{M_s}\begin{bmatrix} \frac{\omega}{\gamma} & iH_1 \\ -iH_2 & \frac{\omega}{\gamma} \end{bmatrix} \vec{\vec{g}}(\omega, \mathbf{Q}) = \vec{\vec{\mathbf{1}}} \tag{5.104}$$

The solution to this inhomogeneous equation is easily shown to be

$$\vec{\vec{g}}(\omega, \mathbf{Q}) = M_s \sum_{\Omega} \frac{\mathbf{m}_r(\Omega)\mathbf{m}_l(\Omega)}{\omega/\gamma - \Omega} \tag{5.105}$$

where the sum is over all the allowed eigenfrequencies Ω. For the monolayer, there are only two eigenfrequencies given by $\pm(H_1H_2)^{1/2}$. As we will see below, both the positive and the negative frequencies are necessary to properly describe the response of the system.

First, however, it must be noted that these response functions do **not** directly represent the response of the spin system to a driving magnetic **field**, as did the Green's functions of the previous sections. This is because in writing the equations of motion as an eigenvalue problem, we did not insure that our driving terms corresponded to a physical driving field. In order to obtain the Green's functions which measure the response of the spin system to driving magnetic fields, we must perform the transformation [8]:

$$g_{xx} \to +i\, g_{yx} \tag{5.106}$$
$$g_{xy} \to +i\, g_{yy} \tag{5.107}$$
$$g_{yx} \to -i\, g_{xx} \tag{5.108}$$
$$g_{yy} \to -i\, g_{xy} \tag{5.109}$$

This transformation makes **g** correspond to the Green's functions in our previous examples where the equations of motion were solved in the presence of a driving dipolar field. The transformed **g** satisfies

$$\frac{1}{M_s}\begin{bmatrix} H_2 & i\dfrac{\omega}{\gamma} \\ -i\dfrac{\omega}{\gamma} & H_1 \end{bmatrix} \ddot{g}(\omega,\mathbf{Q}) = \ddot{\mathbf{1}} \tag{5.110}$$

Solving explicitly for the left and right eigenvectors, the result takes on a very simple form:

$$\ddot{g}(\omega,\mathbf{Q}) = \frac{M_s}{2}\left(\frac{1}{\omega/\gamma-\sqrt{H_1H_2}} - \frac{1}{\omega/\gamma+\sqrt{H_1H_2}}\right)\begin{bmatrix} \sqrt{\dfrac{H_2}{H_1}} & 0 \\ 0 & \sqrt{\dfrac{H_1}{H_2}} \end{bmatrix} + \frac{M_s}{2}\left(\frac{1}{\omega/\gamma-\sqrt{H_1H_2}} + \frac{1}{\omega/\gamma+\sqrt{H_1H_2}}\right)\begin{bmatrix} 0 & i \\ -i & 0 \end{bmatrix} \tag{5.111}$$

It can easily be verified that both the positive and negative frequency solutions are required in order for this Green's function to satisfy Eq. (5.110). Also, it is interesting to note that for **Q**=0, **g** reduces to the usual magnetic susceptibility defined by Eq. (5.14).

Finally, the magnitude of the eigenfrequency is given by

$$|\Omega| = \sqrt{H_1H_2} = \sqrt{\left(H_o+\varepsilon_{\mathbf{Q}}+4\pi M_s-2\pi M_s Qa\right)\left(H_o+\varepsilon_{\mathbf{Q}}+2\pi M_s (Qa)\frac{q_x^2}{Q^2}-\frac{K}{M_s}\right)} \tag{5.112}$$

Here it is easy to see how dipolar interactions modify the long wavelength dispersion relation. In particular, as **Q** goes toward 0, $\Omega \propto Q^{1/2}$, which is very different from the $\Omega \propto Q^2$ behavior of a three dimensional ferromagnetic lattice.

By means of the fluctuation-dissapation theorem mentioned at the end of section 5.2.3, we can now construct the temperature dependent spin correlation functions. As an example, suppose that a ferromagnetic system is at some temperature T so that a number of spin waves are thermally excited. In our geometry the saturation magnetization lies along the z axis, so the spin waves correspond to fluctuations in the xy plane. Neglecting the zero point fluctuations, the temperature dependent part of the net reduction of the saturation magnetization is proportional to [2]

$$\Delta = \int d\omega \sum_{\mathbf{Q}} \left[S_{xx}(\mathbf{Q};\omega) + S_{yy}(\mathbf{Q};\omega) \right] \tag{5.113}$$

Converting to an integral over **Q** and using Eq. (5.69) and **g** given above, this is

$$\Delta = \int dQ\, Q\, n(\Omega(\mathbf{Q})) \left[\frac{H_1 + H_2}{\Omega(\mathbf{Q})} \right] \tag{5.114}$$

This result agrees with previous calculations [37,41] for the temperature dependence of two dimensional ferromagnets in the presence of dipolar coupling. Dipolar interactions as well as the uniaxial anisotropy make the spin wave modes highly elliptical and the term in brackets measures this ellipticity. While a nonzero anisotropy stabilizes a two dimensional ferromagnet by creating an energy gap at **Q**=0, we note that dipolar interactions alone will also stabilize two dimensional ferromagnetic order. In particular, the behavior of $\omega(\mathbf{Q})$ at long wavelengths removes the two dimensional instability even with zero anisotropy (K=0) [37,41].

Since the correlation functions have such a simple form for the planar ferromagnet, it is interesting to examine the form of the Brillouin light scattering intensity. Proceeding exactly as in Section 5.3.2 except using the Green's function **g**, the intensity of the scattered light for the planar ferromagnet at frequency Ω is given by

$$I(\pm\Omega)=r_{xx}\sqrt{\frac{H_2}{H_1}}+r_{yy}\sqrt{\frac{H_1}{H_2}}\pm 2\,\mathrm{Im}\left(r_{xy}\right) \tag{5.115}$$

The $\pm$ signs refer to Stokes and anti-Stokes, respectively. The r_{ij} are chosen appropriate to a monolayer on a substrate and are pure real for i=j but complex for i≠j in an absorbing medium.

Thus the relative intensities of the Stoke's and anti-Stoke's peaks will, in an absorptive media, depend on a competition between the diagonal and off-diagonal elements of the magnetic correlation function [42]. Referring to the explicit expressions for H_1 and H_2 given by Eqs. (5.98) and (5.99), we see that this ratio depends sensitively on the ellipticity of the spin wave mode [42]. At long wavelengths, this ellipticity is governed in large part by the strength of the uniaxial anisotropy.

So far in this section we have considered only monolayer ferromagnetic structures. It is a simple matter to extend these results to multiple layer films, and one can show that the basic features of a quasi-two dimensional ferromagnet are displayed by ultra-thin ferromagnetic films consisting of three or four atomic layers. In Fig. 5.8, we plot the ratio of Stoke's to anti-Stoke's intensities for a number of different anisotropies as functions of applied field strength for a three atomic layer thick cobalt film. The ellipticity of the modes becomes more circular as the applied field strength is increased since H_1 and H_2 become nearly equal at large field strengths. For small applied field strengths, however, the modes are highly elliptical due to the dipolar interactions and the uniaxial anisotropy. For these small applied fields, the Stokes/anti-Stokes ratio first decreases and then slowly increases with increasing applied field strength.

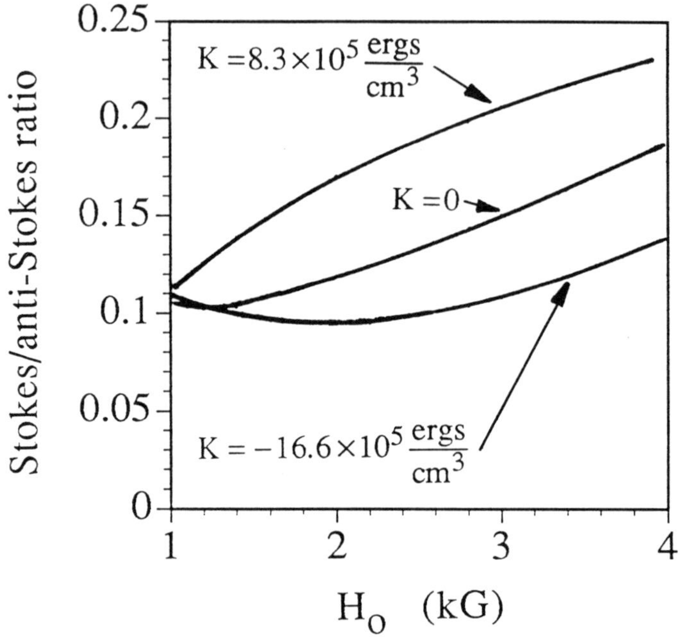

Fig. 5.8. The Stokes/anti-Stokes ratio for light scattering from a three layer film as a function of applied field for different values of an out-of-plane anistropy.

This can be understood in terms of the in-plane and out-of-plane fluctuations of the magnetization. The ratios H_1/H_2 and H_2/H_1 represent the average out-of-plane and average in-plane fluctuations of the magnetization, respectively. As the field strength is increased, the ratio H_1/H_2 will also increase whereas the ratio H_2/H_1 will decrease. The rate of change with applied field differs for these two terms, and is strongly modified by the magnitude of the anisotropy. For monolayers and ultra-thin ferromagnetic films, the Stokes/anti-Stokes ratio will display a minimum at some applied field strength which will depend on the strength of any uniaxial anisotropies present in the system.

5.4. ADDITIONAL APPLICATIONS FOR GREEN'S FUNCTIONS

There are of course far more applications for Green's functions than the few examples that we have presented here. To give the reader some feeling for the broad range of problems which can be attacked with a Green's function technique, in this final section we list some other interesting works that we have not had the opportunity to mention. We emphasize that the following list is by no means complete and represents only a small fraction of the existing work. Furthermore, we concentrate on studies with surfaces or interfaces.

Magnetic excitations in antiferromagnetic materials have been studied using Green's functions for semi-infinite structures [9], films [43], and layered systems[16]. Also excitations in antiferromagnetically coupled layered have been discussed using Green's functions [44]. In addition Green's functions have also been used to discuss spin wave excitations in ferrimagnetic thin films [45].

The models we have discussed throughout this chapter have been macroscopic in the sense that we have said nothing about the electronic band structure which governs the magnetic properties of the material. Various approaches have been developed which attempt to provide a microscopic picture of magnetism that make use of spin dependent Green's functions for the electrons in a crystal structure. These include tight binding methods for ferromagnetic films [46], and magnetic semi-conductors [47].

Problems involving impurities and inhomogeneities are often formulated as driven equations of motion where the impurities or inhomogeneities supply the driving terms. These problems can thus be examined with a Green's function approach . Examples include the propagation of surface excitations on rough surfaces [9], and impurities in a semi-infinite structure [48-49].

For most of our applications we have used the magnetic correlation functions which we derived from magnetic response functions. These functions were useful in discussing the magnetic behavior at temperatures far away from phase transition points. On the other hand, methods have been developed for using correlation functions to examine magnetic behavior in the critical regions using forms of the scaling hypothesis. Some notable applications of this technique have been made for ferromagnetic monolayers by Prokovsky, et. al. [50].

Besides being used to study purely magnetic excitations, Green's functions have also been used to examine a variety of other collective excitations. Plasmon excitations of an electron gas, for example, can be described by Green's functions which are formally identical to the magnetostatic Green's functions discussed here [51]. There are also excitations that arise from the coupling between the magnetic spins in a lattice structure and the elastic forces acting between the atoms at each lattice site [2]. Magnetic polaritons, which are excitations in the coupled magnon and photon field, have also been studied with a response function technique in order to examine the effects of dissapation mechanisms [9].

The generation of magnetostatic surface waves and bulk waves by a strip line transducer is a common experimental technique [52]. A very direct application of Green's functions in finite geometries addresses the excitation of spin waves by strip line transducers. [53]

Acknowledgements

We would like to thank Professor A. S. Carriço for a careful reading of this manuscript. This work was supported by the U.S. Army Research Office through grant #DAAL03-91-G-0229. We would like to thank Professor G. Güntherodt and Dr. B. Hillebrands for their hospitality during our stay at the RWTH Aachen where some of this work was perfomed. One of us (R.L.S.) also thanks the Alexander von Humbolt foundation for support.

REFERENCES

1. A.A. Abrikosov, L.P. Gorkov, and I.E. Dzyaloshinski, *Methods of Quantum Field Theory in Statistical Physics* (Dover, New York, 1975).

2. M. G. Cottam and A. A. Maradudin, in *Surface Excitations*, eds V. M. Agranovich and R. Loudon (North-Holland, Amsterdam, 1984) p. 1.
3. L. D. Landau and E. M. Lifshitz, *Statistical Physics*, (Pergamon Press, Oxford, 1962).
4. M. G. Cottam and D. L. Lockwood, *Light Scattering in Magnetic Solids* (John Wiley & Sons, New York, 1986).
5. A.A. Maradudin and D.L. Mills, *J. Phys. Chem. Solids* **28**, (1967) 1855; erratum, *J. Phys. Chem. Solids* **30** (1969) 784.
6. R. C. Moul and M. G. Cottam, *J. Phys. C.* **16**, (1983) 1307.
7. M. G. Cottam, *J. Phys. C* **12**, (1979) 1709.
8. R. E. Camley and D. L. Mills, *Phys. Rev. B* **18**, (1978) 4821.
9. R. L. Stamps and R. E. Camley, *Phys. Rev. B* **40**, (1989) 596; *Phys. Rev. B* **40**, (1989) 609.
10. T. Wolfram and R. E. DeWames, *Prog. Surf. Sci.* **2**, (1972) 233.
11. E. A. Turov, *Physical Properties of Magnetically Ordered Crystals* (Academic Press, New York, 1965) p. 156
12. D. L. Mills and E. Burstein, *Rep. Prog. Phys.* **37**, 817 (1974)
13. L. R. Walker, *Phys. Rev.* **105** (1957) 390; *J. Appl. Phys.* **29** (1958) 318
14. R. E. Camley, *Surf. Sci. Rep.* **7**, (1987) 103.
15. R. E. Camley, T. S. Rahman, and D. L. Mills, *Phys. Rev. B* **27**, (1983) 261.
16. H. T. Diep, *Phys. Rev. B* **40**, (1989) 4818.
17. N. Raj and D. R. Tilley, *Physica Status Solidi b* **152**, (1989) 135.
18. L. Dobrzynski, *Surf. Sci. Rep.* **6**, (1986) 119.
19. H. Puskarski and L. Dobrzynski, *Phys. Rev. B* **39** (1989) 1819
20. R. E. Camley, T. S. Rahman and D. L. Mills, *Phys. Rev. B* **23**, (1981) 1226.
21. R. E. Camley and A. A. Maradudin, *Phys. Rev. Lett.* **49**, (1982) 168.
22. R. E. Camley, *Solid State Comm.* **28**, (1978) 321.
23. W. Wettling, M. G. Cottam, and J. R. Sandercock, *J. Phys. C.* **8**, (1975) 211.
24. P. Grünberg and F. Metawe, *Phys. Rev. Lett.* **39** (1977) 1561.
25. J. R. Sandercock and W. Wettling, *J. Appl. Phys.* **50** (1979) 7784.
26. R. E. Camley and M. Grimsditch, *Phys. Rev. B* **22** (1980) 5420.
27. P. Grünberg and K. Mika, *Phys. Rev. B* **27** (1983) 2955.
28. M. Grimsditch, M. R. Kahn, A. Kueny and I. K. Schuller, *Phys. Rev. Lett.* **51** (1983) 498.
29. B. Hillebrands, A. Boufelfel, C. M. Falco, P. Baumgart, G. Güntherodt, E. Zirngiebl, and J. D. Thompson, *J. Appl. Phys.* **63** (1988) 3880.
30. J.F. Cochran and J.R. Dutcher, *J. Magn. Mag. Mat* **73** (1988) 299.
31. N.D. Mermin and H. Wagner, *Phys. Rev. Lett.* **17**, (1966) 1133.
32. D. Kerkmann, J.A. Wolf, D. Pescia, Th. Woike, and P. Gruenberg, *Solid State Commun.* **72** (1989) 963.
33. U. Gradmann, *Phys. Rev. Lett.* **52** (1984) 771.
34. C.M. Schneider, P. Bressler, P. Schuster, J. Kirschner, J.J. de Miguel, and R. Miranda, *Phys. Rev. Lett.* **64** (1990) 1059.

35. S.O. Demokritov, N.M. Kreines, V.I. Kudinov, and S.V. Petrov, *Zh. Eksp. Teor Fiz.* **95**, (1989) 2211 [Sov. Phys.--JETP **68** (1989) 1277].
36. V.L. Pokrovskii and M.V. Feigel'man, *Zh. Eksp. Teor. Fiz.* **72** (1977) 557 [*Sov. Phys.--JETP* **45** (1977) 291].
37. Y. Yafet, J. Kwo, and E.M. Gyorgy, *Phys. Rev. B* **33** (1986) 6519.
38. R.L. Stamps and B. Hillebrands, *Phys. Rev. B* **44** (1991) 12417.
39. B.A. Kalinikos and A.N. Slavin, *Acta Physica Polonica* **A75** (1989) 541.
40. R.P. Erickson and D.L. Mills, *Phys. Rev. B* **44** (1991) 11825.
41. P. Bruno, *Phys. Rev. B* **43** (1991) 6015.
42. R. E. Camley, P. Grünberg, and C. M. Mayr, *Phys. Rev. B* **26** (1982) 2609.
43. H. T. Diep, *Phys. Rev. B* **43** (1991) 8509.
44. N. S. Almeida and D. L. Mills, *Phys. Rev. B* **31** (1988) 6698.
45. Ze-Nong Ding, Yu Xia, Li-Bin Lin, Hao Che and Xin-Chuan Zeng, *Solid State Comm.* **79**, (1991) 273.
46. J. Mathon, *Phys. Rev. B.* **24** (1981) 6588.
47. Sudha Gopalan and M. G. Cottam, *Phys. Rev. B***42** (1990) 10311
48. Niu-Niu Chen, and M. G. Cottam, *Solid State Comm.* **76** (1990) 437.
49. Niu-Niu Chen, and M. G. Cottam, *Phys. Rev. B* **44** (1991) 7466.
50. V. L. Prokovsky, M. V. Feigel'man, and A. M. Tsvelick, in *Spin Waves and Magnetic Excitations 2*, ed. A. S. Borovik-Romanov and S. K. Sinha (Elsevier Science Publishers, B. V., 1988).
51. M. S. Kushwaha, B. Djafari Rouhani, *Phys. Rev. B* **43** (1991) 902.
52. L. K. Brundle and N. J. Freedman, *Electron Lett.* **4** (1968) 2521
53. B. A. Kalinikos, *IEE Proc. Pt. H.* **127** (1980) 4.

CHAPTER 6

LIGHT SCATTERING AND MICROWAVE RESONANCE STUDIES OF SPIN WAVES IN METALLIC FILMS AND MULTILAYERS

J.R. DUTCHER

Department of Physics, University of Guelph,
Guelph, Ontario N1G 2W1, Canada

Contents

6.1 INTRODUCTION

6.1.1 General Introduction and Historical Perspective

Thin films of magnetic materials have long been of both scientific and technological interest. Recently, there have been dramatic developments in the techniques used for depositing magnetic thin films, such as the use of molecular beam epitaxy (MBE) and sputtering. These techniques, together with sophisticated surface science tools for the evaluation of the film structure, have been used not only to grow very high quality epitaxial films of known magnetic materials, such as body centered cubic (bcc) Fe [1], but also to synthesize new, metastable structures by forcing the magnetic film to grow on an atomic template which is lattice-mismatched with the equilibrium structure of the magnetic material, such as bcc Co [2] and face centered cubic (fcc) Fe [3]. In addition, artificial layering of magnetic films with other magnetic and nonmagnetic films have allowed the production of magnetic multilayer films [4, 5]. Concurrently, sophisticated calculations of the magnetic properties of ultrathin and multilayer films have been performed which have allowed the prediction of the properties of new magnetic structures [6, 7, 8]. Because of these film growth and calculational developments, there was an urgency to measure the magnetic properties of these newly available structures. The resulting experimental thrust has not only confirmed some of the theoretical predictions, such as the existence of large perpendicular magnetic anisotropies [1] for ultrathin films and collective spin-wave excitations for multilayers [9, 10], but has also revealed new, unexpected results, such as the observation of oscillatory exchange coupling between thin magnetic films [11] and giant magnetoresistance [12].

In this chapter we will discuss recent experimental results obtained from studies of spin-waves in magnetic films and multilayers using Brillouin light scattering (BLS) and the microwave resonance techniques of ferromagnetic resonance (FMR) or spin-wave resonance (SWR), ferromagnetic antiresonance (FMAR) and standing spin-wave resonance (SSWR). Although neutron scattering has been used extensively as a probe of bulk spin-waves in magnetic materials, recent experiments involving grazing incidence of the neutron beam have allowed the study of spin-waves in thin magnetic films and multilayers [13,14]. The neutron scattering measurements involve the study of spin-waves with wave-vectors comparable to that of the Brillouin zone boundary which are much larger than the small wavevector (long wavelength) spin-waves studied using BLS and FMR. The reader is referred to the above references for more details. The magnetic properties of thin magnetic films have also been measured using static magnetometry techniques such as surface magneto-optic Kerr effect (SMOKE)[15, 16] and torsion oscillating magnetometry (TOM)[17]. Except for certain cases for which both the dynamic and static properties were measured for the same samples, the static measurements will not be discussed further.

Spin-waves in opaque ferrmagnets in a variety of thin film geometries have been investigated recently using both light scattering and microwave resonance techniques. Three different film geometries will be discussed in this chapter: 1) single magnetic layer films, from ultrathin films to bulk crystals; 2) double magnetic layer structures, which consist of two magnetic films with and without an intervening layer; and 3) magnetic multilayer structures, which consist of alternating layers of two different materials of which at least one is magnetic.

In the remainder of this section, a brief history of both experimental and theoretical results will be given. A summary is given of the experimental results prior to those presented in section 6.3. Theoretical work is summarized up to the present day and will be referred to, where appropriate, in section 6.3.

The study of spin-waves in ferromagnetic materials began with the microwave resonance experiments of Griffiths in 1946 [18]. In microwave absorption studies of thin (film thickness L = 25 μ), electrodeposited films of Ni, Fe and Co, Griffiths observed a maximum in the microwave power absorbed by the samples as the external d.c. magnetic field was changed. His experimental results were correctly interpreted by Kittel [19] in terms of a resonant absorption of microwave energy by spin-waves within the materials using a simplified theory of spin-wave resonance (SWR) or ferromagnetic resonance (FMR). A detailed interpretation of SWR results were worked out by Ament and Rado [20] and Rado and Weertman [21], who recognized the importance of the influence of the exchange interaction at the surfaces of a magnetic film, which can lead to pinning of the surface spins. The excitation of spin-waves by microwaves requires a nonuniformity of the microwave field inside the material. In SWR measurements of metals, the field nonuniformity is provided by the skin depth effect due to the large electrical conductivity σ.

The term spin-wave resonance has also been used to describe a different type of microwave resonance experiment. In 1958, Kittel [22] predicted that standing spin waves across the thickness of a thin magnetic film could be excited if the surface spins were pinned. In this case, the nonuniformity in the microwave field is provided by the pinning of the surface spins. Later that year Seavy and Tannenwald [23] observed this phenomenon which we will call standing spin wave resonance (SSWR). In their measurements of thin (film thickness $L \sim 4000$ Å) permalloy films, they observed resonant microwave absorption for as many as seven discrete standing spin-waves.

In another type of microwave resonance experiment, called ferromagnetic antiresonance (FMAR), a minimum was observed in the microwave absorption for bulk supermalloy samples by Yager [24] in 1949, corresponding to an increase in the penetration depth of the microwaves into the metal at FMAR.

Cochran et al. [25] have worked out the details of FMR in a system composed of a thin ferromagnetic film which is exchange-coupled to a ferromagnetic substrate. The FMR theory for an exchange-coupled double magnetic layer system has been presented by Heinrich et al. [26].

There have been excellent reviews of SWR, SSWR and FMAR by Bhagat [27], in which he has emphasized experimental aspects, and Frait and Fraitová [28].

The inelastic scattering of light from acoustic sound waves was predicted by Brillouin [29] in 1922 and, independently, by Mandelstam [30] in 1926. This effect was first observed experimentally in transparent liquids by Gross [31] in 1930 and it is now commonly known as Brillouin light scattering (BLS). Although the prediction and observation of scattering of light from sound waves in transparent condensed matter occured prior to the microwave resonance experiments described above, it was not until recent improvements in experimental techniques that light scattering from spin-waves in opaque magnetic materials could be observed using BLS. In 1970, Sandercock [32] demonstrated that dramatic increases in experimental sensitivity could be obtained by multipassing of the scattered light through a single Fabry-Perot interferometer (FPI). This improvement led to the observation of thermally excited spin-waves in the ferrimagnet yttrium iron garnet (YIG) by Sandercock and Wettling [33]. Thermally excited surface-guided modes were observed using BLS in the opaque ferromagnetic semiconductor EuO by Grünberg and Metawe [34] in 1977 using three passes through a single FPI. Sandercock and Wettling [35] successfully observed surface-guided and bulk spin-waves in polycrystalline samples of the opaque ferromagnetic metals Fe and Ni using tandem operation of two FPIs. Grünberg et al. [36] studied the behavior of the surface mode for Fe films as thin as 30 Å deposited on sapphire, and observed crossover between the surface mode and standing spin wave modes by using films of different thicknesses. Crossover between the modes was observed directly by Kabos et al. [37] for individual Fe films by varying the angle of incidence of the laser light. Grünberg [38, 39] has pioneered experiments on exchange-coupled double layer films, in which the exchange coupling between two 100 Å thick Fe films was varied by varying the thickness of intervening nonmagnetic and antiferromagnetic spacer layers. The first BLS studies of magnetic superlattice films, consisting of magnetic layers separated by nonmagnetic layers, were performed by Grimsditch et al. [9, 40], in which they observed collective modes of the superlattice. A subsequent comprehensive study of the collective modes of superlattices was performed by Hillebrands et al. [10].

The calculation of BLS magnetic mode frequencies and scattering intensities for thin single magnetic films for in-plane magnetization and including exchange effects was first carried out by Camley et al. [41] using Green's function techniques. Recently, the normal modes approach of Damon and Eshbach [42] and Wolfram and DeWames [43] to the case of single magnetic films has been extended by Cochran and Dutcher [44], including the effects of exchange, spin-pinning boundary conditions, magnetic damping and metallic conductivity. Rado and Hicken [45] have presented a calculation very similar to that of Cochran and Dutcher, except that magnetic damping and conductivity were neglected. The light scattering intensity from magnetic modes in single films has been calculated by Cochran and Dutcher [46] for in-plane magnetization including the effects of exchange, and by Dutcher et al. [47] for large perpendicular anisotropy allowing the magnetization to tilt out of the film plane in the absence of exchange. With the neglect of exchange, only the surface magnetic mode is calculated; this calculation has been generalized to include the effects of exchange by Stamps and Hillebrands [48].

Grünberg [49, 50] first calculated the spin-wave frequencies for a magnetic double layer system including the effects of dipolar coupling, but neglecting the effects of exchange and exchange-coupling between the magnetic layers. Cochran and Dutcher [51] extended their single magnetic film calculation [44, 46] to double layer films, with and without an intervening nonmagnetic spacer layer, including exchange and exchange-coupling of the magnetic films. They found excellent agreement between the calculated magnetic mode frequencies and scattering intensities and the BLS data of Grünberg et al. [39]. Vayhinger and Kronmüller [52], Barnaś and Grünberg [53] and Vohl et al. [54] have also calculated the magnetic mode frequencies for exchange-coupled double layer films. Each of these calculations assumed parallel or antiparallel alignment of the two film magnetizations which are parallel to the applied magnetic field. Cochran et al. [55] have generalized the calculation to allow for equilibrium magnetizations which are nonparallel to each other, and to the applied magnetic field.

Inspired by the measurements of Grimsditch et al. [9], Camley et al. [8] used a Bloch theorem to extend their single film calculation to the case of multilayered films, in the absence of exchange and exchange coupling between layers, for which they calculated the properties of collective modes of the multilayered film as a result of dipolar coupling between the individual magnetic layers. Using an extension of the Rado and Hicken/Cochran and Dutcher calculation for single magnetic films, Hillebrands [56, 57] has calculated the mode frequencies for magnetic multilayer films composed of a finite number of film layers. The effects of exchange, accounting for the spin-pinning boundary conditions at each magnetic interface, and exchange coupling between layers were included for a superlattice film consisting of up to 10 magnetic layers. In the absence of exchange, the calculation can be extended up to 90 magnetic layers. Stamps and Hillebrands [58] have recently calculated the properties of multilayer films having perpendicular anisotropies large enough to tilt the magnetization out of the sample plane. The concept of representing a magnetic multilayer as a single film with effective magnetic properties has been demonstrated by Raj and Tilley [59] and Almeida and Mills [60].

Recent reviews on BLS studies of magnetic materials include those by Patton [61], Grünberg [38, 62], Borovik-Romanov and Kreines [63], Cochran [64] and Hillebrands and Güntherodt [65].

6.1.2 Relationship Between Spin-Waves and Magnetic Properties

In BLS and FMR calculations, the magnetic media are generally treated using a continuum approach, in which the magnetic properties of the media are characterized by macropscopic parameters such as the magnetization, the Lande g-factor and magnetic anisotropies, cf. section 1.2.3 of this book. The microscopic theory of spin-waves is presented in sections 1.2.2 and 1.3.5 of this book. In this section, a brief summary of the continuum theory related to BLS and FMR measurements is presented.

In the continuum approach, the motion of the spins in a magnetic material is determined by torques acting on the spins due to various interaction energies [66, 46]. One must consider the interaction energies between the magnetization and the applied d.c. magnetic field H_0, and the magnetization and the static shape demagnetizing field H_{demag}. In addition, there are interaction energies due to the magnetocrystalline anisotropy E_{mca}; the exchange energy E_{ex}; and the surface anisotropy E_s. Equivalently, one can describe the various interaction energies in terms of the effective fields H_{eff} acting on the magnetization:

$$H_{eff} = -\frac{\partial E}{\partial M} \tag{6.1}$$

where E represents each one of the above energy terms and M is the total magnetization, e.g. $M = (m_x, m_y, M_0)$ with the saturation magnetization M_0 lying in-plane along the z-direction, see Fig. 6.1. In general, H_0 and M_0 lie in the x-z plane. The effect of magnetic damping can also be included as an effective field H_d. The dynamic behavior of the spins is described by the Landau-Lifshitz equation of motion or torque equation for the motion of the magnetization (Eq. (1.36))

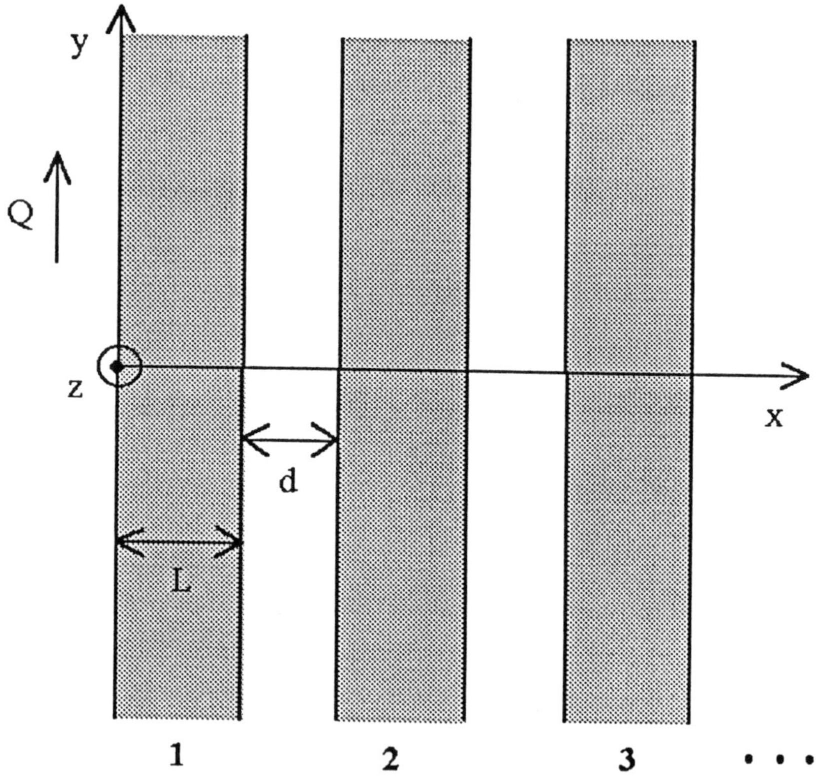

Fig. 6.1. Coordinate system used for spin-wave calculation discussion.

$$\frac{\partial \mathbf{M}}{\partial t} = -\gamma\left(\mathbf{M}\times\mathbf{H}_{\mathbf{eff}}\right) \tag{6.2}$$

where $\gamma = g|e|/(2m_ec)$ is the gyromagnetic ratio, g is the Lande g-factor, and m_e is the electron mass.

Equation (6.2) can be used to derive expressions for the effective magnetic permeability μ_{eff} where $B = \mu_{eff}H = (1 + 4\pi\chi_{eff})H$, which can be used to illustrate the basic phenomena of FMR and FMAR. For simplicity, we will consider only the cases for which the applied field is parallel and perpendicular to the sample plane, which are the two most common configurations for magnetic resonance experiments. If one neglects the effects of exchange, magnetic anisotropy and magnetic damping, one can use Eq. (6.2) to derive expressions for μ_{eff} for the parallel configuration [67]:

$$\mu_{\mathrm{eff}} = \frac{\mathbf{B}_0^2 - \left(\frac{\omega}{\gamma}\right)^2}{\mathbf{B}_0\mathbf{H}_0 - \left(\frac{\omega}{\gamma}\right)^2} \tag{6.3}$$

and for the perpendicular configuration:

$$\mu_{\mathrm{eff}} = \frac{\mathbf{H}_0 \pm \frac{\omega}{\gamma}}{\mathbf{H}_0 - 4\pi\mathbf{M}_0 \pm \frac{\omega}{\gamma}} \tag{6.4}$$

In Eqs. (6.3) and (6.4), H_0 is the applied d.c. magnetic field, $B_0 = H_0 + 4\pi M_0$, and we have taken into account the static shape demagnetizing field H_{demag} for a thin film, which is zero for the parallel configuration and equal to $-4\pi M_0$ along the sample normal for the perpendicular configuration. The + and - signs in Eq. (6.4) refer to the circularly polarized electromagnetic modes that can be used to describe the perpendicular configuration. FMR occurs for the H_0 values for which

μ_{eff} becomes very large, corresponding to the vanishing of the resonant denominators of Eqs. (6.3) and (6.4):

$$H_0^2 + (4\pi M_0)H_0 - \left(\frac{\omega}{\gamma}\right)^2 = 0 \tag{6.5}$$

for the parallel configuration, and

$$H_0 = 4\pi M_0 \mp \frac{\omega}{\gamma} \tag{6.6}$$

for the perpendicular configuration. FMAR occurs for the H_0 values for which μ_{eff} becomes very small, corresponding to the vanishing of the numerators of Eqs. (6.3) and (6.4):

$$H_0 = \frac{\omega}{\gamma} - 4\pi M_0 \tag{6.7}$$

for the parallel configuration, and

$$H_0 = \frac{\omega}{\gamma} \tag{6.8}$$

for the perpendicular configuration. Since the effective penetration depth δ_{eff} of the microwaves into the magnetic metal is $\delta_{eff} = \delta/\sqrt{(i\mu_{eff})}$, where $\delta = c/\sqrt{(4\pi\omega\sigma)}$ is the electromagnetic skin depth, δ_{eff} increases relative to its off-resonance value δ for the H_0 value corresponding to FMAR and allows the observation of the transmission of microwaves through magnetic films [67]. For FMR, δ_{eff} decreases relative to δ. The inclusion of magnetic damping prevents μ_{eff} from becoming infinite or zero.

For thin films, the calculation of mode frequencies for BLS and FMR can be presented within a common framework. The two calculations differ with respect to the in-plane component of wavevector Q: for BLS, $Q \sim 10^5$ cm^{-1}, whereas for FMR, $Q \sim 0$. In the FMR experiment, we can assume that the waves vary as

$\exp[i(q_x x-\omega t)]$, i.e. the spatial variation of the waves occur across the thickness of the films (x-direction). The waves studied in the BLS experiment have an additional wavevector Q dependence in-plane in, e.g., the y-direction perpendicular to the applied field such that the waves can be assumed to vary as $\exp[i(Qy + q_x x - \omega t)]$. The major effect introduced by a nonzero value of Q is that of dipole-dipole interactions. In addition, smaller Q-dependent exchange interactions are present.

By combining Eq. (6.2) with Maxwell's equations, which also relate the magnetization components m_i to the magnetic field components h_i, and neglecting terms that are second or higher order in the r.f. quantities, one can solve for the allowed values of the wavevector components q_x across the thickness L of the film. To obtain values of the spin-wave frequencies for thin film and multilayer geometries, it is necessary to consider the proper boundary conditions at each interface between the magnetic and nonmagnetic layers.

Neglecting exchange, Damon and Eshbach [42] derived the frequency of the magnetic mode propagating parallel to the surface of a single magnetic film surrounded on both sides by a dielectric by requiring that the tangential components of the electric and magnetic fields be continuous across both film surfaces. For propagation perpendicular to the in-plane applied magnetic field, the frequency of the Damon-Eshbach mode is given by:

$$\left(\frac{\omega}{\gamma}\right)^2 = H_0(H_0 + 4\pi M_0) + \left(2\pi M_0^2\right)\left(1 - e^{-2QL}\right) \qquad (6.9)$$

In the limit of Q = 0, corresponding to FMR, the mode frequency is equivalent to that given by Eq. (6.5).

If the effects of exchange are included, it is also necessary to account for the exchange boundary conditions [21] at each magnetic film surface:

$$A\frac{\partial m_x}{\partial x} \pm K_x^{\pm} m_x = 0 \tag{6.10}$$

$$A\frac{\partial m_y}{\partial x} \pm K_y^{\pm} m_y = 0 \tag{6.11}$$

where A is the exchange constant, + refers to x = j(L + d), and - refers to x = L + j(L + d), and j is the bilayer index for the case of a multilayer film. Positive values of the anisotropy constants K_x and K_y correspond to spin pinning which tends to orient the magnetization along the sample normal. Inclusion of exchange also leads to exchange-dominated standing spin-wave modes [41], with mode frequencies which increase as $1/L^2$. These are the modes that are observed in a standing spin-wave resonance (SSWR) experiment.

Rado [68] has shown that for very thin single films of thickness L that the effect of spin pinning is equivalent to adding an additional surface anisotropy field H_s:

$$H_s = \frac{2K_s}{M_0 L} \tag{6.12}$$

where $K_s = K_x^+ + K_x^-$, since it includes the effects due to both film surfaces. It is convenient to define the effective magnetization of the single magnetic film $(4\pi M_0)_{eff}$ as the difference between the saturation magnetization $4\pi M_0$ and the surface anisotropy field H_s:

$$(4\pi M_0)_{eff} = 4\pi M_0 - \frac{2K_s}{M_0 L} \tag{6.13}$$

For large spin pinning, $(4\pi M_0)_{eff}$ becomes small, and can even become negative corresponding to perpendicular magnetization in zero applied magnetic field.

For double magnetic layer and magnetic multilayer films, the calculation of the mode frequencies proceeds in a manner similar to that outlined above, taking into account the electromagnetic and exchange boundary conditions at each interface. It can be important to consider the coupling between adjacent magnetic films due to an interlayer exchange interaction of the form used by Hoffmann, Stankoff and Pascard [69]:

$$\mathbf{E_{ex}} = -\mathbf{A_{AB}}\, \mathbf{M_A} \cdot \mathbf{M_B} \tag{6.14}$$

where M_A and M_B are the saturation magnetizations of the two adjacent magnetic films, evaluated at the facing surfaces of the films, and A_{AB} is the interlayer exchange-constant. In this case, the exchange boundary conditions of Eqs. (6.10) and (6.11) are replaced by the Hoffmann boundary conditions [57].

In BLS and FMR studies of double magnetic layer films, there are two modes of interest [26]: an acoustic mode, in which the magnetizations of the two films precess in phase; and an optical mode, in which the magnetizations precess 180° out of phase. In the limit of large separation between the magnetic layers, only the effect of dipolar coupling between the layers has to be considered. Dipolar coupling of the magnetic layers results from the nonzero stray magnetic fields produced by the in-plane variation of the transverse magnetization components for the modes measured in a BLS experiment; dipolar coupling is negligible for FMR experiments since $Q \sim 0$. For spacer layer thicknesses ≤ 50 Å, the effect of interlayer exchange can dominate and the dipolar coupling is essentially saturated at its zero spacer layer thickness value. Because the interlayer exchange-coupling of Eq. (6.14) depends on the angle between the two magnetizations, the frequency of the optical mode depends strongly on the interlayer exchange-coupling constant A_{AB}, whereas the frequency of the acoustic mode is only weakly dependent on A_{AB}. Measurement of the optical mode allows one to obtain a value for A_{AB}. The direction of change of the optical mode frequency, an increase or a decrease, as the magnitude of the exchange-coupling is increased tells one whether the interlayer exchange-coupling is ferromagnetic or antiferromagnetic, respectively. For relatively thick magnetic films, as used by Grünberg et al. [39] ($L \sim 200$ Å), the intensity of the optical mode is large enough to be observed for identical magnetic films. Heinrich et al. [26] have pointed out for ultrathin magnetic films, the optical mode is observable in FMR and BLS studies only if the two magnetic films have different magnetic properties. One can observe the optical mode by using ultrathin magnetic films made of the same material if the films have a large perpendicular anisotropy, since choosing different film thicknesses produces different perpendicular anisotropy fields (cf. Eq. (6.12)).

Magnetic superlattice films consist of N magnetic layers of thickness L, with each pair separated by a nonmagnetic spacer layer of thickness d. As in the case of a magnetic double layer, each magnetic film has a Damon-Eshbach surface mode which can couple to the Damon-Eshbach modes in other layers via their dipolar fields. The result of including dipolar interactions between the individual magnetic films is to form a set of N collective superlattice modes [8, 9]. If the magnetic layer thickness L is greater than or equal to the spacer layer thickness d, one of the modes corresponds to the collective surface mode of the superlattice film, i.e. the magnetization components for each of the magnetic layers is such that the overall magnetization profile for the superlattice decays exponentially from the free surface of the film. This collective superlattice surface mode is analogous to the Damon-Eshbach mode for a single magnetic film. For small spacer layer thickness, interlayer exchange can be important and can give rise to exchange-dominated collective superlattice modes [56, 57, 65].

6.2 EXPERIMENTAL TECHNIQUES

6.2.1 Brillouin Light Scattering

In the BLS experiment, monochromatic light is directed onto the surface of a magnetic film. Most of the light is scattered from the static roughness on the sample surface and is unshifted in frequency. A very small fraction of the light ($\sim 10^{-8}$) is scattered from the thermally excited spin-waves travelling inside the sample. In very simple terms, the BLS process can be understood as a Doppler shift of the frequency of the light as it scatters from a propagating spin-wave [70]. Alternatively, one can think of the precessing magnetization of the spin-wave as modulating the index of refraction of the film through the magneto-optical interaction. The frequency shift of the scattered light occurs because of the interaction between the modulated index of refraction and the electric field of the incident optical wave, which generates a distribution of electric dipoles that give rise to the scattered light wave. The light does not excite the spin-waves; it serves only as a probe of the magnetic modes which exist already inside the film.

The scattering process for materials that are opaque to the incident radiation such as metallic ferromagnets is described by conservation laws for energy and momentum:

$$\omega_s - \omega_i = \pm\Omega \tag{6.15}$$

$$k_s - k_i = \pm Q \tag{6.16}$$

where ω_i, k_i, ω_s and k_s are the frequencies and in-plane wavevector components of the incident and scattered light, and Ω and Q are the spin-wave frequency and in-plane wavevector component. For metals, the incident light is strongly absorbed within a very small distance $\delta_{opt} \sim 150$ Å from the free surface of the film. Since δ_{opt} is small compared with the wavelength of the light $\lambda \sim 5000$ Å, the wavevector components along the sample normal are not conserved [71]; wavevector conservation applies only to the in-plane wavevector components. The value of Q is determined by the choice of scattering geometry, i.e. k_i and k_s, and measurement of the scattered light frequency using a high-contrast Fabry-Perot interferometer [71] allows the determination of Ω. In BLS studies of spin-waves, the spin-wave frequencies Ω and relative scattering intensities are measured for ranges of experimental parameters, e.g. spin-wave wavevector Q and applied magnetic field H_0. These measured quantities are compared to those calculated using a phenomenological theory to obtain the macroscopic magnetic parameters, such as the Lande g-factor and the magnetic anisotropies, using a least squares fitting procedure. Although δ_{opt} is small for opaque materials, the light scatters from modes which are determined by the magnetic properties and geometry of the entire sample. The light will scatter efficiently only from those modes that have substantial transverse magnetization components, and therefore substantially alter the index of refraction, within a distance δ_{opt} from the sample surface.

The terms Brillouin light scattering (BLS) and Raman light scattering (RLS) are both used to describe the inelastic scattering of light. Both scattering processes involve excitation wavevectors that are small (~ 10^5 cm^{-1}) compared with the wavevector value at the Brillouin zone edge (~ 10^8 cm^{-1}). The two processes are distinguished by the size of the frequency shift between the incident and scattered light. For the unit of measure of frequency shift, we will use GHz, where 30 GHz = 1 cm^{-1} = 0.12 meV. BLS frequency shifts range from approximately 1 GHz to 500 GHz, whereas RLS frequency shifts range from approximately 500 GHz to 10^5 GHz. Because of the large difference in the size of the frequency shifts for BLS and RLS, different experimental techniques are used. As described below, in a BLS experiment, a Fabry-Perot interferometer is used to separate the inelastically scattered light from the elastically scattered light. In a RLS experiment, a grating spectrometer is used for the same purpose. For more details concerning RLS studies of magnetic materials, see reference [72].

A typical experimental setup for BLS studies of spin-waves is shown schematically in Fig. 6.2. Typically several hundred milliwatts of linearly-polarized light from a single-mode, continuous wave (cw) laser, such as an Ar^+ laser which can be operated with either λ = 5145 Å or 4880 Å, is focussed onto the sample surface. Single mode operation of the laser is achieved by using a temperature-stabilized, intra-cavity etalon [73], such that the laser linewidth (~ 30 MHz) is small compared with the Fabry-Perot instrumental linewidth (~ 0.5 GHz). A 180° backscattering geometry is commonly used, in which the lens that is used to focus the laser light onto the sample surface is also used to collect the diffusely scattered light. The diameter of the focussed laser light spot is typically 10 μ. Because the laser light is focussed into such a small spot size, one has to be careful not to overheat the sample during the measurement. Dutcher et al. [74] have shown that the laser heating is not too severe for metallic samples: the local temperature increase for 140 mW of laser light focussed onto a bulk Fe whisker was about 25°C.

The sample is placed between the poles of an electromagnet used to produce a d.c. magnetic field H_0 of between 0 and 15 kOe at the sample position. Typically, H_0 is applied parallel to the film surface and perpendicular to the plane of incidence of the light. Because the effective area of the sample is so small for BLS measurements, of the order of $(10\ \mu)^2$, the requirements of the d.c. magnetic field uniformity are not very stringent; one can use a smaller electromagnet (3 or 4 inch diameter pole pieces) which can be mounted directly on the surface of an optical table.

The scattered light contains a range of scattering wavevectors centred about that corresponding to 180° backscattering due to the nonzero aperture of the lens. This range of scattering wavevectors parallel to the film surface can be written as [75]

$$\frac{\Delta Q}{Q} = \frac{1}{2}\cot an\,\theta \cdot \sin\left[\tan^{-1}\left(\frac{1}{2f}\right)\right] \tag{6.17}$$

where $\Delta Q = Q_{max} - Q_{min}$ is the spread in scattering wavevectors parallel to the film

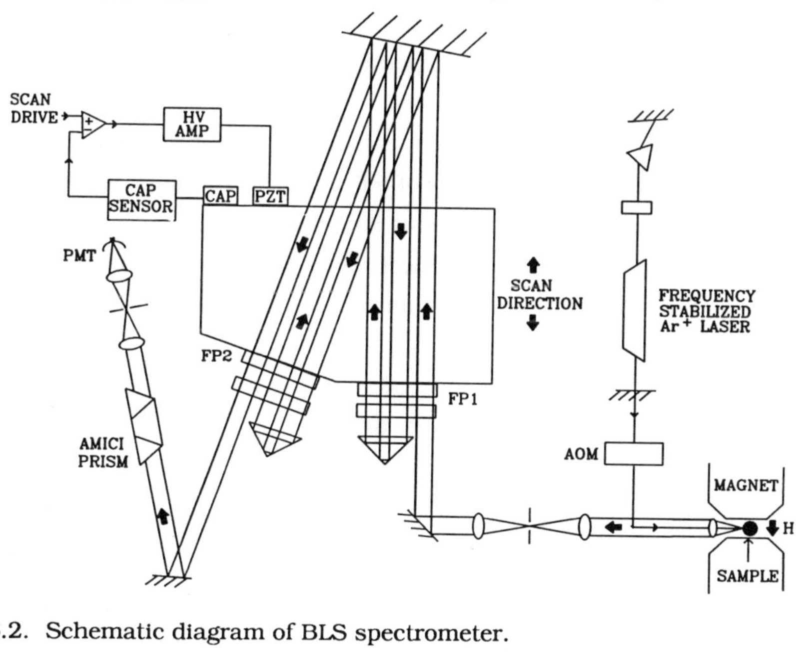

Fig. 6.2. Schematic diagram of BLS spectrometer.

surface, θ is the angle of incidence and f is the f-number of the lens. $\Delta Q/Q$ can be quite large; for example, for a lens of f-number 1.4, $\Delta Q/Q$ ranges from 60% for $\theta = 30^o$ to 12% for $\theta = 70^o$. If necessary, ΔQ can be reduced by the use of a slit in the scattered light beam [76]. Unwanted reflections from e.g. the lens surfaces are removed by spatially filtering the scattered light using a lens-pinhole-lens arrangement such that only light which has been collected from the focal spot of the sample lens is passed through the pinhole and is recollimated by the second lens of the spatial filter. The beam is then sent through a high contrast, tandem Fabry-Perot interferometer to separate the weak, frequency shifted signal from the very large unshifted background signal. In the system of Fig. 6.2, the light is passed six times through the two Fabry-Perot interferometers (FPIs). A discussion of the multipassed, tandem FPI is given below. The light transmitted through the FPIs is again filtered using an Amici dispersing or direct-vision prism [77] and a second lens-pinhole-lens arrangement, such that only collimated light of the desired wavelength is sent to the detector. The detector is a photomultiplier tube, which is cooled thermoelectrically to about -20 °C to obtain a low dark count rate of < 1 count/s. Data collection and experimental control can be performed using a personal computer.

The heart of the BLS experiment is the Fabry-Perot interferometer, which consists of two plane, parallel mirrors of reflectivity R [78]. It is necessary to use mirrors that are very flat to within about 30 Å over their 50 mm aperture, with a reflectivity value of R ~ 0.9. For monochromatic light, there is a maximum in transmission through the Fabry-Perot mirrors if the separation of the mirrors d_0 satisfies the relationship

$$d_0 = m\frac{\lambda}{2} \tag{6.18}$$

where λ is the wavelength of the light and m is an integer which specifies the order of the interferometer. The very small component of light which has been shifted in frequency (or wavelength) by scattering from the thermally-excited spin-waves satisfies the constructive interference relation of Eq. (6.18) for a value of d_0 which differs slightly from that for the elastically scattered light. The resulting spectrum of transmitted intensity versus mirror separation consists of very large peaks due to the elastically scattered light together with very small, shifted peaks due to the inelastically scattered light. Changes in the mirror separation δd are equivalent to shifts in the frequency of the light δf:

$$\delta f = -\frac{f}{d_0}\delta d \tag{6.19}$$

The frequency range corresponding to a change in mirror separation of $\lambda/2$ is known as the free spectral range (FSR) and is determined by d_0:

$$FSR = \frac{c}{2d_0} \tag{6.20}$$

Typically, d_0 = 3 mm, corresponding to a FSR of 50 GHz. Using Eq. (6.19), the BLS spectra can be interpreted as transmitted intensity versus frequency shift.

A typical BLS spectrum of scattering intensity versus frequency shift for a 7 Å thick film of Fe(001) on Ag(001) [79] is shown in Fig. 6.3, for a free spectral range of 50 GHz and an applied magnetic field of 4.21 kOe. The very large peaks at 0 and ± 50 GHz correspond to light which has scattered elastically from static defects on the film surface. The peaks at ± 20 GHz correspond to light scattered from the surface magnetic mode of the magnetic film.

In principle, the measured linewidth of the magnetic modes can be used to determine the magnetic damping. However, the intrinsic linewidth due to magnetic damping is usually small compared with the instrumental resolution of the Fabry-Perot interferometer (~ 0.5 GHz).

The experimental variables that can be changed in a BLS experiment include: the d.c. magnetic field strength H_0; the angle of incidence of the laser light, which determines Q; the free spectral range of the interferometer, which determines the frequency range which is scanned; the angle between H_0 and the crystalline axes of the sample; the angle between H_0 and the scattering plane; the incident laser light polarization; and the sample temperature.

Recent advances by J.R. Sandercock [71] in both the multipassing optics and the stabilization electronics used in the BLS experiment have enabled BLS studies of spin-waves and sound waves in materials that are opaque to the incident light. A dramatic increase in the contrast of the FPI is obtained by multipassing the light through the interferometer using cube corner prisms. The multipass contrast C_n is given by the nth power of the single pass contrast C_1, where n is the number of passes. For large values of n, the contrast is very large; for $n = 6$ and $C_1 = 100$, $C_n = 10^{12}$. This large value of C_n reduces the contribution of the large, elastically scattered peaks to negligible values for frequency shift values between the large peaks, and allows the measurement of very weak frequency-shifted signals.

Sandercock's [71] arrangement of two FPIs with a common scan direction and their axes inclined at a tandem angle $\alpha \sim 20^{\circ}$, as in Fig. 6.2, is very useful for the unambiguous identification of frequency shifts. The equilibrium mirror

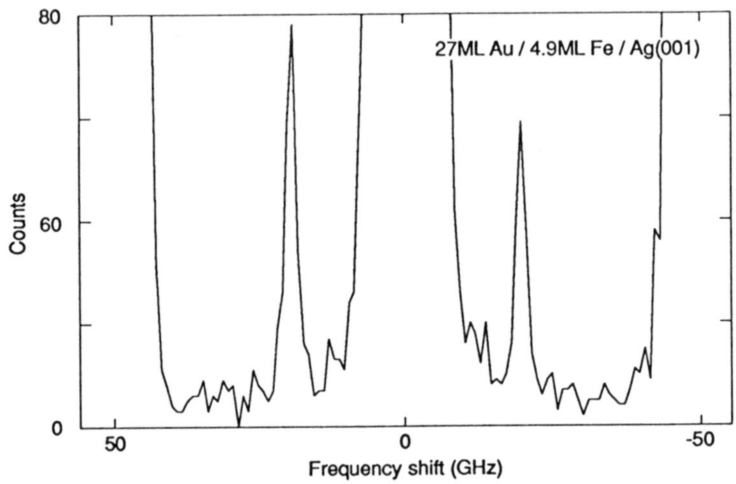

Fig. 6.3. BLS spectrum for a 7 Å thick Fe(001) film on Ag(001) [75] for a field of H0 = 4.21 kOe, a laser power of 35 mW and a counting time for each of the 128 data points of 4 s.

separation d_2 of the second interferometer is related to that of the first interferometer by

$$d_2 = d_1 \cos\alpha \tag{6.21}$$

Scanning both interferometers along the axis of the first interferometer creates a vernier effect. By adjusting the relative spacings of the two FPIs such that the maximum transmissions coincide on a particular, central order of the FPI, the neighbouring orders and their associated frequency-shifted peaks do not coincide and are attenuated. This means that frequency-shifted peaks observed in the tandem FPI spectra correspond to frequency shifts measured from the central order of the FPI.

In the experiment, the Fabry-Perot mirrors are scanned repetitively over two FSRs of the interferometer, corresponding to a small distance approximately equal to the wavelength of the light (~ 0.5 μ). The scanning of the mirrors is accomplished by applying a ramp voltage with a peak-to-peak voltage of ~ 50 V to a piezoelectric transducer (PZT) which moves one mirror relative to the other. The scanning rate is typically 1 Hz, and the data collection time can range from several minutes to several hours, depending on the strength of the signal of interest. Because the displacement of a PZT is not linearly related to its motion, the scanning of the mirrors must be linearized using a capacitance displacement transducer [79].

It is also important to maintain the parallelism of the mirrors during the BLS experiment to within a tolerance of about 30 Å. This is accomplished by adjusting the voltages supplied to piezoelectric transducers (PZTs) attached to one of the two mirrors in each FPI. Throughout the experiment, the PZT voltages, and therefore the tilts of the mirrors, are continuously adjusted to maximize the strength of the elastically scattered light transmitted through the system, which corresponds to maximum parallelism of the mirrors. A similar feedback technique is used to centre the spectrum to prevent unwanted drifts in the equilibrium mirror separation. It is necessary to use a dynamic feedback system which can be controlled by analog electronics or by computer.

Because of the stringent alignment requirements for high contrast Fabry-Perot interferometers, it is necessary to isolate the interferometer from vibrations. This can be done using a passive vibration isolation optical table, which consists of a rigid table top supported by pneumatically controlled legs. This type of vibration isolation is most effective in damping high frequency vibrations but is less effective in damping building vibrations of the order of 10 to 30 Hz. Sandercock [80, 81] has recently developed an dynamic vibration isolation system jsut for the interferometer in which accelerometers and PZTs are used to isolate the experiment from vibrations with frequencies greater than 1 Hz.

In addition to light scattering from the spin-waves, light is also scattered from thermally-excited phonons or sound waves within the sample [71]. There are two ways to distinguish the scattering due to spin-waves from that due to sound waves. First of all, because the spin-wave energy depends strongly on the value of the applied d.c. magnetic field, as in e.g. Eq. (6.9), the spin-wave peaks shift with a change in the applied magnetic field H_0, whereas the sound wave peaks do not shift with H_0. Also, for simple crystal symmetries such as cubic, the light scattered from sound waves is typically polarized scattering (scattered polarization is the same as incident polarization), whereas the light scattered from spin-waves is typically depolarized (scattered polarization is rotated by 90° with respect to the incident polarization). The relative strengths of the scattering from sound waves and spin-waves depends on the properties of a given material. For example, for BLS measurements of the bulk Fe whisker crystal discussed in section 6.3.1 for an angle of incidence of 45°, the scattering due to the surface spin-wave (Damon-Eshbach wave) is about five times stronger than that due to the surface sound wave (Rayleigh wave).

6.2.2 Microwave Resonance

In a microwave resonance experiment, a radio frequency (r.f.) magnetic field at fixed frequency ω_0 is directed onto a magnetic sample as the magnitude of an applied d.c. magnetic field H_0 is varied. The natural resonance frequency ω of the spins depends strongly on H_0, e.g. for a disk shaped magnetic sample having gyromagnetic ratio γ and saturation magnetization $4\pi M_0$ with H_0 applied in the plane of the sample,

$$\left(\frac{\omega}{\gamma}\right)^2 = H_0 \left(H_0 + 4\pi M_0\right) \tag{6.5}$$

where the effects of magnetic anisotropy, exchange and magnetic damping have been neglected. For values of H_0 for which ω is very nearly equal to ω_0, the precessing spins extract energy from the r.f. field. For example, for a sample with the same saturation magnetization and g-factor as Fe ($4\pi M_0$ = 21.55 kG and g = 2.09) and a microwave frequency of 36 GHz, Eq. (6.5) yields a resonance field of H_0 = 5.6 kOe. Because there is a resonant absorption of energy from the incident electromagnetic field, the microwave resonance experiment is qualitatively different from the BLS experiment, for which the magnetic excitations are thermally excited. The microwave apparatus described below is discussed in detail in several standard textbooks [82, 83, 84]. Rudd [85] also gives an excellent description of FMR theory and experimental details.

To increase the size of the microwave field at the sample position, the sample is placed inside, or forms the end wall, of a resonant microwave cavity. The cavity is designed such that the Poynting vector be zero inside the cavity, i.e. there is no net energy flow into the cavity. This means that the r.f. magnetic field h and r.f. electric field e form standing wave patterns inside the cavity, characteristic of the different cavity modes. The basic requirements of the cavity for the observation of FMR are that, at the sample position, h be perpendicular to H_0, h be a maximum, and e be a minimum [84]. H_0 is typically either parallel to or perpendicular to the sample plane. Because of the large index of refraction of metals for microwave radiation [86], the microwaves are refracted dramatically such that they travel essentially along the sample normal. This means that the in-plane wavevector component $Q \sim 0$ to a very good approximation. Typically, the area of sample which is exposed to the microwaves is of the order of 1 cm^2.

The cavity is characterized by its quality factor Q_L in the presence of loading due to the sample. The enhancement of the r.f. magnetic field h due to the use of a resonant cavity is discussed by Poole [84]. For a rectangular cavity, the ratio of the average field in the cavity h_c to that in the waveguide h_w is roughly $\sqrt{(Q_L)}$; for $Q_L = 4000$, $h_c/h_w \sim 63$. For a microwave power of 1 W incident on the cavity, h_c is several Oe. The coupling of the cavity to the attached waveguide is important for determining the sensitivity of the measurement of the reflected power [85]. In general, for reflection measurements it is best to adjust the coupling using an iris diaphragm between the cavity and the waveguide such that 1/3 of the incident power is reflected.

In most resonant cavities, it is necessary to have good electrical contact between the sample and the metallic walls of the cavity to ensure that the required microwave currents can flow. In the mounting of the sample it is very important to avoid introducing stresses in the sample which can shift and distort the resonance signal. MacDonald [87] has calculated the effect of a uniform in-plane stress on FMR due to magnetostriction. Sample mounting for studies in which the temperature is varied is very difficult because of differential thermal expansions between the sample and the cavity wall material. In such cases, spring loaded contacts have been found to be successful [85].

A typical experimental setup for the measurement of microwave absorption is shown schematically in Fig. 6.4. The resonant absorption of microwave power by the sample is observed by measuring the power reflected from the sample, as described below. Other types of spectrometers have been designed to measure the transmission of microwaves through magnetic samples [86]. The microwaves are produced typically by using a klystron with a continuous wave (cw) output power of several hundred mW; for these low microwave powers, the behavior of the spin-waves can be described by the linear theory presented in section 6.1.2. The operating frequency of klystrons, typically between several GHz and 150 GHz, can be tuned both mechanically, by about ± 20%, and electronically, by about ± 1%. The frequency of the microwaves is measured using a wavemeter which is typically accurate to within 0.05%. Microwaves reflected from the rest of the spectrometer are prevented from entering the klystron by using an isolator. Waveguides, having dimensions specific to the particular frequency of the klystron, are used to guide the microwaves to the sample, which is placed in a resonant cavity as discussed above, and to the detector, which is typically a diode that is sensitive to microwave radiation. The microwaves reflected from the cavity are sent to the detector using either a magic tee or directional coupler. Use of a small quantity of diphenylpicryl-hydrazyl (DPPH) within the resonant cavity allows the use of its EPR signal for calibration of signal strengths [1].

Because the resonant frequency of the cavity can shift slightly with changes in the microwave absorption of the sample near resonance, the klystron frequency is locked to that of the resonant cavity using automatic frequency control (AFC) [84, 86]. The resonant frequency of the cavity can be lowered easily by inserting a dielectric, e.g. a micrometer-driven quartz rod, into the cavity.

Usually an a.c. modulation technique, in which a small amplitude (tens of Oe) a.c. magnetic field is applied to the sample parallel to the large d.c. magnetic field H_0, is used to increase the sensitivity of the measurement. A lock-in amplifier is used to measure the component of the reflected signal at the modulation field frequency. Since the modulation amplitude is small compared with the FMR

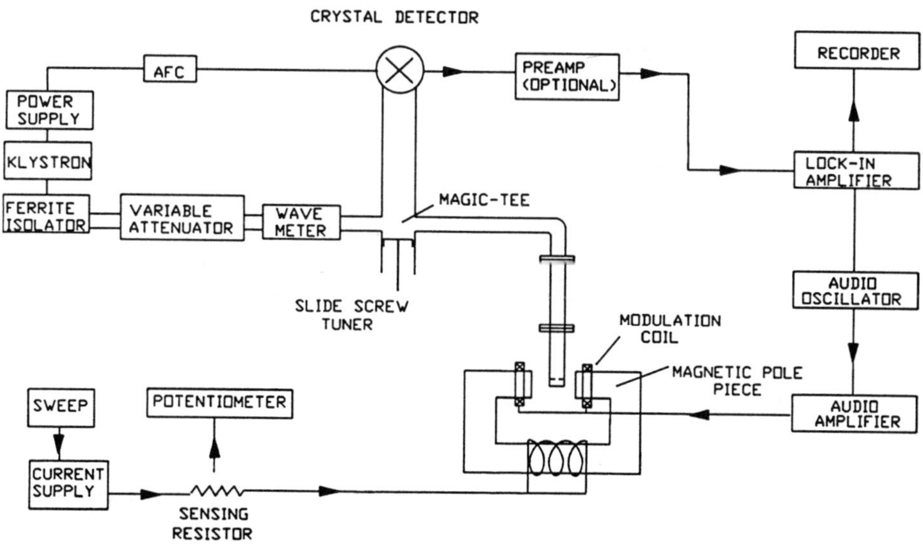

Fig. 6.4. Schematic diagram of microwave absorption spectrometer [27].

linewidth, the lock-in amplifier signal is the derivative of the absorption with respect to H_0. Field modulation at audio frequencies is desirable to maximize the signal-to-noise ratio. However, audio frequency modulation can cause problems because of the decreased value of the microwave skin depth, especially at cryogenic temperatures, which gives limited penetration of the microwaves into the cavity [85]. A low frequency (~ 130 Hz) modulation works best for low temperature studies.

It is necessary to measure the d.c. magnetic field strength H_0 accurately (to within about 0.1%). Use of nuclear magnetic resonance (NMR) probes [88] is one of the most reliable techniques to calibrate the H_0 values. It is also necessary to be able to sweep H_0 slowly over a field range which is large enough to observe the resonance of interest.

A typical microwave resonance spectrum, which consists of the derivative of the microwave absorption with respect to the d.c. magnetic field H_0 versus H_0, is shown for a 7 Å thick Fe film [1] in Fig. 6.5. Measurement of the derivative

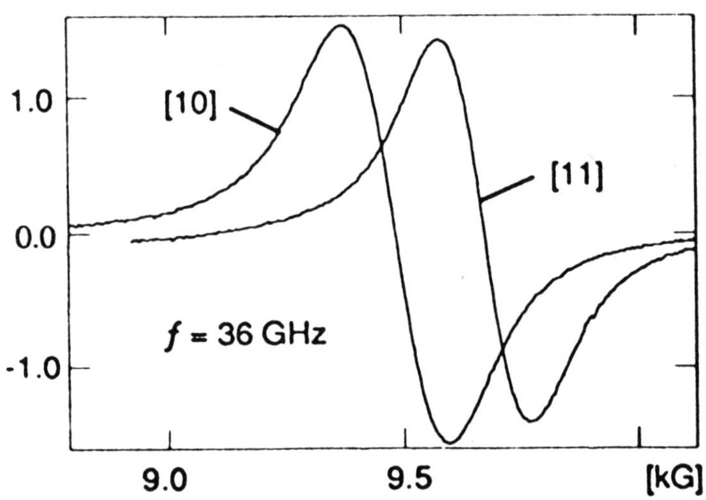

Fig. 6.5. Microwave absorption spectrum for the same 7 Å thick Fe(001) film on Ag(001) as in Fig. 6.3 [1]. The curves for the [10] easy and the [11] hard axes are shown.

signal is especially useful for determining the resonance field (for which the derivative is equal to zero) and the resonance linewidth (field range between the two extrema of the derivative). The linewidth of the microwave resonance signal is related to the magnetic damping and magnetic inhomogeneities in a sample and can be used to judge the quality of the sample [89].

The experimental variables that can be changed in a microwave resonance experiment include: the angle between the d.c. magnetic field H_0 and the crystalline axes of the sample; the angle between H_0 and the sample normal (typically 0° or 90°); the microwave polarization relative to the sample crystalline axes; and the sample temperature. Changing the microwave frequency requires the use of another complete microwave spectrometer designed for use at the new frequency.

6.2.3 Ultrahigh vacuum techniques

Typically, BLS and FMR measurements are performed in air after the film growth procedure is complete. To protect the magnetic films from oxidation in air, it is necessary to cover the films with a protective capping layer which is usually tens of MLs of nonmagnetic material. The choice of the capping layer material is usually determined by the lattice matching condition with the magnetic film crystal and its ability to limit oxidation of the magnetic film; typically the substrate material is also used for the capping layer [3]. Using BLS and FMR to study such capped films allows the measurement of the magnetic properties of the magnetic films with substrate/film and film/capping layer interfaces.

It is also very interesting to measure the properties of bare magnetic films. To do this, it is necessary to perform the film growth and magnetic measurements in an ultra-high vacuum (UHV) environment so that the bare film surface does not become contaminated. Under such conditions, contamination of the surface is limited, e.g. with a base pressure of 10^{-10} Torr, the time taken for the formation of one monolayer (ML) of impurities is about 5 hours, and careful deposition of the films can be performed using deposition rates that can range from 1 Å/s to 1 Å/min. For such an "in-situ" apparatus, one not only has the ability to study the magnetic properties of the film/vacuum interface, but one can also use sophisticated surface science tools to perform simultaneously sensitive measurements of the structure of the film and its surface during the film growth procedure. Both "in-situ" BLS [90] and FMR [91, 92] systems have been developed for use in an UHV chamber, in combination with molecular beam epitaxy (MBE) film growth facilities.

In the UHV studies using BLS by Hillebrands et al. [90] and Baumgart et al. [93], the magnetic film is placed in the gap of an electromagnet (maximum field of 2 kOe) which is inside the UHV chamber. The laser light is focussed onto the sample through a glass window on an optical port of the UHV chamber, with the focussing lens placed outside the chamber. The scattered light is collected by the same lens using a 180° backscattering geometry. To collect BLS data, the film growth was interrupted and the sample surface was directed at the optical port. Potential difficulties for the BLS UHV studies include the effect of the vibration of UHV system vacuum pumps on the optical alignment of the BLS system, and modification of the polarization of the light and an increase in the background signal due to the passage of the laser light through the glass window. Despite these potential difficulties, scattering from the magnetic modes of films within the UHV chamber has been observed [90, 93].

For the FMR UHV studies by Urquhart et al. [91], the sample was placed in the gap of an electromagnet (maximum field of 5.5 kOe) which is outside the UHV chamber. The sample formed the end wall of a TE_{012} mode microwave resonant cavity with a resonant frequency of 16.88 GHz, and microwaves were directed onto the sample through a waveguide-coupling flange on the UHV system. As with the BLS UHV studies, to perform the FMR UHV measurements the film growth was interrupted and the sample was moved onto the end of the resonant cavity. A distinct advantage of using the TE_{102} cavity in a UHV system was that physical contact between the sample and the rest of the cavity was not required. In addition, the system was insensitive to tilts of up to 4° between the sample and the cavity end wall. Vacuum pump vibrations did not significantly degrade the performance of the FMR UHV spectrometer. UHV FMR measurements have also been reported by Li et al. [92].

For both BLS and FMR UHV spectrometers, the maximum magnetic field was been limited to several kOe because of strict limitations on magnet materials that are compatible with a UHV system for UHV magnets, or because of limitations on the minimum gap size for exterior magnets. Also, it was necessary to move the sample away from the growth configuration to measure the magnetic properties. A discussion of the unique results obtained using the UHV techniques is given in section 6.3.1.

6.2.4 Discussion and Comparison

Both BLS and FMR involve measuring magnetic resonances which occur at particular values of frequency and applied d.c. magnetic field H_0. However, there are basic differences between the BLS and FMR experiments:

1) In a BLS measurement, H_0 is fixed and the spin-wave modes that occur at different frequencies are measured. In a FMR measurement, the microwave frequency is fixed and H_0 is swept to observe the spin-wave modes that occur at different values of H_0. It should be noted that it is much easier to choose a different value of H_0 for the BLS experiment than to choose a different microwave frequency for the FMR experiment.

2) The spin-wave excitations are thermally excited for BLS, and are driven by the incident microwave fields for FMR. An important consequence is that the signal intensity for low temperature studies is much larger for FMR than for BLS.

3) The area of the sample which is probed differs dramatically for the two techniques. For BLS, the effective sample area is about $(10\ \mu)^2$, corresponding to the focal spot size of the optical lens which directs the incident light onto and collects the scattered light from the sample. For FMR, the effective sample area is much larger, typically about 1 cm^2, corresponding to the surface area of the sample exposed to the microwave radiation. As a result, BLS can be used to measure samples with much smaller dimensions, e.g. Fe whiskers which have widths of several hundred microns.

4) The in-plane wavevector $Q \sim 0$ for FMR, but is of the order of 10^5 cm^{-1} for BLS. The additional Q-dependence for BLS experiments allows the investigation

of dipolar coupling between the magnetic films in double magnetic layer and magnetic multilayer films. Both techniques can be used to study interlayer exchange-coupling. However, the presence of dipolar coupling for BLS measurements interferes somewhat in the measurement of the onset of interlayer exchange-coupling as the spacer layer thickness is reduced.

5) In both experiments, it is important to measure to measure both the d.c. magnetic field H_0 and the mode frequencies precisely. The measurement of H_0 has the same precision for FMR and BLS, but the measurement of the resonant mode frequencies can be made much more precisely for FMR (~ 0.01 GHz) than for BLS (~ 0.1 GHz).

Despite these fundamental differences, both BLS and FMR standard apparatus have sufficient sensitivity to allow the measurement of the magnetic spin-waves of ultrathin films. Cochran [64] has given a rule-of-thumb criterion for the measurement of equivalent signal-to-noise ratios for BLS and FMR for ultrathin magnetic films.

6.3 RECENT EXPERIMENTAL RESULTS

In this section, some of the highlights of recent experimental results are presented.

6.3.1 Single Magnetic Layer Films

The unambiguous identification of surface anisotropy was first made by Heinrich et al. [1] in an FMR study of ultrathin Fe(001) films grown on a bulk Ag(001) single crystal substrate. By measuring the in-plane resonance field values for films of different thicknesses at two different microwave frequencies (36.5 and 73 GHz), they were able to show that the results were consistent with a surface anisotropy which scaled as 1/L, where L is the film thickness. In Fig. 6.6 is shown the strong dependence of the measured FMR fields on the film thickness d = L. As the film thickness is decreased, passage of the data points above the horizontal lines indicates perpendicular magnetization. For the smallest film thickness (4 Å ~ 3 monolayers), the magnetization is on the verge of turning out of the plane. BLS measurements on the same set of films gave good agreement with the FMR measurements [75, 94]. FMR and BLS data was shown above in Figs. 6.3 and 6.5, respectively, for the 7 Å thick Fe film. BLS measurements on an Fe(001)/Ag(001) structure grown on a Fe(001) whisker substrate, chosen because of its very smooth surface, showed that the increase in the perfection of the substrate surface resulted in a larger perpendicular anisotropy than for Fe films grown on bulk Ag(001) crystals, such that a perpendicular magnetization in zero applied magnetic field was obtained ($(4\pi M_0)_{eff}$ = - 2 kG) [26].

Another magnetic system which has been studied using both BLS and FMR is metastable fcc Fe(001)/Cu(001) [47, 95, 96], using 3 ML thick Fe(001) films grown by Egelhoff et al. [3]. The films studied using BLS and FMR, and also

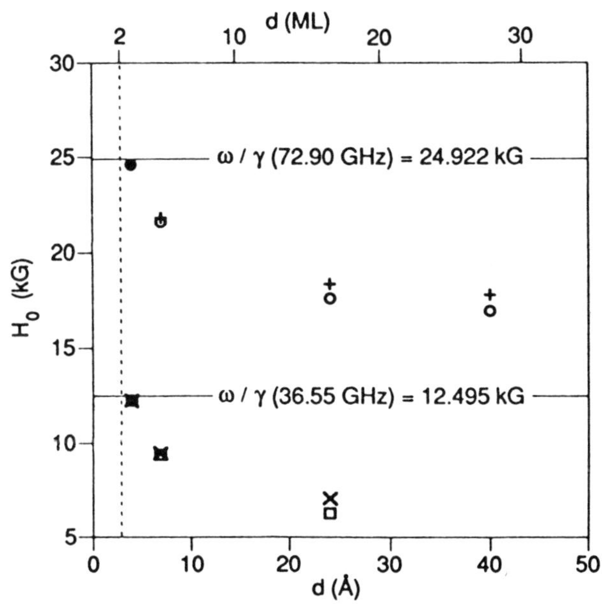

Fig. 6.6. FMR fields for two microwave frequencies (36.5 and 73 GHz) as a function of Fe film thickness for Fe(001) films on Ag(001) for applied fields along the [10] easy (o, □) and the [11] hard (+, ×) axes [1].

SMOKE [15], were clearly ferromagnetic with very large perpendicular anisotropy such that the magnetization was strongly perpendicular to the film plane in zero applied magnetic field. In Fig. 6.7 is shown the measured surface mode frequency versus H_0 applied in the plane of the film. Instead of a monotonic increase in the mode frequency with increasing H_0 as given by Eq. (6.9), the surface mode frequency initially decreases to a minimum value and then increases monotonically [95]. This behavior is the signature of films that have perpendicular magnetization in zero applied magnetic field. As H_0 is increased from zero, the magnetization is pulled down into the sample plane, and it locks into the plane for a critical field $H_c = -(4\pi M_0)_{eff} \sim 4$ kOe, corresponding to the minimum in the mode frequency. The mode frequency did not decrease to zero, probably because of slight variations in the magnetic film thickness, and therefore the surface anisotropy field (Eq. (6.12)), which would smear out the minimum. A dramatic

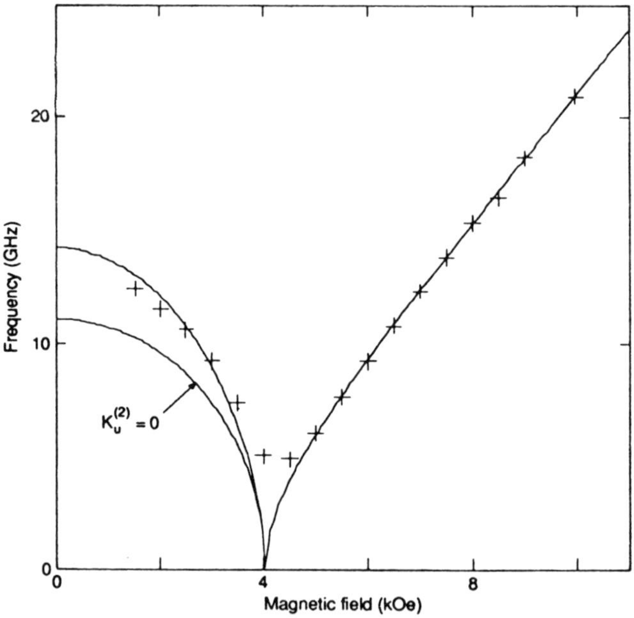

Fig. 6.7. Dependence of the surface magnetic mode frequency on the applied in-plane magnetic field for a 5.5 Å thick film of fcc Fe(001) on Cu(001) [47].

increase in the scattering intensity from the surface magnetic mode was also observed near $H_0 = H_c$ [47]. The effective field acting on the spins goes to zero for this value of H_0, allowing the spins to deviate substantially from their equilibrium direction. Since the scattering intensity is proportional to the square of the transverse magnetization, a large increase in the scattering intensity is observed. Cochran et al. [96] have studied the variation in $(4\pi M_0)_{eff}$ with substrate preparation. They found that for smoother Cu(001) substrates that had been polished chemically and covered with a few layers of epitaxial Cu before the Fe growth, there was a slight increase in the perpendicular anisotropy.

Hillebrands et al. [90, 93, 97] have used a UHV BLS apparatus to study the magnetic properties of Fe(110) films grown on W(110) with and without a covering layer of Pd(111). It is particularly interesting to grow Pd on magnetic materials because it is believed that small changes in the lattice parameter may make Pd ferromagnetic [98] (cf. section 6.3.2). As the Pd layer thickness is increased from zero, the surface spin-wave frequency decreases to a minimum value near a Pd thickness of 1 monolayer. Further increases in the Pd thickness give a slight increase in the mode frequency, which saturates for Pd thicknesses > 3 monolayers. Hillebrands et al. [97] suggest that this behavior is due to the replacement of the Fe/vacuum interface with an Fe/Pd interface, with a corresponding change in the perpendicular anisotropy, as well as a polarization of the Pd layer.

The scattering from the surface and bulk spin waves of an Fe(001) whisker single crystal has been measured by Dutcher et al. [74]. The whisker was cleaned and covered with both epitaxial Au(001) and Ag(001) under ultrahigh vacuum conditions. A substantial increase in the precision of the magnetic surface mode frequency measurements was achieved by focussing the laser light onto the 180° domain wall of the whisker and using four passes through a single Fabry-Perot interferometer. This allowed accurate measurements of small changes δf in the surface magnetic mode frequency as the angle of incidence of the laser light was varied between 32° and 68°. For semi-infinite magnetic media such as the Fe whisker, δf is approximately a linear function of the surface spin pinning coefficient K_x, and is very insensitive to the value of the applied field and the magnetic damping. For Fe(001) along the (100) direction in zero applied magnetic field,

$$\delta f = (0.456 - 0.300 K_x)\,\mathrm{GHz} \tag{6.22}$$

Values of K_x = 0.54 +/- 0.05 erg/cm^2 and 0.79 +/- 0.05 erg/cm^2 were obtained for the Fe(001)/Au(001) and Fe(001)/Ag(001) surfaces, respectively. These values are in excellent agreement with those obtained from FMAR measurements on UHV-prepared bulk Fe crystal surfaces [99], which yielded K_x = 0.4 to 0.5 erg/cm^2 and 0.7 erg/cm^2 for Fe(001)/Au(001) and Fe(001)/Ag(001) surfaces, respectively. The insensitivity of δf to the magnetic damping offers a distinct advantage over FMAR measurements of K_x, which depend sensitively on the magnitude of the magnetic damping. The most remarkable feature of the bulk Fe(001) crystal work is the absolute agreement obtained between the K_x values for bulk and ultrathin films for Fe(001)/Ag(001) measured using FMR and BLS in air [1, 94] and UHV FMR [91]. This is strong evidence that the pinning of the surface spins is related only to the magnetic surface because of the abrupt change in the symmetry of the magnetic atoms at the surface.

6.3.2 Double Magnetic Layer Films

Heinrich et al. [26] have used FMR and BLS to study the magnetic properties of bilayers of ultrathin Ni films grown directly on ultrathin bcc Fe(001) films using Ag(001) as a substrate. Because of the intimate contact between the Ni and Fe layers, the interlayer coupling is very large. It was found that for Ni thicknesses less than 6 monolayers, corresponding to "bcc" Ni, the magnetic properties of the underlying Fe films were not modified. For larger Ni thicknesses, reconstruction of the Ni layers was obtained, which introduced a large in-plane magnetic anisotropy with four-fold symmetry due to the formation of misfit dislocations created during the Ni reconstruction.

Heinrich et al. [100] have also grown a new phase of Cu, bcc Cu, in between ultrathin layers of bcc Fe(001). They used FMR, BLS and SMOKE to study the coupling between the Fe layers as the thickness of the Cu layer was varied. As mentioned in section 6.1.2, it is necessary to choose the two ultrathin Fe film thicknesses to be different to give the two films different perpendicular anisotropies and allow the observation of the optical mode. Despite choosing very different Fe film thicknesses, for Cu layers thinner than 9 monolayers, the Fe films were coupled so strongly that only the acoustic modes were observed. They found that the interlayer exchange changed from ferromagnetic to antiferromagnetic as the Cu layer thickness was increased above 9 monolayers. Both the acoustic and optical modes can be seen in Fig. 6.8 which is a FMR spectrum for a Fe/Cu/Fe film consisting of 5 monolayers of Fe/9.5 monolayers of Cu/10 monolayers of Fe. The only fitting parameter used to calculate the solid curve was the interlayer exchange. Further details of the BLS experiments on the Fe/Cu/Fe films have been given by Cochran et al. [55].

Another interesting experiment by Celinski et al. [101] involved the study

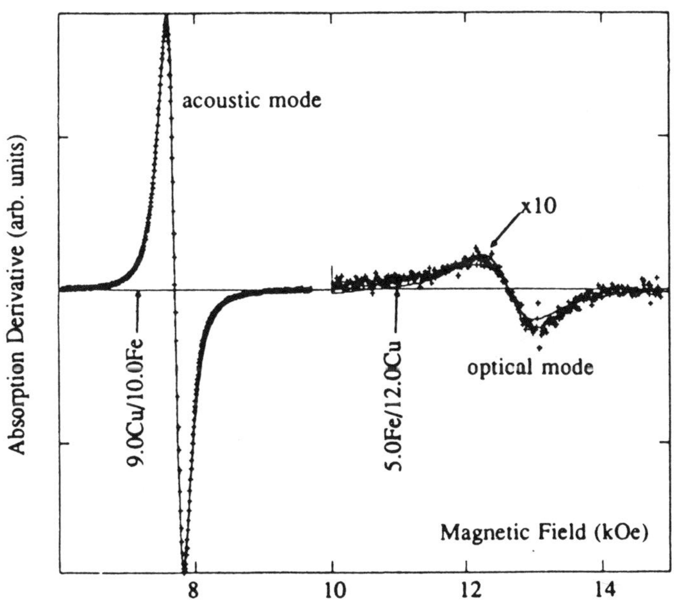

Fig. 6.8. Magnetic field derivative of FMR absorption for a 5 monolayer Fe/9.5 monolayer Cu/10 monolayer Fe film for a microwave frequency of 36 GHz [100].

of Fe/Pd/Fe films using FMR and BLS. The epitaxial growth of Pd on Fe(001) stretches the Pd lattice spacing by 5%, for which ferromagnetic order in Pd has been predicted [98]. Celinski et al. found that Pd is not easily polarized. Although 4 monolayers of Pd were found to order ferromagnetically throughout the entire Pd film, 5 monolayers of Pd (and larger Pd thicknesses) were not ferromagnetic. Ferromagnetism was absent in thicker Pd films because the exchange-coupling measured between the Fe films was only weakly ferromagnetic. In Fig. 6.9 is shown a BLS spectrum for a 5 monolayer Fe/5.5 monolayer Pd/10 monolayer Fe film. It can be seen that the coupling is weak because the optical mode is observed for this small separation of the Fe films. It was found that the exchange-coupling between the two Fe films was ferromagnetic for all Pd thicknesses (up to 11 monolayers). It is worth noting that the exchange coupling found for Fe/Pd/Fe

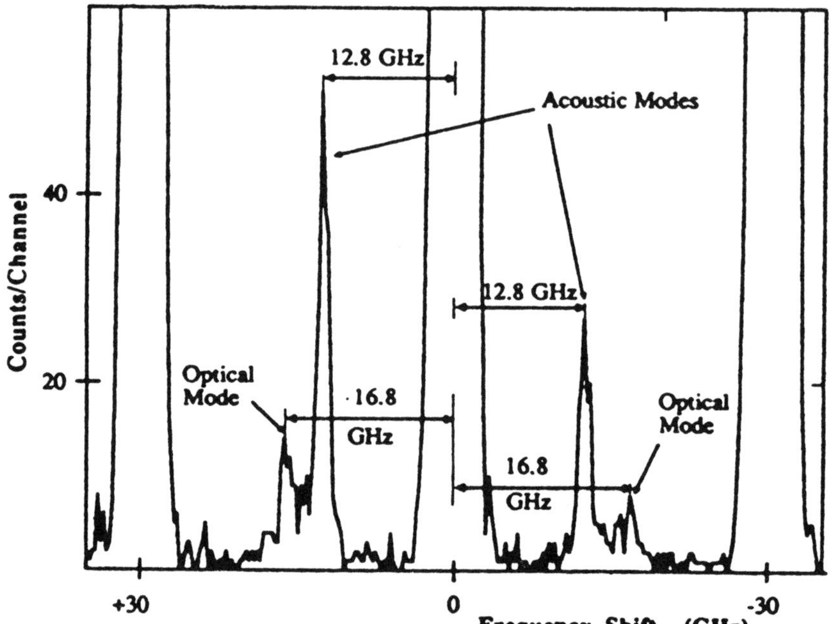

Fig. 6.9. BLS spectrum of scattered intensity versus frequency shift for a 5 monolayer Fe/5.5 monolayer Pd/10 monolayer Fe film with an applied field of 0.82 kOe [101]. The counting time for each of the 256 channels was 1 s.

films was considerably smaller than that measured for Fe/Cu/Fe [100]. This is quite surprising since Pd is thought to be on the verge of becoming ferromagnetic, whereas Cu is not.

The oscillation of the exchange-coupling of two films between ferromagnetic and antiferromagnetic was discovered by Parkin et al. [11] from magnetic hysteresis loop measurements. Grünberg et al. [102, 103, 104] have used a different sample geometry to study this phenomenon using BLS and SMOKE, and have observed the ferromagnetic portion of the oscillation curves which cannot be obtained from magnetic hysteresis loops. By using wedge-shaped spacer layers between the two magnetic layers, the full range of exchange-coupling could be studied on a single sample. In Fig. 6.10 is shown a plot of the exchange-coupling parameter A_{12} versus the thickness of the Cr spacer layer. Positive (negative)

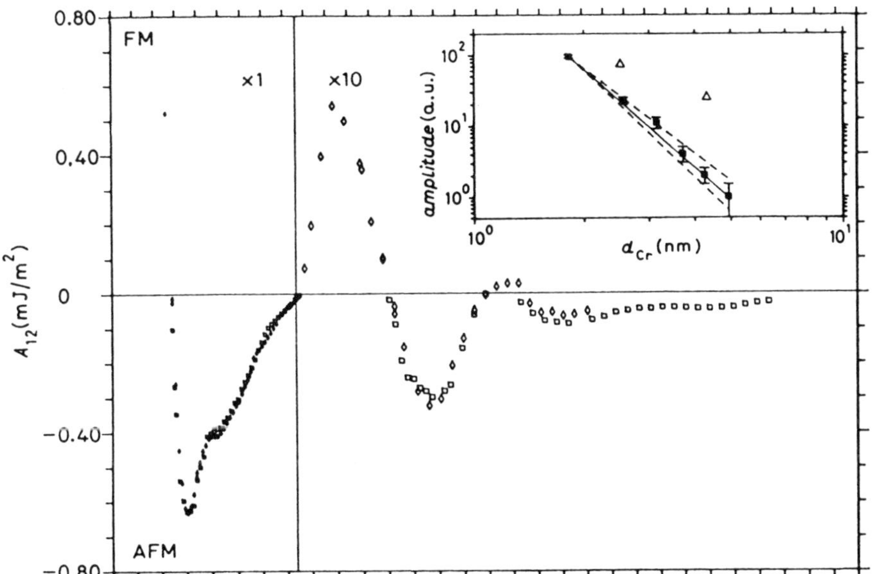

Fig. 6.10. Exchange-coupling constant A_{12} as a function of Cr spacer layer thickness between two Fe films obtained from BLS (◊) and SMOKE (□) measurements using a wedge-shaped spacer layer [103].

values of A_{12} correspond to ferromagnetic (antiferromagnetic) coupling. For Cr thicknesses > 15 Å the vertical scale has been expanded by a factor of 10. The inset of Fig. 6.10 shows the A_{12} amplitude as a function of Cr thickness. The damping of the oscillations is very large, and three times larger than that observed for sputtered films [11]. The feature which occurs at a Cr thickness of 8 Å is indicative of an additional oscillation with a shorter period which has also been observed using other techniques [105, 106]. The use of a very flat substrate enhances the strength of the short period oscillations [104].

6.3.3 Magnetic Multilayer Films

Harzer et al. [107] have measured the magnetic properties of Co/Pd superlattice films. Co/Pd and Co/Pt superlattices are of technological interest because films become perpendicularly magnetized for small Co layer thicknesses of several monolayers [108, 109]. In Fig. 6.11 typical BLS spectra of 4 different Co/Pd films are shown for an applied field of H_0 = 1 kOe. The numbers of atomic layers n_{Co}/n_{Pd} are indicated for each spectrum. For large Co thicknesses the surface collective mode is observed on only one side of the spectra, and the band of bulk-like collective modes is observed on both sides. For small Co thicknesses, the surface and bulk-like modes become degenerate because of the strong coupling across the Pd layers.

The measured spin-wave frequency as a function of magnetic field applied in the plane of the films is shown in Fig. 6.12 for a film consisting of 10 bilayers made up of 2 monolayers of Co and 5 monolayers of Pd. The open squares correspond to the centre frequency of the collective spin-wave band, and the solid curves have been calculated using the theory of Stamps and Hillebrands [58], allowing the magnetization to turn out of the sample plane, using the best fit parameters obtained from the large H_0 data. A sharp minimum in the spin-wave frequencies appears in the calculated curves at a critical field H_c, but a correspon-ding minimum is not observed in the data. Harzer et al. attribute the absence of a well-defined minimum in the data to spatially varying film thicknesses, and therefore anisotropies, and magnetization within the Co layers.

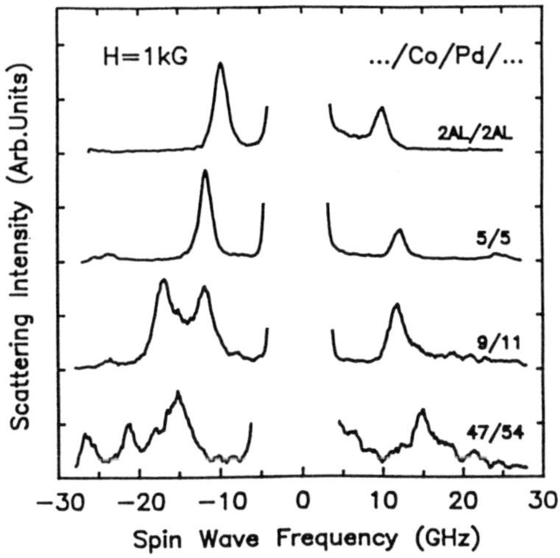

Fig. 6.11. BLS spectra of Co/Pd multilayer films with an applied field of 1 kOe parallel to the film plane [107]. All multilayers consist of 10 bilayers.

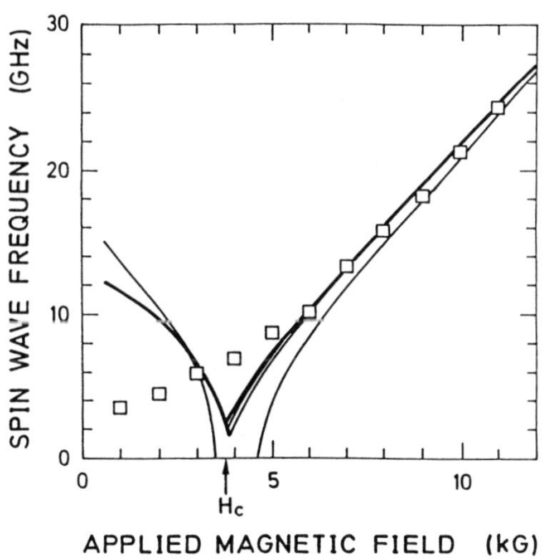

Fig. 6.12. Spin-wave frequencies versus magnetic field for a Co/Pd multilayer with 10 bilayers consisting of 2 monolayers Co and 5 monolayers Pd [107].

The interfaces of Co/Au superlattice films were shown by den Broeder et al. [110] to sharpen as a result of annealing the films. This unusual behavior occurs because of the mutual insolubility of Co and Au at elevated temperatures, and can increase the perpendicular anisotropy sufficiently to turn the magnetization out of the sample plane. Krams et al. [111] have investigated the effect of annealing on the BLS spectra for films consisting of 70 bilayers of 9 Å of Co and 9 Å of Au. After annealing, they found that the perpendicular anisotropy did increase, but they also observed an unexpected broadening of the spin-wave peaks. They concluded that the spin-wave broadening was due to interface roughness on a length scale comparable to that of the in-plane wavevector Q (~ 2000 Å), which did not significantly affect the perpendicular anisotropy.

Hillebrands et al. [112] have discovered evidence for the existence of collective superlattice modes that are dominated by the effects of interlayer exchange-coupling for sputtered Co/Pd superlattice films. Each superlattice contained 30 Co and Pd layers. As the Pd spacer layer thickness was decreased, the frequencies of a band of collective spin-waves was observed to increase. This behavior was in agreement with a simple theory presented by Hillebrands et al. [112] in which the exchange-coupling was assumed to decay exponentially with increasing spacer layer thickness. Unfortunately, the complete calculation of superlattice collective spin-wave modes, including the effects of intralayer and interlayer exchange, could not be used for quantitative comparison with experiment because of numerical instabilities for > 10 magnetic layers in the superlattice. A series of superlattice films containing 8 Co and Pd layers did not show evidence for the exchange-dominated modes.

In a subsequent study of e^--beam evaporated (111)-oriented Co/Pt superlattice films [113], exchange-dominated collective modes were observed and the value of the interlayer exchange constant was estimated. In Fig. 6.13 a series of BLS spectra is shown for films having different Pt thicknesses. As with the Co/Pd superlattices, as the Pt layer thickness decreases, a band of modes (indicated by vertical arrows) shifts to larger frequency which is indicative of exchange-dominated collective modes. The additional band of modes at ± 20 GHz does not shift with decreasing Pt thickness, and is dominated by dipolar coupling. The observed widths of both collective spin-wave bands was very large. Harzer et al. [113] attributed the broadening of the bands to spatial variations in the perpendicular anisotropy, caused by thickness variations within the Co layers of ~ ± 1 Å.

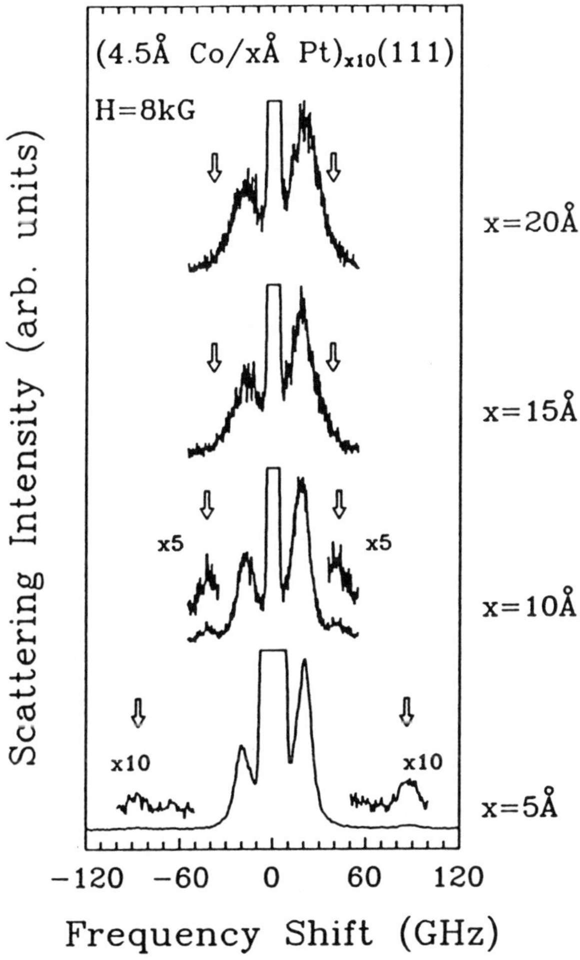

Fig. 6.13. BLS spectra of (111)-oriented Co/Pt multilayer films with 10 bilayers and an in-plane applied magnetic field of 8 kOe [113].

6.4 ACKNOWLEDGEMENTS

I thank A. Basso for preparing Figs. 6.2 and 6.4. Financial support from the Natural Sciences and Engineering Research Council of Canada is gratefully acknowledged.

REFERENCES

1. B. Heinrich, K.B. Urquhart, A.S. Arrott, J.F. Cochran, K. Myrtle and S.T. Purcell, Phys. Rev. Lett. **59**, 1756 (1987).
2. G. Prinz, Phys. Rev. Lett. **54**, 1051 (1985).
3. D.A. Steigerwald and W.F. Egelhoff, Jr., Surf. Sci.
4. M.R. Khan, C.S.L. Chun, G.P. Felcher, M. Grimsditch, A. Kueny, C.M. Falco and I.K. Schuller, Phys. Rev. B **27**, 7186 (1983).
5. J. Kwo, E.M. Gyorgy, D.B. McWhan, M. Hong, F.J. DiSalvo, C. Vettier and J.W. Bower, Phys. Rev. Lett. **55**, 1402 (1985).
6. for example, C.L. Fu and A.J. Freeman, Phys. Rev. B 35, 925 (1987).
7. J.G. Gay and R. Richter, Phys. Rev. Lett. **56**, 2728 (1986).
8. R.E. Camley, T.S. Rahman and D.L. Mills, Phys. Rev. B **27**, 261 (1983).
9. M. Grimsditch, M.R. Khan, A. Kueny and I.K. Schuller, Phys. Rev. Lett. **51**, 498 (1983).
10. B. Hillebrands, A. Boufelfel, C.M. Falco, P. Baumgart, G. Güntherodt, E. Zirngiebl and J.D. Thompson, J. Appl. Phys. **63**, 3880 (1988).
11. S.S.P. Parkin, N. More and K.P. Roche, Phys. Rev. Lett. **64**, 2304 (1990); S.S.P. Parkin, Phys. Rev. Lett. **67**, 3598 (1991).
12. M.N. Baibich, J.M. Broto, A. Fert, F. Nguyen Van Dau, F. Petroff, P. Etienne, G. Creuzet, A. Friedrich and J. Chazelas, Phys. Rev. Lett. **61**, 2472 (1988); G. Binasch, P. Grünberg, F. Saurenbach and W. Zinn, Phys. Rev. B **39**, 4828 (1989).
13. G.P. Felcher, in *Dynamical Phenomena at Surfaces, Interfaces and Superlattices*, edited by F. Nizzoli, K.-H. Rieder and R.F. Willis (Springer-Verlag, Berlin, 1985), p. 316.

14. R.F. Willis, J.A.C. Bland and W. Schwarzacher, J. Appl. Phys. **63**, 4051 (1988).
15. C. Liu, E.R. Moog and S.D. Bader, Phys. Rev. Lett. **60**, 2422 (1988).
16. C.A. Ballentine, R.L. Fink, J. Araya-Pochet and J.L. Erskine, Appl. Phys. A **49**, 459 (1989).
17. H.J. Elmers, T. Furubayashi, M. Albrecht and U. Gradmann, J. Appl. Phys. **70**, 5764 (1991).
18. J.H.E. Griffiths, Nature **158**, 670 (1946).
19. C. Kittel, Phys. Rev. **71**, 270 (1947); C. Kittel, Phys. Rev. **73**, 155 (1948).
20. W.S. Ament and G.T. Rado, Phys. Rev. **97**, 1558 (1955).
21. G.T. Rado and J.R. Weertman, J. Phys. Chem. Solids **11**, 315 (1959).
22. C. Kittel, Phys. Rev. **110**, 1295 (1958).
23. M.H. Seavey and P.E. Tannenwald, Phys. Rev. Lett. **1**, 168 (1958).
24. W.A. Yager, Phys. Rev. **75**, 316 (1949).
25. J.F. Cochran, B. Heinrich and A.S. Arrott, Phys. Rev. B **34**, 7788 (1986).
26. B. Heinrich, S.T. Purcell, J.R. Dutcher, K.B. Urquhart, J.F. Cochran and A.S. Arrott, Phys. Rev. B **38**, 12879 (1988).
27. S.M. Bhagat, in *Techniques of Metals Research VI*, Part 2, edited by E. Passaglia (Wiley, New York, 1973), p. 79.
28. Z. Frait and D. Fraitová, in *Spin Waves and Magnetic Excitations 2*, edited by A.S. Borovik-Romanov and S.K. Sinha (Elsevier Science, 1988), p. 1.
29. L. Brillouin, Ann. Phys. (Paris) **17**, 88 (1922).
30. L.I. Mandelstam, Zh. Russ. Fiz.-Khim. Ova. **58**, 381 (1926).
31. E.F. Gross, Nature **126**, 201 (1930).
32. J.R. Sandercock, Optics Commun. **2**, 73 (1970); J.R. Sandercock, *2nd Int. Conf. on Light Scattering in Solids*, ed. M. Balkanski, Flammarion, Paris, 1971.
33. J.R. Sandercock and W. Wettling, Solid State Commun. **13**, 1729 (1973).
34. P. Grünberg and F. Metawe, Phys. Rev. Lett. **39**, 1561 (1977).
35. J.R. Sandercock and W. Wettling, IEEE Trans. Magn. MAG-**14**, 442 (1978); J.R. Sandercock and W. Wettling, J. Appl. Phys. **50**, 7784 (1979).

36. P. Grünberg, M.G. Cottam, W. Vach, C. Mayr and R.E. Camley, J. Appl. Phys. **53**, 2078 (1982).

37. P. Kabos, C.E. Patton and W.D. Wilber, Phys. Rev. Lett. **53**, 1962 (1984).

38. P. Grünberg, Prog. Surf. Sci. **18**, 1 (1985).

39. P. Grünberg, R. Schreiber, Y. Pang, M.B. Brodsky and H. Sowers, Phys. Rev. Lett. **57**, 2442 (1986).

40. A. Kueny, M.R. Khan, I.K. Schuller and M. Grimsditch, Phys. Rev. B **29**, 2879 (1984).

41. R.E. Camley, T.S. Rahman and D.L. Mills, Phys. Rev. B **23**, 1226 (1981).

42. R.W. Damon and J.R. Eshbach, J. Phys. Chem. Solids **19**, 308 (1961).

43. T. Wolfram and R.E. DeWames, Prog. Surf. Sci. **2**, 233 (1972).

44. J.F. Cochran and J.R. Dutcher, J. Appl. Phys. **63**, 3814 (1988).

45. G.T. Rado and R.J. Hicken, J. Appl. Phys. **63**, 3885 (1988).

46. J.F. Cochran and J.R. Dutcher, J. Magn. Magn. Mater. **73**, 299 (1988).

47. J.R. Dutcher, J.F. Cochran, I. Jacob and W.F. Egelhoff, Jr., Phys. Rev. B **39**, 10430 (1989).

48. R.L. Stamps and B. Hillebrands, Phys. Rev. B **43**, 3532 (1991).

49. P. Grünberg, J. Appl. Phys. **51**, 4338 (1980).

50. P. Grünberg, J. Appl. Phys. **52**, 6824 (1981).

51. J.F. Cochran and J.R. Dutcher, J. Appl. Phys. **64**, 6092 (1988).

52. K. Vayhinger and H. Kronmüller, J. Magn. Magn. Mater. **72**, 307 (1988).

53. J. Barnaś and P. Grünberg, J. Magn. Magn. Mater. **82**, 186 (1989).

54. M. Vohl, J. Barnaś and P. Grünberg, Phys. Rev. B **39**, 12003 (1989).

55. J.F. Cochran, J. Rudd, W.B. Muir, B. Heinrich and Z. Celinski, Phys. Rev. B **42**, 508 (1990).

56. B. Hillebrands, Phys. Rev. B **37**, 9885 (1988)

57. B. Hillebrands, Phys. Rev. B **41**, 530 (1990).

58. R.L. Stamps and B. Hillebrands, Phys. Rev. B **44**, 5095 (1991).

59. N. Raj and D.R. Tilley, Phys. Rev. B **36**, 7003 (1987).

60. N.S. Almeida and D.L. Mills, Phys. Rev. B **38**, 6698 (1988).

61. C.E. Patton, Phys. Rep. **103**, 251 (1984).

62. P. Grünberg, in *Light Scattering in Solids V*, edited by M. Cardona and G. Güntherodt (Springer-Verlag, Berlin, 1989), p. 303.

63. A.S. Borovik-Romanov and N.M. Kreines, in *Spin Waves and Magnetic Excitations 1*, edited by A.S. Borovik-Romanov and S.K. Sinha (Elsevier Science, 1988), p. 81.

64. J.F. Cochran, in *Ultrathin Magnetic Films*, edited by B. Heinrich and J.A.C. Bland (Springer-Verlag, Berlin, 1992), to be published.

65. B. Hillebrands and G. Güntherodt, in *Ultrathin Magnetic Films*, edited by B. Heinrich and J.A.C. Bland (Springer-Verlag, Berlin, 1992), to be published.

66. W.F. Brown, Jr., *Micromagnetics*, Krieger, New York, 1978.

67. J.F. Cochran and B. Heinrich, IEEE Trans. Magn. **Mag-16**, 660 (1980).

68. G.T. Rado, Phys. Rev. B **26**, 295 (1982).

69. F. Hoffmann, A. Stankoff and H. Pascard, J. Appl. Phys. **41**, 1022 (1970); F. Hoffmann, Phys. Status Solidi **41**, 807 (1970).

70. J.C. Slater, Rev. Mod. Phys. **30**, 197 (1958).

71. J.R. Sandercock, in *Light Scattering in Solids III*, edited by M. Cardona and G. Güntherodt (Springer, Berlin, 1982), p. 173.

72. M.G. Cottam and D.J. Lockwood, *Light Scattering in Magnetic Solids* (John Wiley and Sons, New York, 1986).

73. M. Hercher, Appl. Opt. **8**, 1103 (1969).

74. J.R. Dutcher, J.F. Cochran, B. Heinrich and A.S. Arrott, J. Appl. Phys. **64**, 6095 (1988).

75. J.R. Dutcher, Ph.D. dissertation, Simon Fraser University, 1988.

76. J.R. Dutcher, S. Lee, B. Hillebrands, G.J. McLaughlin, B.G. Nickel and G.I. Stegeman, Phys. Rev. Lett. **68**, 2464 (1992).

77. F.A. Jenkins and H.E. White, *Fundamentals of Optics*, McGraw-Hill, New York, 1957, chapter 2.

78. M. Born and E. Wolf, *Principles of Optics* (Pergamon, Oxford, 1975).

79. R.V. Jones and J.C.S. Richards, J. Phys. E **6**, 589 (1973).

80. J.R. Sandercock, in *Vibration Control in Optics and Metrology*, Soc. Photo-optical Instrumentation Engineers 732, 157 (1987).

81. F. Nizzoli and J.R. Sandercock, in *Dynamical Properties of Solids*, Vol. 6, edited by G.K. Horton and A.A. Maradudin (North-Holland, Amsterdam, 1990), p. 281.
82. C.G. Montgomery, editor, *Technique of Microwave Measurements*, Radiation Laboratory Series 11 (McGraw-Hill, New York, 1947).
83. E.L. Ginzton, *Microwave Measurements* (McGraw-Hill, New York, 1957).
84. C.P. Poole, *Electron Spin Resonance, A Comprehensive Treatise on Experimental Techniques*, (Interscience, New York, 1967).
85. J.M. Rudd, M.Sc. thesis, Simon Fraser University, 1985.
86. J.F. Cochran, B. Heinrich and G. Dewar, Can. J. Phys. **55**, 834 (1977).
87. J.R. MacDonald, Proc. Phys. Soc. London A **64**, 968 (1951).
88. K. Borer and G. Fremont, Nucl. Instr. and Meth. **154**, 61 (1978).
89. Z. Celinski and B. Heinrich, J. Appl. Phys. **70**, 5935 (1991).
90. B. Hillebrands, P. Baumgart and G. Güntherodt, Phys. Rev. B **36**, 2450 (1987).
91. K.B. Urquhart, B. Heinrich, J.F. Cochran, A.S. Arrott and K. Myrtle, J. Appl. Phys. **64**, 5334 (1988).
92. Y. Li, M. Farle and K. Baberschke, Phys. Rev. **41**, 9596 (1990).
93. P. Baumgart, B. Hillebrands and G. Güntherodt, J. Magn. Magn. Mater. **93**, 225 (1991).
94. J.F. Cochran, B. Heinrich, A.S. Arrott, K.B. Urquhart, J.R. Dutcher and S.T. Purcell, J. de Physique, Colloque C8, **49**, C8-1671 (1988).
95. J.R. Dutcher, B. Heinrich, J.F. Cochran, D.A. Steigerwald and W.F. Egelhoff, Jr., J. Appl. Phys. **63**, 3464 (1988).
96. J.F. Cochran, W.B. Muir, J.M. Rudd, B. Heinrich, Z. Celinski, T.-T. Le-Tran, W. Schwarzacher, W. Bennett and W.F. Egelhoff, Jr., J. Appl. Phys. **69**, 5206 (1991).
97. B. Hillebrands, P. Baumgart and G. Güntherodt, Appl. Phys. A **49**, 589 (1989).
98. V.L. Moruzzi and P.M. Marcus, Phys. Rev. B **39**, 471 (1989).
99. S.T. Purcell, B. Heinrich and A.S. Arrott, J. Appl. Phys. **64**, 5337 (1988).

100. B. Heinrich, Z. Celinski, J.F. Cochran, W.B. Muir, J. Rudd, Q.M. Zhong, A.S. Arrott, K. Myrtle and J. Kirschner, Phys. Rev. Lett. **64**, 673 (1990).

101. Z. Celinski, B. Heinrich, J.F. Cochran, W.B. Muir, A.S. Arrott and J. Kirschner, Phys. Rev. Lett. **65**, 1156 (1990).

102. P. Grünberg, S. Demokritov, A. Fuss, M. Vohl and J.A. Wolf, J. Appl. Phys. **69**, 4789 (1991).

103. S. Demokritov, J.A. Wolf and P. Grünberg, Europhys. Lett. **15**, 881 (1991).

104. P. Grünberg, S. Demokritov, A. Fuss, R. Schreiber, J.A. Wolf and S.T. Purcell, Proc. Int. Conf. Magn. 1991, J. Magn. Magn. Mater., to be published.

105. J. Unguris, R.J. Celotta and D.T. Pierce, Phys. Rev. Lett. **67**, 140 (1991).

106. S.T. Purcell, W. Folkerts, M.T. Johnson, N.W.E. McGee, K. Jager, J. aan de Stegge, W.B. Zeper, W. Hoving and P. Grünberg, Phys. Rev. Lett. **67**, 903 (1991).

107. J.V. Harzer, B. Hillebrands, R.L. Stamps, G. Güntherodt, C.D. England and C.M. Falco, J. Appl. Phys. **69**, 2448 (1991).

108. P.F. Carcia, A.D. Meinholdt and A. Suna, Appl. Phys. Lett. **47**, 178 (1985).

109. F.J.A. den Broeder, D. Kuiper, H.C. Dondersloot and W. Hoving, Appl. Phys. A **49**, 507 (1989).

110. F.J.A. den Broeder, D. Kuiper, A.P. van de Mosselaer and W. Hoving, Phys. Rev. Lett. **60**, 2769 (1988).

111. P. Krams, B. Hillebrands, G. Güntherodt, K. Spörl and D. Weller, J. Appl. Phys. **69**, 5307 (1991).

112. B. Hillebrands, J.V. Harzer, G. Güntherodt, C.D. England and C.M. Falco, Phys. Rev. B **42**, 6839 (1990).

113. J.V. Harzer, B. Hillebrands, R.L. Stamps, G. Güntherodt, D. Weller, Ch. Lee, R.F.C. Farrow and E.E. Marinero, Proc. Int. Conf. Magn. 1991, J. Magn. Magn. Mater., to be published.

CHAPTER 7

SPIN-WAVE INSTABILITIES, AUTO-OSCILLATIONS, CHAOS, AND CONTROL OF CHAOS IN YIG SPHERES

S.M. REZENDE, A. AZEVEDO, and F.M. de AGUIAR

Departamento de Física, Universidade Federal de Pernambuco,
50739 Recife, Brazil

Contents

7.1 INTRODUCTION

A great deal of work has been done in recent years on the dynamics of nonlinear systems in various branches of science and engineering [1]. A remarkable feature of nonlinear dynamics is that large classes of quite different physical systems display very similar qualitative and quantitative behavior. Among the rich variety of nonlinear phenomena the most common are self-oscillations, spiking, period multiplication, intermittency and chaos. Microwave pumped spin-wave instabilities were found to display nonlinear phenomena over 30 years ago. Recently they have been under intense investigation in the light of the modern ideas of chaotic dynamics of nonlinear dissipative systems.

Nonlinear behavior in ferromagnets was first observed in microwave magnetic resonance experiments in the early 50s. Damon [2] and Bloembergen and Wang [3] observed a premature saturation of the main resonance absorption and the appearance of a subsidiary absorption when the microwave power exceeded certain threshold levels. A few years later Suhl [4] produced a theory to explain those intriguing observations based on the parametric generation of spin waves. Later it was predicted [5,6] that spin waves could also be parametrically pumped by a microwave field parallel to the dc magnetic field $\vec{h}//\vec{H}_0$. This was called the parallel pumping process to distinguish from the perpendicular pumping Suhl process observed in the magnetic resonance configuration ($\vec{h} \perp \vec{H}_0$). In all those parametric pumping processes it was soon observed [7] that the radiation absorbed by the sample displayed low-frequency ($10kHz - 1MHz$) auto-oscillations or turbulent behavior when the microwave intensity was increased beyond the Suhl instability (SI) threshold. Over the years several explanations were given for these phenomena [8,9] until it was realized that their origin lied in the nonlinear dynamics of spin waves [10]. Today we know that spin-wave instabilities exhibit universal nonlinear behavior such as turbulent fluids [11], lasers [12], plasmas [13], electric discharges [14], electric circuits and several solid state systems [15]. Realization of this fact has revived the interest in spin-wave instabilities [16]. Bifurcations and chaos have been studied in ferromagnetic [17-25] and antiferromagnetic [26-29] materials, in bulk form and thin filmes [30-32] as well, and several theoretical papers have been published on the subject [16,33-39].

In this chapter we review the experimental situation and the present understanding of nonlinear phenomena associated with spin-wave instabilities in bulk samples. We begin with the concept of spin-wave parametric instabilities, showing a simple derivation of the threshold fields for the parallel pumping and subsidiary resonance processes, with a formalism that will be useful later. Next we present experimental results on spin-wave instabilities and auto-oscillations obtained with these two pumping processes in spheres of yttrium iron garnet (YIG). YIG is very convenient for the experiments because it has very low magnetic losses at room temperature so that spin waves can be driven parametrically with low microwave power. The experiments are then extended to the chaotic regime. The routes to chaos and the strange attractors are characterized quantitatively. The interacting spin-wave equations that govern the dynamics are derived from microscopic Hamiltonians and their numerical solutions are shown to explain the experimental observations. Finally we show how the chaotic state can be controlled by means of a small modulation in the applied magnetic field.

7.2 SPIN-WAVE INSTABILITIES

In an insulating ferromagnet, a spin wave with wave vector $\vec{k}$ has all spins precessing with frequency ω_k about the equilibrium direction with phase planes determined by $\vec{k}$. The spin-wave dispersion relation depends on the interactions between the spins which arise mainly from the Zeeman (interaction with the applied field H_0), exchange, crystalline anisotropy and dipolar energies. If H_0 is applied along a crystal direction of high symmetry the dispersion relation for wavelengths much shorter than the sample size and much large than the lattice parameter is [40-43]

$$\omega_k = \gamma(H_Z + H_A + Dk^2)^{1/2}\,(H_Z + H_A + Dk^2 + 4\pi M sin^2\theta_k)^{1/2}, \qquad (7.1)$$

where $H_z = H_0 - 4\pi M N_z$, N_z is the demagnetizing factor in the (z) direction of the external dc field $\vec{H}_0$, $\gamma = g\mu_B/\hbar$ is the gyromagnetic ratio ($2.8GHz/kOe$ or $17.6 \times 10^6 sec^{-1}Oe^{-1}$ for $g = 2$), M is the saturation magnetization, θ_k is the angle between $\vec{k}$ and $\vec{H}_0$, D is the exchange stiffness, and H_A is the effective magnetic anisotropy. In pure YIG [111] is the easy axis, $H_A \simeq 60Oe$, $D = 5.4 \times 10^{-9} Oe.cm^2$ and $4\pi M = 1760Oe$ at room temperature. By doping YIG with Gallium, $4\pi M$ can be reduced to $\sim 200Oe$ without much change in the other parameters. For fields of a few kOe the spin-wave frequency for $k \leq 10^6 cm^{-1}$ lies in the microwave range.

At room temperature spin waves in this frequency range are thermally excited with magnon population $n_k \sim 10^3$. However, they can be pumped by microwave radiation to levels as high as 10^{17} by means of several parametric processes.

In the parallel pumping process a microwave magnetic field h at a frequency ω_p is applied parallel to the static field H_0 [5,6,44]. Because of the ellipticity in the spin precession (due to the dipolar interaction), the oscillating field couples to spin waves with frequency $\omega_k \simeq \omega_p/2$. As a result, a uniform radiation ($k = 0$) photon can drive parametrically two magnons with opposite wavevectors $\vec{k}$ and $-\vec{k}$, thus conserving energy and momentum as illustrated in Fig.7.1a. When the driving field h is larger than a certain threshold value h_c, the rate at which energy is pumped into the spin wave system exceeds that lost to the lattice through various relaxation mechanisms. This causes the n_k magnon population to increase exponentially until it reaches saturation produced by the nonlinear interactions.

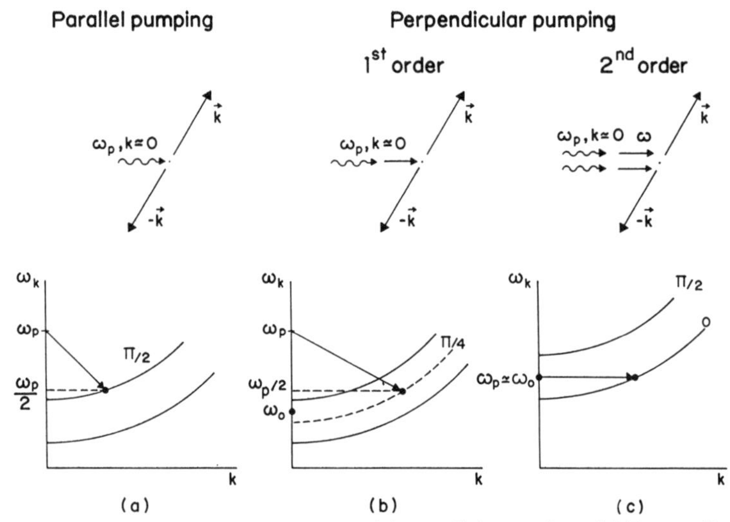

Fig.7.1: Illustration of spin-wave processes: (a) Parallel pumping; (b) Perpendicular pumping, subsidiary resonance (first order Suhl processes); (c) Perpendicular pumping, premature saturation of the main resonance (second order Suhl process).

In the perpendicular pumping configuration, the microwave field couples linearly to the uniform mode ($k = 0$) and one observes a ferromagnetic resonance (FMR) absorption at $H_0 \simeq \omega_p/\gamma$. Although spin waves with $k \neq 0$ do not couple directly to the radiation they can be excited via magnon-magnon interactions. One can distinguish two main processes [4,44]. In the first-order Suhl process, spin waves with $\omega_k \simeq \omega_p/2$ in a $\vec{k}$, $-\vec{k}$ pair can be driven parametrically by the uniform mode by means of the three-magnon interaction as shown in Fig.7.1b. In this case the uniform mode is driven far off-resonance with $\omega_0 \approx \omega_p/2$, so when the pumping exceeds a certain threshold one observes a *subsidiary resonance* at a field H_0 roughly half the value for the main resonance. In the second order Suhl process a spin-wave pair $\vec{k}$, $-\vec{k}$ is driven by two uniform mode magnons with frequency $\omega_0 \approx \omega_p \approx \omega_k$ pumped by the microwave radiation. This is made possible by the four-magnon interaction as illustrated in Fig.7.1c. This process limits the growth of the FMR absorption with increasing microwave power and is refered to as the premature saturation of the main resonance. In this chapter we will restrict attention to the parallel pumping and subsidiary resonance processes.

The instability processes can be treated quantitatively by means of several different approaches [4-6,44]. We use here the method of second quantization based on the creation and annihilation magnon operators $c_k^\dagger$ and c_k, obtained from the spin operators with the Holstein-Primakoff transformations [40-43]. Notice that the expectation value of the magnon operator $< c_k^\dagger >$ is proportional to the transverse value of the precessing magnetization $M^+ = M_x + iM_y$ and thus represents a spin-wave amplitude. Using a quantum mechanical formalism or a semi-classical approach one can show that the Hamiltonian for a spin-wave system pumped by a microwave field can be written as [40-44]

$$\mathcal{H} = \mathcal{H}^{(2)} + \mathcal{H}^{(3)} + \mathcal{H}^{(4)} + \cdots + \mathcal{H}'(t) \tag{7.2}$$

where

$$\mathcal{H}^{(2)} = \sum_k \hbar\omega_k c_k^\dagger c_k \tag{7.3}$$

is the Hamiltonian for a system of independent harmonic oscillators (magnons) with frequency ω_k, $\mathcal{H}^{(3)}$ and $\mathcal{H}^{(4)}$ represent three- and four-magnon interactions and $\mathcal{H}'(t)$ describes the interaction with the microwave field. In order to understand the onset of the spin-wave instability we consider initially the parallel pumping configuration neglecting the magnon-magnon interactions.

7.2.1 Parallel-pumping

If a uniform microwave field $h\ cos\omega_p t$ is applied in the (z) direction of $\vec{H}_0$, one can show that in an isotropic infinite medium the driving term of the Hamiltonian becomes [40]

$$\mathcal{H}'_{\shortparallel}(t) \simeq \frac{1}{2} h \sum_k \hbar \rho_k\ e^{-i\omega_p t}\ c_k^\dagger c_{-k}^\dagger + h.c. \tag{7.4}$$

where $\rho_k = \gamma\omega_M sin^2\theta_k\ exp(-i2\varphi_k)/4\omega_k$ represents the coupling of the driving field to the $\vec{k}$, $-\vec{k}$ magnon pair, $\omega_M = \gamma\ 4\pi M$ and φ_k is the azimuthal angle of $\vec{k}$. Notice that the coupling is strongest for spin waves propagating perpendicularly to $\vec{H}_0$. Using the Heisenberg equation with the Hamiltonians (7.3) and (7.4) and introducing dissipation phenomenologically in the usual manner, we obtain the equation of motion for the magnon operator,

$$\frac{dc_k}{dt} = -(i\omega_k + \gamma_k)c_k - ih\rho_k\ e^{-i\omega_p t}\ c_{-k}^\dagger \tag{7.5}$$

where γ_k is the relaxation rate of mode k. Since spin waves are excited in pairs one can assume that $c_{-k} = exp(iq_k)c_k$ where q_k is a real phase [21]. The fast time dependence in (7.5) can be eliminated with the use of the slowly varying spin-wave amplitude

$$\tilde{c}_k = < c_k >\ e^{iq_k/2}\ e^{i(\omega_p/2)t} \tag{7.6}$$

Equation (7.5) then becomes

$$\frac{d\tilde{c}_k}{dt} = -(i\Delta\omega_k + \gamma_k)\tilde{c}_k - ih\rho_k\tilde{c}_k^* \tag{7.7}$$

where $\Delta\omega_k = \omega_k - \omega_p/2$. Eq.(7.7) and its complex conjugate have solutions of the form $\tilde{c}_k$, $\tilde{c}_k^* \sim exp(\alpha_k t)$, where

$$\alpha_k = -\gamma_k + \left(|h\rho_k|^2 - \Delta\omega_k^2\right)^{1/2} \tag{7.8}$$

Clearly the spin-wave amplitude will grow exponentially in time when $\alpha_k > 0$. This happens when the pumping field exceeds the threshold value

$$h_c = \left(\gamma_k^2 + \Delta\omega_k^2\right)^{1/2} / |\rho_k| \tag{7.9}$$

For waves propagating at $\theta_k = \pi/2$ and with $\omega_k = \omega_p/2$ the threshold attains a minimum value,

$$h_{c_{min}} = 2\omega_p \gamma_k / \gamma \omega_M . \tag{7.10}$$

At fixed microwave frequency one can selectively excite modes with different k by changing the applied field H_0, as given by (7.1). For this reason the parallel pumping process is a very useful technique for measuring the relaxation rate γ_k as a function of k. In YIG at room remperature, $\omega_M \simeq 3 \times 10^{10} sec^{-1}$, $\gamma_k \sim 10^6 sec^{-1}$ for $k \approx 0$ so that $h_c \sim 10^{-1} Oe$ at X-band microwave frequencies $\omega_p \sim 6 \times 10^{10} sec^{-1}$. This driving field can be obtained in typical microwave cavities with $Q \sim 10^3$ with incident power of order $10 mW$.

7.2.2 Subsidiary resonance

In the perpendicular pumping configuration [4] the microwave field $h \; cos\omega_o t$ couples directly only to the $k = 0$ mode. The driving term in the Hamiltonian (7.2) can be written in this case as [40]

$$\mathcal{H}_{\perp}(t) = \hbar\gamma(SN/2)^{1/2} \, h(c_0^{\dagger} \, e^{-i\omega_p t} + h.c.) \tag{7.11}$$

where N is the number of spins S in the sample. In the first-order Suhl process, or subsidiary resonance [4], the $k = 0$ mode drives a $\vec{k}$, $-\vec{k}$ magnon pair via the three-magnon interaction

$$\mathcal{H}^{(3)} = \frac{\hbar}{2} \sum_{k \neq 0} F_k c_0 c_k^{\dagger} c_{-k}^{\dagger} + h.c. \tag{7.12}$$

arising from the dipolar interaction . The interaction coefficient is given approximately by [44]

$$F_k \simeq (2SN)^{1/2} \, \omega_M \; sin 2\theta_k \; e^{-i\varphi_k} \tag{7.13}$$

and is largest for spin waves propagating at an angle $\theta_k = \pi/4$. Since it is driven far off-resonance, the uniform mode behaves essentially like a virtual mode intermediary between the driving field and the spin wave pairs. Its amplitude acts like a classical variable [21] which can be obtained from the Heisenberg equation with the Hamiltonian (7.3) and (7.11),

$$c_0 \simeq \gamma \frac{(SN/2)^{1/2}}{(\omega_p - \omega_0) - i\gamma_0} h\, e^{-i\omega_p t} \tag{7.14}$$

With this approximation interaction (7.12) becomes identical to the parallel pumping driving Hamiltonian (7.4) with the coupling coefficient given by (with $\omega_p \approx 2\omega_0 \approx 2\omega_k$, $\omega_0 - \omega \gg \gamma_0$)

$$\rho_k \simeq \gamma \omega_M sin 2\theta_k e^{-i\varphi_k}/4\omega_k \tag{7.15}$$

This result shows that the subsidiary resonance instability is described by the same equation as the parallel-pumping process. The only difference to the parallel pumping is that there the modes with minimum threshold have $\theta_k = \pi/2$ whereas here they have $\theta_k = \pi/4$. The minimum threshold for subsidiary resonance given by (7.9) and (7.15) with $\omega_p = 2\omega_k$, $\theta_k = \pi/4$, is

$$h_{c_{min}} \simeq 2\omega_k \gamma_k / \gamma \omega_M \tag{7.16}$$

which is the same as the parallel pumping value (7.10). Note that when other effects are taken into account the mode with minimum threshold is not exactly the one with $\theta_k = \pi/4$ [45].

7.3 EXPERIMENTAL OBSERVATION OF INSTABILITIES AND AUTO-OSCILLATIONS

7.3.1 Experimental set-up

The experiments described here were performed at room temperature with polished single crystal YIG spheres using the experimental set-up shown schematically in Figure 7.2. The sample is located at the center of a critically coupled rectangular TE_{102}

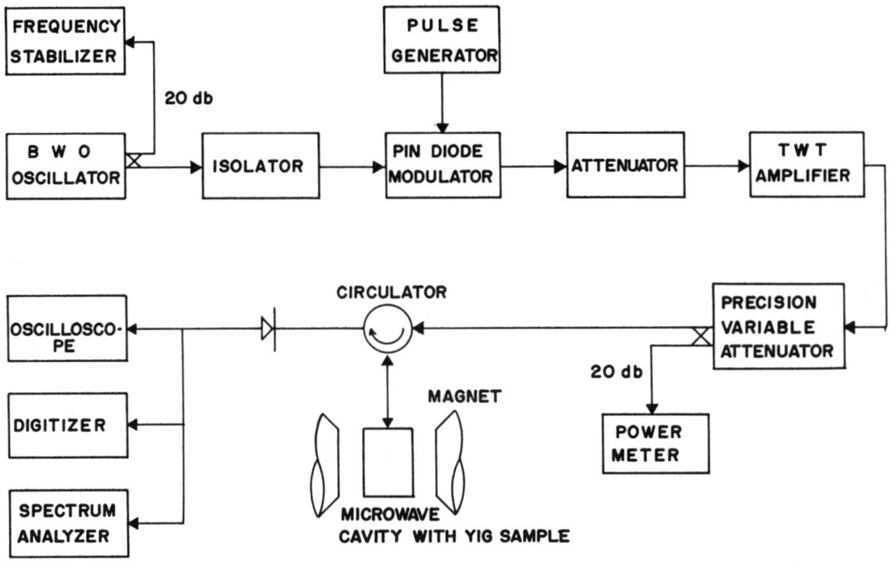

Fig.7.2: Schematic diagram of the microwave system used to excite spin-wave instabilites.

microwave cavity with $Q \simeq 2000$, placed between the poles of an electromagnet that provides the static field $\vec{H}_0$. The cavity can be oriented so that the microwave magnetic field $\vec{h}$ is either parallel or perpendicular (subsidiary resonance) to $\vec{H}_0$. The microwave radiation is generated by a X-band backward-wave oscillator (BWO) and amplified to 1.8 Watt by a traveling-wave tube (TWT). The BWO frequency is stabilized by an external crystal oscillator and manually adjusted to the center of the cavity resonance. The radiation power is then controlled with a variable precision attenuator and directed by a circulator to the resonant cavity, where it drives spin waves in the sample. The reflected microwave signal is detected with a sensitive Schottky-barrier diode at the output port of the circulator and recorded in a digital storage oscilloscope or observed with a spectrum analyzer. In order to avoid sample heating the microwave radiation is pulsed (duration $100\mu sec$ at 100 $pulses/sec$) with a PIN modulator placed before the TWT amplifier. This causes no difficulty for recording the steady-state regime of the sample response because the transient time is normally much shorter than the pulse length.

7.3.2 Spin-wave instabilities

At low power levels the pulse reflected from the cavity has essentially the same shape as the incoming microwave pulse. As the power is increased, an abrupt change in the pulse shape occurs when the driving field h reaches the minimum threshold h_c^{min} for spin-wave parametric excitation. Spin-waves are excited in pairs with wavevectors $\pm\vec{k}$ and nearly half the microwave pumping frequency, $\omega_k \simeq \omega_p/2$. Further increase in the power usually leads to the spontaneous appearance of low-frequency $(10kHz - 1MHz)$ amplitude modulations in the microwave reflection, the so-called auto-oscillation, above a second threshold h'_c. The auto-oscillations can be observed directly on the oscilloscope screen or studied with a spectrum analyzer. Fig.7.3(a) shows data for the parallel pumping spin-wave SI threshold h_c and auto-oscillation critical field h'_c versus H_0 for a $1mm$ diameter YIG sphere magnetized in the [111] direction, obtained with $f_p = \omega_p/2\pi = 9.5GHz$. The shape of the $h_c \times H_0$ curve, the so-called "butterfly curve" can be explained as follows. To the left of the minimum, the mode with $\theta_k = \pi/2$ can be excited with $\omega_k = \omega_p/2$, and since it has the lowest

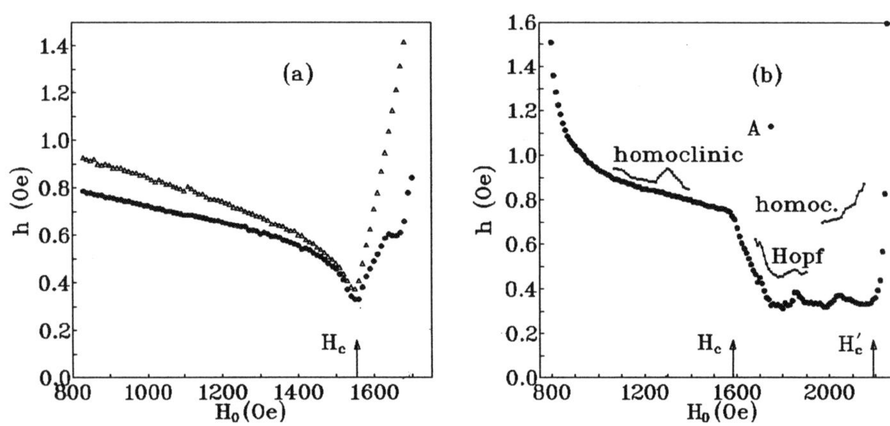

Fig.7.3: Threshold microwave fields versus dc field for spin-wave instability and auto-oscillation in a $1mm$ YIG sphere for (a) parallel pumping at $9.5GHz$, H_0 along [111]; (b) subsidiary resonance at $8.87GHz$, H_0 along [110].

threshold, the variation of h_c with H_0 reflects the linear dependence of γ_k with k for this mode. The minimum occurs at the field value $H_0 = H_c$ for which ω_k ($k = 0$, $\theta_k = \pi/2) = \omega_p/2$. Above this value the $\pi/2$ mode is not available for pumping and so h_c increases because the coupling decreases with decreasing θ_k. This is expressed by the factor $sin^2\theta_k$, of ρ_k in the denominator of (7.9).

The gross features of the spin-wave instabilities driven in subsidiary resonance are similar to those in parallel pumping. Fig.7.3(b) shows the phase boundaries for spin wave excitation and auto-oscillation measured in the same $1mm$ YIG sphere of Fig.7.3(a) with H_0 along the [110] axis and microwave frequency $8.87GHz$. Instead of a sharp minimum, the "butterfly curve" for subsidiary resonance has a broad minimum and three characteristic regions in field H_0. For $H_0 < H_c \simeq 1580Oe$ the modes with lowest threshold are plane waves with $k > 0$ propagation at an angle $\theta_k = \pi/4$ with $\vec{H}_0$ (or $\theta_k \simeq 35°$ depending on the effect of the relaxation rate [45]). For $H_c < H_0 < H_c' = 2190Oe$ the modes with minimum threshold are magnetostatic waves with $k \approx 0$. For $H_0 > H_c'$ the threshold rises sharply with H_0 as $\theta_k \rightarrow 0$. Since in the various field regions the modes excited have different nonlinear coupling parameters, a variety of bifurcations leading to self-oscillations can be observed at different field values, as indicated by the boundary lines in Fig.7.3(b).

7.3.3 Auto-Oscillations

The auto-oscillations that appear at $h_c > h_c' > h_c$ have shape and frequency which depend on the pumping level, dc field value, crystal axis orientation with respect to $\vec{H}_0$ and also on sample shape, dimensions, surface polishing and temperature. Over the years several explanations were proposed for these so-called auto-oscillations. The most noticeable prenonlinear dynamics model was proposed by Wang and co-workers [8,9]. According to their model, due to the finite linewidth of the spin waves, several modes with spread Δk and $\Delta\omega_k$ near $\omega_p/2$ are excited above the SI threshold. The beat between these modes would be the source of the auto-oscillation with frequency $\Delta\omega = v_g\Delta k$ which is determined by the boundary conditions via $\Delta k \simeq \pi/d$, where v_g is the group velocity and d the sample dimension along the propagation direction. This model explained some features of the experimental data, like the dependence of the frequency on H_0 and on sample dimensions. However it completely failed in explaining the nature of the second threshold and the dependence of the frequency on the pump power. This was noted by Zakharov, L'vov, Starobinets and co-workers [10] who were the first to use a nonlinear dynamics approach to the problem. They established that the auto-oscillations arise from the back-and-forth dynamic power exchange between parametric modes due to the nonlinear interaction between them. The research in this field has intensified in recent years and we now have a good understanding of the phenomena in the light of the theory of bifurcations.

The auto-oscillations arise naturally in the solutions of the nonlinear differential equations that govern the interacting spin-wave modes excited above the SI threshold. Their onset is associated to a bifurcation of the spin-wave equations. A bifurcation is the analog of a phase transition in equilibrium thermodynamics and corresponds to a qualitative change in the dynamic state of the system. Since the nonlinear parameters change markedly with k and θ_k, one expects the nature of the bifurcation to change with the conditions of the experiment. This is indeed the case as illustrated in Figure 7.4.

Figure 7.4(a) shows the oscilloscope traces of the self-oscillations with increasing pumping in the parallel pumping configuration of Fig.7.3(a) at $H_0 = 1570 Oe$. The oscillations appear at a field h'_c just above the SI threshold $h'_c \leq h_c$, with a finite frequency $f_0 \simeq 220 kHz$ and vanishing amplitude A. As h increases the amplitude increases smoothly while the frequency stays constant. The detailed data reveals that A varies with the reduced driving field $r = h/h'_c - 1$ in the critical region as $A \sim r^\beta$, where $\beta = 0.5$. This is typical of a Hopf bifurcation and is analogous to a second-order phase transition [46,47].

Quite different behavior is shown in Fig.7.4(b) by the oscilloscope traces obtained in subsidiary resonance pumping with $\vec{H}_0$ along the [110] axis of the YIG sphere, as in Fig.7.3(b), with $H_0 = 1950 Oe$, $f_p = 9.4 GHz$. In this case the auto-oscillation sets in with finite amplitude and vanishing frequency, in analogy with a first-order phase transition and in marked contrast with the scenario of Fig.7.4(a). This has been identified as a homoclinic bifurcation [47]. Notice that the subsidiary resonance configuration presents a wide variety of behavior, as indicated by the boundary lines in Fig.7.3(b). This results from the fact that modes in different regions in

k-space, and consequently different parameters, are excited as H_0 changes.

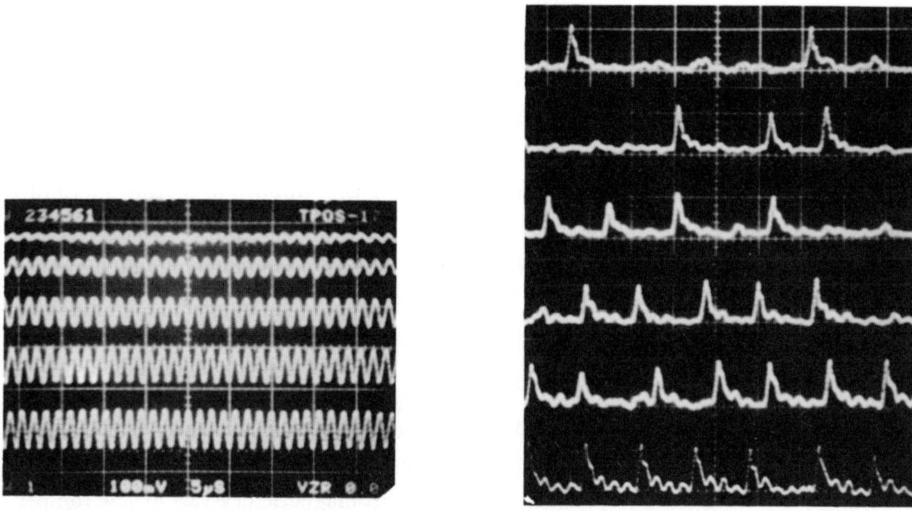

(a) (b)

Fig.7.4: Oscilloscope traces showing the behavior of the auto- oscillation with increasing microwave field $R = h/h_c$. Figure (a) was obtained with parallel-pumping of a YIG sphere aligned along [111], with R = 1.224, 1.255,1.293, 1.321, 1.342 from top to bottom. Figure (b) was obtained in subsidiary-resonance with the same sphere along [110] for R = 1.862, 1.872, 1.883, 1.894, 1.905, 1.916 from top to bottom. Two cases correspond to completely different bifurcations [Ref.47].

7.4 PERIOD DOUBLING AND CHAOS

The auto-oscillations described in the previous section are very easily observed in spin-wave pumping experiments. Samples with almost any shape and orientation in the magnetic field display auto-oscillations when the pumping is increased beyond the Suhl threshold. On the other hand, period doubling, intermittency, chaos and other manifestations of nonlinear dynamic behavior are not so easy to observe. They appear only under certain conditions of crystal orientation, dc field value and sample shape. This is probably the reason why it took so many years for these phenomena to be observed in spin-wave instabilities.

The first observation of a complete period doubling route to chaos was made by Gibson and Jeffries [17] with the second order Suhl process in a Ga-YIG sphere pumped at $1.3GHz$. They employed an arrangement used in YIG filters where the sample is placed between two orthogonal one-turn coils, one for driving and the other for detection. Auto-oscillations were observed in the amplitude of the microwave transmission signal when the incident power exceeded the threshold for the premature saturation. Auto-oscillations were observed with frequencies $250kHz$ and $16kHz$. As the driving field was increased, the oscillation at $f_0 \approx 16kHz$ showed a typical Feigenbaum route to chaos, with successive period doublings leading to chaos.

The first studies of chaotic behavior in spin-wave pumping experiments had been done earlier with the parallel pumping configuration. Waldner, Barberis and Yamazaki [27] observed a route to chaos by irregular periods and Yamazaki [26] observed one period doubling but no chaos in antiferromagnetic crystals at low temperatures. The first observation of period-doubling cascades in routes to chaos in parallel pumping was made by the present authors in pure YIG sphere [18,19]. The experiments were done with a $1mm$ diameter sphere pumped by X-band radiation. For some crystal directions and values of H_0, up to three period doublings were observed as in a typical Feigenbaum cascade. However, for other conditions a new scenario was observed. This is shown by the data of Fig.7.5 obtained with the YIG sphere oriented with the crystal axis [111] parallel to the dc field H_0 in the same configuration which gave the data of Fig.7.3(a). Chaotic behavior was observed for $H_0 = 1540Oe$, a value very close to the minimum of the butterfly curve H_c. Figs.7.5(a)-(d) show the oscilloscope traces of the auto-oscillation for increasing microwave pumping field h, represented by the control parameter $R = h/h_{c_{min}}$. The auto-oscillation first appears at $R_c' \simeq 1.50$ with frequency $f_p \simeq 100kHz$. In (a) one sees the nearly sinusoidal signal at $R = 1.52$. The first period doubling occurs at $R = 1.62$ as shown in (b). Immediately above this value, at $R = 1.64$, chaotic behavior develops with no evidence of other period doublings. This scenario, which we call C to distinguish from the Feigenbaum sequence [48], was also observed by Mino and Yamazaki [20] in parallel pumping. It is shown later that it is also present in the subsidiary resonance configuration. Increasing R further leads to another sequence of more pronounced

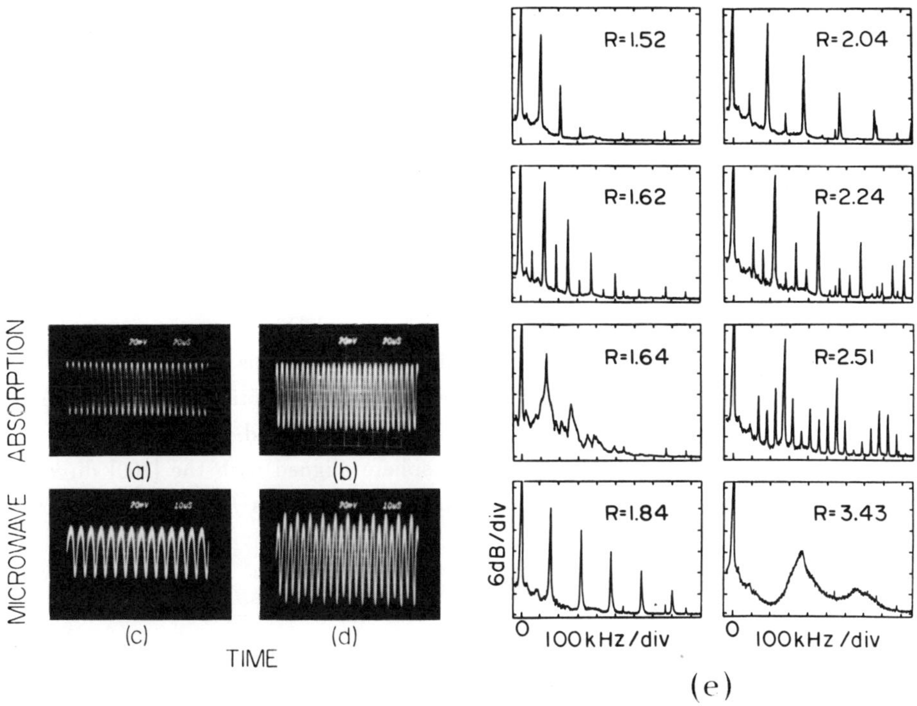

Fig.7.5: Oscilloscope (a)-(d) traces showing oscillations and chaos observed in parallel-pumping experiments with $f_p = 9.4GHz$ and $H_0 = 1540Oe$, along [111]. Power spectra of the oscillations in the same configuration [Ref.19] shown in (e).

period doublings, as shown in (c) and (d).

The two sequences of period multiplications are more clearly demonstrated with the power spectra of Fig.7.5(e), obtained directly with a spectrum analyzer. The spectrum with $R = 1.52$ is that of a nearly sinusoidal signal with fundamental frequency $f \simeq 100kHz$ and several weak harmonics. In the first cascade one period doubling is observed at $R = 1.62$ and there is no evidence of further bifurcations before the onset of chaos at $R = 1.64$. When R is increased further a coherent modulation reappears at $R = 1.84$ and another sequence of bifurcations develops distinctly from a window of the Feigenbaum scenario. Bifurcations with periods $2T$ and $4T$ occur at $R = 2.04$ and 2.24 and then the period-doubling cascade is interrupted and a series of periods $3T$, $6T$, and $12T$ sets in leading to chaos at $R = 3.43$. A similar route to chaos in which a period-doubling cascade crosses over to an odd-period bifurcation series has been observed in a Rayleigh-Bénard experiment [49].

The first observations of period doubling and chaos in subsidiary resonance experiments were made by Jeffries and co-workers [21] and independently by the present authors [22]. The experiments use the same arrangement as for the parallel pumping except that the cavity is rotated by 90° to provide a driving field $\vec{h}$ perpendicular to $\vec{H}_0$. Like in the parallel pumping experiments, two clearly distinct scenarios of period multiplication were also observed in the subsidiary resonance process. The first scenario, which we call F henceforth, is a typical Feigenbaum route to chaos. The original self-oscillation with period T bifurcates to periods $2T$, $4T$, $8T$, and $16T$ with increasing control parameter R before it becomes chaotic. This is shown by the reproductions of the digitized signals and the corresponding Fourier spectra of Fig.7.6, obtained with the same $1mm$ YIG sphere aligned with the [100] direction along $\vec{H}_0$. The pumping frequency for the data of Fig.7.6 was $f_p = 9.4GHz$ and the

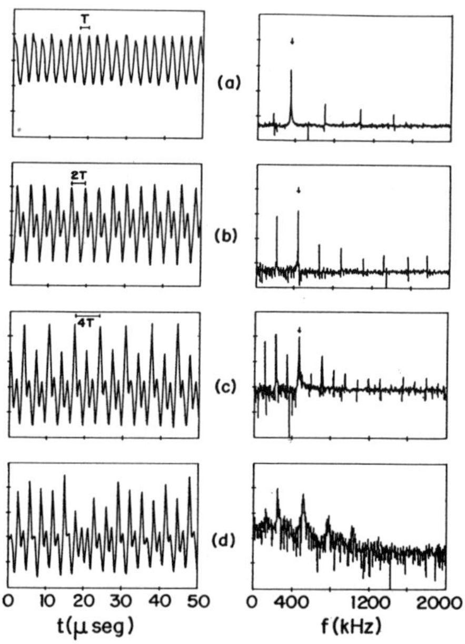

Fig.7.6: Auto-Oscillations and corresponding power spectra observed in subsidiary resonance, with $f_p = 9.4GHz$, $H_0 \simeq 2000Oe$ along [100]. Period doubling bifurcations as in a Feigenbaum route to chaos occur at $R = 1.93$, 1.99, 2.09 and 2.114, from top to bottom.

field was $H_0 = 1950kOe$. The self-oscillation appears initially with $400kHz$ at $h = 0.81Oe$ ($R = 1.93$), a value higher than the thresholds shown in Fig.3(b) for the [110] direction. Period-doubling bifurcations to $2T$ and $4T$ occur at $R = 1.99$ and $R = 2.09$ respectively, leading to chaos at $R = 2.31$. Period $8T$ was also clearly observed but is not shown in the figure.

Scenario C was also observed in subsidiary resonance pumping. Fig.7.7 shows results obtained with the sphere aligned along the [110] direction at $H_0 = 1980Oe$, with $f_p = 9.4GHz$. Period T oscillation ($f_0 \simeq 400kHz$) with $R = 2.29$ is shown at the top. Increasing the driving field leads to period $2T$ at $R = 2.40$, noisy period $2T$ oscillation at $R = 2.45$ and chaos at $R = 2.51$, with no evidence of other period multiplications in the sequence.

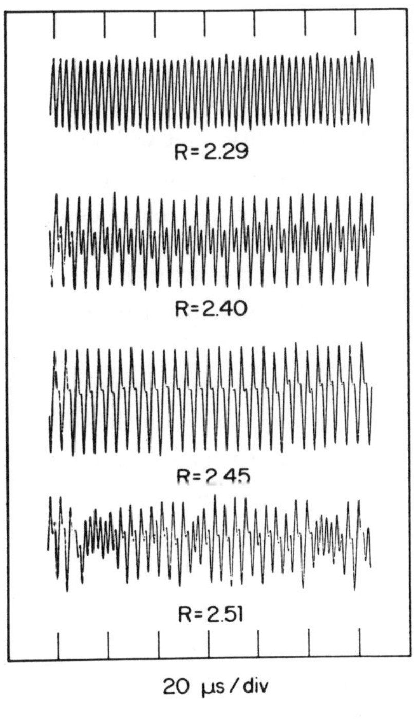

Fig.7.7: Illustration of scenario C route to chaos observed is subsidiary resonance with $f_p = 9.4GHz$, $H_0 = 1980kOe$ along [110].

7.5 CHARACTERIZATION OF CHAOS

The measurement of the power spectra of the amplitude modulation signal $v(t)$ presented in the previous section provides very limited information about the system. However, with a more detailed analysis of the signal it is possible to obtain significant information about the system without knowing its equations of motion. This is made possible by methods of nonlinear dynamics. With these methods it is possible to reconstruct the trajectory in a phase space topologically equivalent to the real trajectory in the space of the dynamical variable of the system. The methods of analysis are usually based on a time-series of the signal, $v(t)$, $v(t+\tau)$, $v(t+2\tau)$, ... $v(t+n\tau)$, using the so-called embeddding technique [50,51]. Several important quantities can be extracted from the time-series such as the Lyapunov exponents, the attractor dimensions, metric entropies and singularity spectra $f(\alpha)$. The Lyapunov exponents measure the rate at which the trajectories converge or diverge in time in a phase space. They have been used by Carrol, Pecora and Rachford [52] to characterize chaos in magnetostatic mode instabilities but will not be explored further here. The singularity spectra of the strange attractor have been measured by Mino, Yamazaki and Nakamura [53] in experiments such as the one described here.

The two quantities used here to characterize chaos are the attractor dimensions and the Kolmogorov entropy. The dimension of an attractor is the first level of knowledge necessary to characterize its static properties and is also a lower bound on the number of independent variables needed to model the dynamics. The Kolmogorov entropy K measures the loss of information on the initial conditions per unit time. Both quantities can be used to discriminate between stochastic and deterministic dynamical systems. For instance, the dimension of a limit cycle in phase space is 1, whereas that of random noise is infinite. A regular trajectory has $K = 0$, while white noise has $K = \infty$. For a deterministic chaotic system, the fractal dimension in general holds a noninteger value and the Kolmogorov entropy has a finite positive value. In order to obtain a reliable characterization of the attractors we have used two algorithms to calculate the dimensions, one proposed by Grassberger and Procaccia (GP) several years ago [51] and another introduced more recently by Badii and Politi (BP) [54].

In the GP algorithm one counts points on the attractor in boxes with fixed size r. A convenient quantity to measure experimentally is the correlation integral $C(r)$.

Considering the points $\{X_i\}$ on the attractor ($i = 1, ...N$) in a D-dimensional phase space, one defines

$$C(r) = \lim_{N\to\infty} \frac{1}{N^2} \sum_{i\neq j} \theta(r - |X_i - X_j|) \tag{7.17}$$

where $\theta(r)$ is the Heaviside function. $C(r)$ may be calculated from a time-series of a single physical quantity $x(t)$ (the voltage of the microwave diode detector in our case) with the embedding technique. This consists of constructing E-dimensional vectors $X(t)$ with coordinates $x(t)$, $x(t+\tau)$, ... $x(t+(E-1)\tau)$, where τ is a fixed delay time and E the embedding dimension [51]. For small r the correlation integral scales as

$$C(r) \sim r^{D_2} \tag{7.18}$$

where the correlation exponent $D_2 = \lim_{r\to 0} logC(r)/logr$ is close to the fractal dimension d of the attractor. D_2 is obtained from the measured slope of $logC(r)$ vs $logr$ curves. For a deterministic signal this slope converges for $E \geq 2D+1$.

In the method of Badii and Politi [54] the attractor is covered with boxes of fixed mass. The central quantity is the moment of order γ calculated from the probability distribution $p(\delta, n)$ of nearest-neighbor distances δ among n randomly chosen points on the attractors

$$<\delta^\gamma> \equiv M_\gamma(n) = \int_0^\infty \delta^\gamma p(\delta, n) d\delta \tag{7.19}$$

From the scaling form of this as $n \to \infty$, one defines a γ dependent dimension function

$$D(\gamma) = -\lim_{n\to\infty} \frac{log\ n}{\log[M_\gamma(n)]} \tag{7.20}$$

The dimension function yields, for specific γ values, several dimensions previously defined [54]. In particular, the fixed point $D_0 = D(\gamma) = \gamma$ is the fractal dimension and $D_1 = D(\gamma = 0)$ is the information dimension, typically $D_0 \geq D_1$. The generalized metric entropies $K(\gamma)$ can be obtained from $D(\gamma)$. For $\gamma = 0$ it satisfies the relation [55]

$$logM_0(n, E) \sim \frac{1}{D(0)}[\tau EK(0) - log\ n] \tag{7.21}$$

where $M_0(n, E)$ is the moment calculated from the E-dimensional embedded vectors and τ is the sampling time [55].

Strange attractors obtained from scenarios F and C described previously, both in the parallel pumping and subsidiary resonance processes, have been studied with the GP and BP methods. We present here details of the results obtained with the subsidiary resonance configuration at the onset of chaos ($R = 2.51$) approached with scenario C previously described. The time-series data were obtained from the low-frequency amplitude modulation at intervals $\tau = 0.2\mu sec$, which is about one tenth of the fundamental period at $R = 2.45$. Fig.7.8(a) shows plots of $log_{10}C(r)$ versus $log_{10}r$ obtained wth 2048 data points that are used to calculated the fractal dimension $d \simeq D_2$ with the GP algorithm. As the embedding dimension increases the slope of

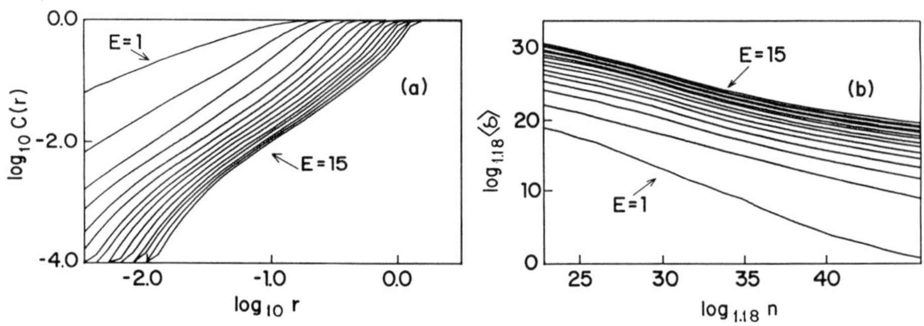

Fig.7.8:(a) Correlation integral $log_{10}C(r)$ vs $log_{10}r$ calculated from the time-series data at the onset of chaos in senario C as in Fig.7.7. (b) Plots of $log_{1.18} < \delta >$ vs $log_{1.18}n$ for $\gamma = 0$ for several embedding dimensions (E) calculated from the same time-series data as (a).

the curve in the range $-3 < logC(r) < -1$ converges to a value $d \simeq 1.6$. In the BP method we determine the dimension from plots of $log_{1.18} < \delta >$ for $\gamma = 0$ versus $log_{1.18}n$ shown in Fig.7.8(b) for $E = 1, 2...15$. For $E > 6$ the slopes at the higher $log_{1.18}n$ end converge to $D_1 \simeq 1.5$, in good agreement with the dimension obtained with the GP method. The same procedure has been used to calculate the dimensions of the attractors with scenarios F in subsidiary resonance and C in parallel pumping. In all cases we find $1.4 < d < 1.8$, which is also close to the value obtained in YIG under parallel pumping at low temperature [20].

From the time-series data we have also calculated the entropy $K(0)$ as a function of the embedding dimension using Eq.(7.21). Fig.7.9 shows the results obtained at the onset of chaos in subsidiary resonance. In both scenarios F and C, for large E the entropy converges to a constant value $K \simeq 0.05$. This positive constant value confirms the deterministic nature of chaos in our experiments.

All of the previous data for the parallel pumping and subsidiary resonance chaotic dynamic experiments presented here have been interpreted with a model for interacting spin waves described next.

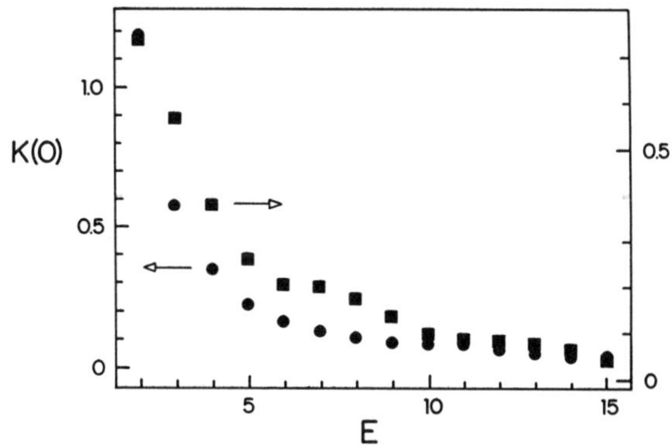

Fig.7.9: Metric entropy *vs* the embedding dimension for the chaotic attractor at the accumulation point of scenarios $F(\bullet)$ and $C(\blacksquare)$ as in Figures 7.6 and 7.7.

7.6 MODEL FOR THE THEORETICAL INTERPRETATION

The nonlinear differential equations that govern the dynamics of large k, plane spin-wave instabilities have been derived by many authors. In all approaches the interacting spin-wave system is considered to be driven by a uniform microwave field in such a way that the pumping term of the Hamiltonian conserves momentum, i.e. magnons are generated in pairs $\pm\vec{k}$. This is strickly correct only for an infinite medium, hence it cannot predict the observed sample size dependence for the self-oscillation frequency and threshold. In a finite sample it has been shown [47] that in addition to the pumping of magnon pairs with opposite and equal wave vectors $\pm k$, it is possible to have excitation of pairs $+k$, $-(k + \Delta k)$. In this case, instead of 7.4, the driving

Hamiltonian is,

$$\mathcal{H}'(t) \simeq \frac{1}{2}h \sum_k \hbar\rho_k e^{-i\omega_p t}\, c_k^\dagger c_{-k'}^\dagger\, (\delta_{kk'} + \alpha_{\Delta k}\delta_{k,k\pm\Delta k} + ...) + h.c. \tag{7.22}$$

where $\alpha_{\Delta k}$ is a factor which depends on the wave vector mismatch Δk_x, Δk_y, Δk_z. For two neighboring standing waves along x (or y or z) with $\Delta k_x = \pi/L_x$, $\alpha_{\Delta k}$ attains its maximum value of $2/\pi$. In an infinite medium $\alpha_{\Delta k} = 0$ so that Eq.(7.22) reduces to the expression used in the usual treatment of spin-wave parametric pumping.

The essential ingredient for the nonlinear dynamics is the coupling between two pairs of parametric magnons provided by the four-magnon interaction. This has relevant terms of the form [10,16,21,35]

$$\mathcal{H}^{(4)} = \hbar \sum_{k,k'} \left(\frac{1}{2} S_{kk'} c_k^\dagger c_{-k}^\dagger c_{k'} c_{-k'} + T_{kk'} c_k^\dagger c_{k'}^\dagger c_k c_{k'} \right) \tag{7.23}$$

In simple ferromagnets this interaction arises from the dipolar, anisotropy and exchange energies, but the latter is negligible for the small k-values excited in the microwave experiments. The interaction (7.23) couples the equations of motion for different k-modes giving rise to a variety of nonlinear behavior. Our model Hamiltonian for a spin-wave system pumped by a microwave field in parallel pumping or subsidiary resonance is then

$$\mathcal{H} = \mathcal{H}^{(2)} + \mathcal{H}^{(4)} + \mathcal{H}'(t) \tag{7.24}$$

where the three terms are given by (7.3), (7.22) and (7.23). Using the Heisenberg equation for the operator c_k and making use of the transformation (7.6) we obtain the following equation of motion for the variable $\tilde{c}_k$ that characterizes the spin-wave pair mode

$$\begin{aligned} \frac{d\tilde{c}_k}{dt} &= -(\gamma_k + i\Delta\omega_k)\tilde{c}_k - ih\rho_k(\tilde{c}_k^* + \alpha_{\Delta k}\tilde{c}_{k'}^* e^{i\beta_{kk'}/2}) - \\ &- i\sum_{k'} \left(S_{kk'}\tilde{c}_{k'}^2 \tilde{c}_k^* + 2T_{kk'}|\tilde{c}_{k'}|^2 \tilde{c}_k \right) \end{aligned} \tag{7.25}$$

where γ_k is the relaxation rate of mode k, $\Delta\omega_k = \omega_k - \omega_p/2$ is its detuning and $\beta_{kk'} = q_k - q_{k'}$ is the phase difference between modes k and k'. Notice that in an infinite medium $\alpha_{\Delta k} = 0$ and Eq.(7.25) reduces to the same equation used in previous works. In this condition we can multiply both sides of Eq.(7.25) by $\tilde{c}_{-k}$ and using the symmetry $\vec{k}$, $-\vec{k}$ we obtain the S-equations [10,16,35] for the Cooper-pair variables σ_k,

$$\frac{1}{2}\frac{d\sigma_k}{dt} = -(\gamma_k + i\Delta\omega_k)\,\sigma_k - ih\rho_k n_k - i\sum_{k'}(S_{kk'}\sigma_{k'}n_k + 2T_{kk'}n_{k'}\sigma_k) \tag{7.26}$$

where $\sigma_k = \tilde{c}_k\tilde{c}_{-k} = n_k exp(i\phi_k)$, n_k being the number of magnons in mode $\vec{k}$ and ϕ_k the phase of the precessing magnetization. The S-equations have quadratic nonlinearities and in the polar form are very convenient for stability analysis. They have been used by many authors to study spin-wave turbulence and nonlinear dynamics. However, they are not suitable for finite samples because of the additional coupling between the spin wave equations introduced by the $\alpha_{\Delta k}$ term.

The spin-wave equations (7.25) represent $2M$ (M is the number of nearly degenerate modes excited by the pump field) nonlinear differential equations of the form $d\vec{x}/dt = F(\vec{x},\vec{\mu})$, where $\vec{x}$ is the dynamical variable vector and $\vec{\mu}$ represents the (time-independent) parameters. The linearization [46] of the equations around the fixed points $\vec{x}_0$ given by $F(\vec{x}_0,\vec{\mu}) = 0$ yields $\vec{x}(t) = \vec{x}_0 + \delta x exp(\lambda t)$, where λ are the eigenvalues of the Jacobian matrix $J_{ij} = \partial F_i/\partial x_j$. Physically there are infinitely many modes involved as soon as the pumping exceeds the Suhl threshold so that the problem of determining the eigenvalues is in principle immensely difficult. However, as Suhl and Zhang [36] have shown, the linearized equations have *collective*, normal mode solutions, formed by linear superpositions of the spin-wave variables δx, each with an associated pair of complex eigenvalues λ. The character of the bifurcation is determined by the eigenvalues whose real part turn positive at the critical field h'_c, all others representing uninteresting damped solutions. In general, at not too intense driving, the dynamics is governed by essentially two modes, each resulting from the collective coupling of many spin-wave pair modes in a certain region of k-space. It appears that each normal mode is described by an equation similar to that of individual k modes with parametes that are suitable averages of the values for the intervenient spin waves. This is justified by the fact that numerical calculations with only two modes have given quite good qualitative agreement with the experimental observations [21,35].

Our analysis of the auto-oscillations is based on the equations for two modes derived from Eq.(7.25),

$$\dot{c}_1 = -(\gamma_1 + i\Delta\omega_1)c_1 - ih\rho_1(c_1^* + \alpha e^{i\beta/2}c_2^*) - i2(S_1 c_1^2 c_1^* + S_{12} c_1^* c_2^2 + 2T_{12} c_2 c_2^* c_1) \tag{7.27}$$

$$\dot{c}_2 = -(\gamma_2 + i\Delta\omega_2)c_2 - ih\rho_2(c_2^* + \alpha e^{-i\beta/2}c_1^*) - i2(S_2 c_2^2 c_2^* + S_{12} c_2^* c_1^2 + 2T_{12} c_1 c_1^* c_2) \tag{7.28}$$

where the subscripts 1 and 2 denote the two collective modes, $S_k \equiv S_{kk} + 2T_{kk}$, $\alpha \equiv \alpha_{\Delta k}$, $\beta \equiv \beta_{12}$, and the tilde in $\tilde{c}_k$ was dropped to simplify the notation. Since c_k is complex, the set (7.27) and (7.28) represents four coupled equations. Numerical solutions of these equations are shown in the next section.

7.7 NUMERICAL SOLUTIONS OF THE SPIN-WAVE EQUATIONS

Although Eqs.(7.27) and (7.28) represent an enormous simplification of the original problem, they are still mathematically very complex. They have twelve independent parameters so that a multitude of bifurcation phenomena can be encountered depending on the parameter values. In order to fully explore these phenomena it is necessary to make use of numerical methods and integrate the equations in a computer. In the numerical studies with an arbitrary set of parameters, one usually finds that the solutions are attracted to stable fixed points. However, in certain regions of the parameter space some fixed points become unstable and the solutions may exhibit a variety of dynamic behavior. Due to the complex nature of the modes involved in the dynamics, it is difficult to relate the parameters with the microscopic ones. However, we have found some sets of parameters that yield results very similar to those observed experimentally. This is done by solving the equations numerically in a VAX 750 computer using a Runge-Kutta subroutine to obtain the steady-state time evolution of c_1, c_2 and the corresponding magnon occupation number $n_k = c_k^* c_k$. Since we want to work with quantities of order 1, the magnon number and the interaction parameters are rescaled as $n_k \to F n_k$, $S_{kk'} \to F^{-1} S_{kk'}$ and $T_{kk'} \to F^{-1} T_{kk'}$ where $F = \omega_M / 4SN\gamma_k$.

In YIG at room temperature, $\gamma_k \sim 2\times10^6 sec^{-1}$, $\omega_M \simeq 3\times10^{10} sec^{-1}$ so that $F \sim 10^{17}$ in a sample with volume $\sim 1mm^3$.

7.7.1 Auto-Oscillations

Let us consider initially the following parameters: $\Delta\omega_1 = \Delta\omega_2 = 0$, $\rho_1/\rho_2 = 0.7$, $\gamma_2/\gamma_1 = 5.0$, $S_1/\gamma_1 = -1.0$, $S_2/\gamma_1 = 0.5$, $S_{21}/\gamma_1 = 2.5$, $T_{21}/\gamma_1 = 1.125$, and $\alpha = 0$. Note that the parameters are quite different for the two modes indicating that they are degenerate modes far in k-space. This justifies making $\alpha = 0$. In this case identical results are obtained with the two-mode equations derived from Eq.(7.26). Fig.7.10 shows the result of the numerical integration for the time evolution of the magnon

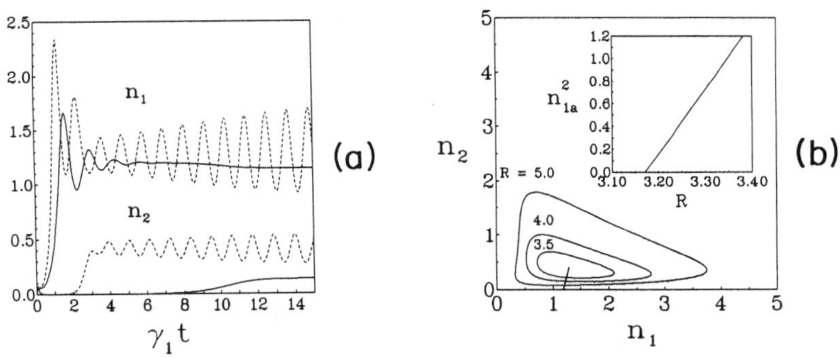

Fig.7.10: Numerical solutions of two-mode equations with parameters: $\Delta\omega_1 = \Delta\omega_2 = 0$, $\rho_2/\rho_1 = 0.7$, $\gamma_2/\gamma_1 = 5.0$, $S_1/\gamma_1 = -1.0$, $S_2/\gamma_1 = 0.5$, $S_{12}/\gamma = S_{21}/\gamma_1 = 2.5$, $T_{12}/\gamma_1 = T_{21}/\gamma_1 = 1.125$. (a) Transient behavior for various pumping levels $R = h/h_c$. The solid lines are for $R = 2.6$ and the dotted lines for $R = 3.3$. (b) Line of fixed points and limit cycles. The inset illustrates the Hopf bifurcation. The amplitude of the oscillation in n varies with $(R - R'_c)^{1/2}$.

population n_1 and the behavior of the projections on the $n_1 \times n_2$ plane for various microwave pumping intensities represented by the control parameter $R \equiv h/h_{c_1}$. For a driving field in the range $1.0 < R \leq 2.4$ only the *strong mode* 1 is excited. Initially n_1 has a transient behavior, but then as $\gamma_1 t \gg 1$ it estabilizes at a steady-state value. For $R \geq 2.4$ the *weak mode* 2 is also excited, even though its threshold field is not reached until $R = 5.0/0.7 \simeq 7$. This happens because sufficient driving is supplied to mode 2 via the nonlinear coupling with mode 1. However, in the range $2.4 \leq R \leq 3.17$ the solutions are attracted to a fixed point, so both n_1 and n_2 estabilize after the initial transient. The heavy line in Fig.10(b) shows the path on the $n_1 \times n_2$ plane of the stable fixed points with varying R. Note that at the critical value $R'_c = 3.17$ the fixed point becomes unstable and a limit cycle develops around it. The limit cycle trajectory is described at a frequency $f_0 \sim \gamma_1$, which bears no relation to the microwave pumping frequency. This corresponds to an oscillation of the magnon numbers n_1 and n_2 and results from the dynamic energy shuffling between the two modes, a consequence of the nonlinear nature of the system. This is the source of the auto-oscillation in the microwave absorption of the sample. The abrupt change in topology of the solutions at $R = 3.17$ characterizes a bifurcation. As shown in the the inset of Fig.7.10(b) the amplitude of the oscillation of n_1 grows with $(R - R'_c)^{1/2}$. This is characteristic of a Hopf bifurcation and is in agreement with the experimental result of Fig.7.4(a).

Quite different types of bifurcations leading to self-oscillations may occur in the spin-wave equations with other sets of parameters [47,59]. The type of bifurcation depends on the stability character of the fixed points, determined by the behavior of the eigenvalues crossing the imaginary axis. The bifurcation leading to the auto-oscillation shown in Fig.7.3(b) has been identified as a homoclinic one [47,59].

7.7.2 Period-doubling, strange-attractors and chaos

Both scenarios C and F observed experimentally have been reproduced by the model presented here. Scenario C is obtained with the same parameters used before, at higher pumping intensities. The results are presented in Fig.7.11. In (a)-(d) we show the trajectories in phase-space $n_1 \times n_2$ and the corresponding Fourier spectra of n_1 for several values of the control parameter R. For $R = 6.0$ we have a stable limit cycle characterizing a self-oscillation with frequency $f_0 \simeq \gamma_1$. Period doubling is observed at $R = 7.7587$. A noisy period-2 attractor is shown for $R = 7.759$ and chaos appears for $R = 7.83$. We varied R up to the fifth decimal place in the range 7.50-7.76 and saw no trace of periods 4 or 8, just like the experimental results. By increasing R further in the chaotic region several windows appears, such as the period-4 window at about $R = 7.925$. Another self-oscillation with period-1 develops at $R = 9.5$ and period-2 occurs at 10.95, but contrary to the experimental results this sequence does

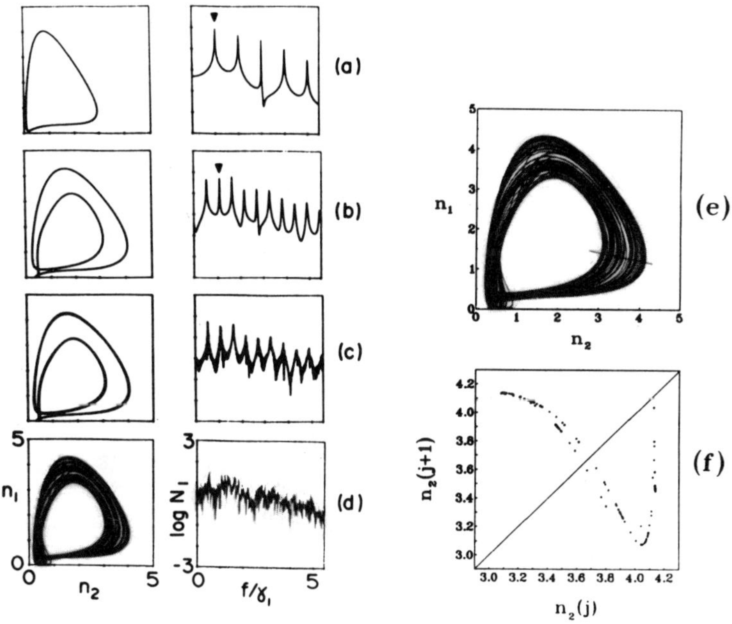

Fig.7.11: Phase-space trajectories, $n_1 \times n_2$ and power spectra, $logN_1$ vs f/γ_1, for two-mode model with the same parameters as in Fig.7.10. (a) period 1 at $R = 6.0$; (b) period 2 at $R = 7.7587$; (c) noisy period 2 attractor at $R = 7.759$; (d) strange attractor at $R = 7.83$; (e) strange attractor for fully developed chaos at $R = 8.0$. (f) shows the Lorenz plot for $R = 8.0$ at the Poincaré section indicated in (e).

not lead to another chaotic state.

Fig.7.11 also shows in (e) the strange attractor in the $n_1 \times n_2$ plane and in (f) the Lorenz plot for fully developed chaos at $R = 8.0$. The Lorenz plot shows the value of the magnon population n_2 at the $(j+1)^{th}$ crossing of the Poincaré section shown in (b) versus the value at the jth crossing. This is in marked contrast to the Lorenz map of strange attractors in Feigenbaum routes to chaos which present a quadratic maximum. Finally Fig.7.12 shows a bifurcation diagram obtained with the values of n_2 at the Poincaré section of Fig.7.11(e) as a function of R. This shows clearly the abrupt change from a period $2T$ oscillation to chaos at $R = 7.759$, just like the experimental observation in scenario C. It also shows the windows of periods $4T$ and $6T$ previously mentioned, as well as inverted Feigenbaum cascades at higher values

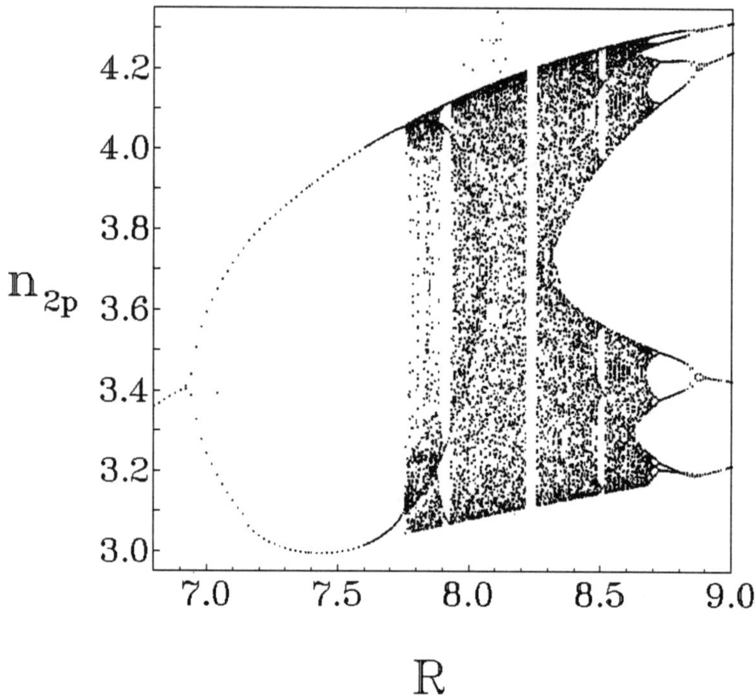

Fig.7.12: Bifurcation diagram $n_{2p} \times R$ at the Poincaré section shown in Fig.7.11(e) obtained with the two-mode model with the parameters as in Fig.7.11.

of R.

The Feigenbaum sequence of period doubling can be obtained from the numerical solution of Eqs.7.27 and 7.28 with many different sets of parameters. In order to demonstrate this we have chosen a set corresponding to two neighboring modes. In this case the parameter α is nonzero since the finiteness of the sample allows the excitation of spin-wave pairs with momenta $+k$, $-(k + \Delta k)$. The results in Fig.7.13 were obtained with the parameters: $\Delta\omega_1 = 0.11$, $\Delta\omega_2 = -0.14$, $\rho_1 = \rho_2$, $\gamma_1 = \gamma_2$, $S_1 = S_2 = 0.8$, $S_{21} = -0.3$, $T_{21} = -0.8$ (all in units of γ_1), $\beta = 0$ and $\alpha = 0.64$. Note that the introduction of the terms in α in Eqs.7.27 and 7.28 sistematically reduces the thresholds for auto-oscillation and for the period doubling bifurcations to the range $R = 1 - 2$, consistent with the experimental observations.

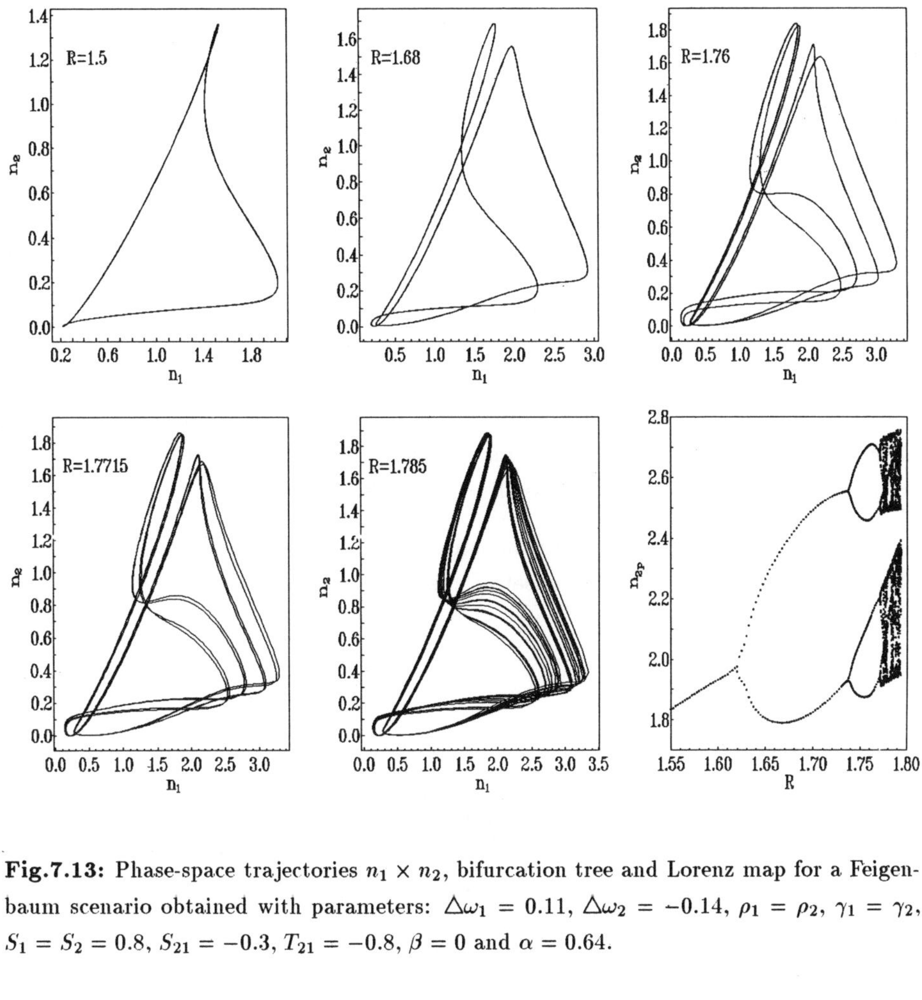

Fig.7.13: Phase-space trajectories $n_1 \times n_2$, bifurcation tree and Lorenz map for a Feigenbaum scenario obtained with parameters: $\triangle\omega_1 = 0.11$, $\triangle\omega_2 = -0.14$, $\rho_1 = \rho_2$, $\gamma_1 = \gamma_2$, $S_1 = S_2 = 0.8$, $S_{21} = -0.3$, $T_{21} = -0.8$, $\beta = 0$ and $\alpha = 0.64$.

7.7.3 Strange attractor dimension

Our further quantitative argument in favor of the two-mode model for the chaotic dynamics of spin waves is provided by the results obtained for the strange attractor dimension. This has been calculated by integrating Eqs.(7.27) and (7.28) numerically over about 10^3 cycles in the *steady state* regime and using the BP method to calculate $D(\gamma)$ with 4000 points ($\sim$ 4 points/cycle) [55]. In this case there is no need to do embedding since one can calculate $< \delta >$ directly using (7.19) in the four-dimensional phase-space. Fig.7.14 shows the variation of the information dimension $D(\gamma = 0)$ with the control parameter R near the onset of chaos, for the same parameters previously used to demonstrate scenario C. As expected, in the range of the period-2 limit cycle, $D(0) \simeq 1$. At $R \simeq 7.76$ the dimension changes abruptly to $D(0) \simeq 1.4$ and then fluctuates with increasing R. This number is quite close to the experimentally measured value at the onset of chaos $1.4 \leq d \leq 1.8$.

The results of this section demonstrate that the auto-oscillations in the pre-chaotic regime as well as at the onset of chaos are accounted for by a model of two interacting spin-wave modes.

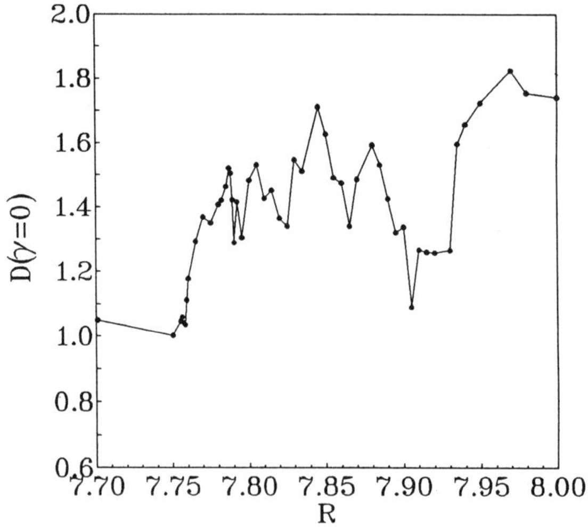

Fig.7.14: Information dimension $D(\gamma = 0)$ versus control parameter $R = h/h_c$ obtained with the two-mode model with the same parameters as in Fig.7.10.

7.8 CONTROLLING CHAOS

It has been proposed by Ott, Grebogi and Yorke [56] that chaos can be controlled in a physical system by means of a small, carefully chosen perturbation in an accessible parameter. This is, in many cases, more feasable or more adequate than the trivial solution of making a large variation in the reverse direction in the system parameter which led to chaos. The feasability of controlling chaos by a small time-dependent perturbation of the biasing magnetic field applied to the sample was recently demonstrated [57]. The method is based on the fact that a chaotic attractor usually has embedded within it an infinite number of unstable periodic orbits. By applying a small time dependent modulation to a conveniently chosen parameter one can stabilize some of these unstable orbits to achieve control of the chaotic state, without having to make large variations in the system parameters. In order to understand the basic principles of the method, consider a dissipative dynamical system described by N nonlinear differential equations of the form $d\vec{x}/dt = F(\vec{x}, \vec{\mu})$, where $\vec{x}$ is the dynamical variable vector and $\vec{\mu}$ represents the various system parameters. When the system is excited by some external source (the microwave radiation in the case of the spin-wave instability), $\vec{x}$ describes a continuous-time orbit which appears as a discrete-times series of points in a conveniently chosen Poincaré surface in the N-dimensional space. If the attractor is chaotic and chaos has been approached by a cascade of period doubling bifurcations, the time interval between two consecutive piercings of the surface is nearly constant and given by $T_0 = f_0^{-1}$, where f_0 is the fundamental frequency. Assume now that some system parameter $p = \mu_i$ can be modulated periodically about its mean value, with period $T_1 \simeq nT_0/m$, where n and m are integers. The periodic change in p can be tailored in such a manner that the unstable orbit is synchronously pushed into describing the same path as in the previous cycles during the time interval mT_1, thus resulting in a controlled trajectory. Ott, Grebogi and Yorke have given a prescription for finding the amplitude p^* of the parameter variation necessary to achieve control of chaos. Their approach relies on the knowledge of the system equations or the time series of an experimental variable. Here we determine p^* experimentally by observing the behavior of the chaotic system response with increasing parameter modulation.

The experiments to control spin-wave chaos were made in the subsidiary resonance configuration with the $1mm$ YIG sphere aligned along the [110] direction. We used the same arrangement of Fig.7.2 with the microwave magnetic field $\vec{h}$ perpendicular to the biasing field $\vec{H}$. However, we have added inside the cavity a loop of diameter $1.5cm$ made with a $0.5mm$ cooper wire to allow the modulation of the sample biasing field $H = H_0 + \delta H cos(2\pi f_1 t)$ over a broad frequency range (0-10)MHz, typically with $\delta H/H_0 \sim 10^{-4}$. By applying a small modulation to the magnetic field H with appropriate wave-shape, frequency and amplitude we have been able to control the chaotic states in most regions of Fig.7.3(b). The field modulation results in a corresponding modulation in the spin-wave frequency ω_k, and so in the detuning parameter $\Delta\omega_k$ of Eqs.(7.27) and (7.28), thus providing a handle to control the orbits. Figures 7.15-7.17 show the results obtained with a sinusoidal field modulation applied after the system is driven to a fully chaotic regime with $h = 1.1Oe$ and $H_0 = 1750Oe$, represented by the point A in Fig.7.3(b). Fig.7.15 shows the evolution of the power spectra, obtained from the digitally stored signals, with incresing field modulation δH at a frequency $f_1 = 1480kHz$. For $\delta H \leq 200mOe$ the spectra display broadband features characteristic of chaos as shown in Fig.7.15(a) for $\delta H = 0$. As δH increases

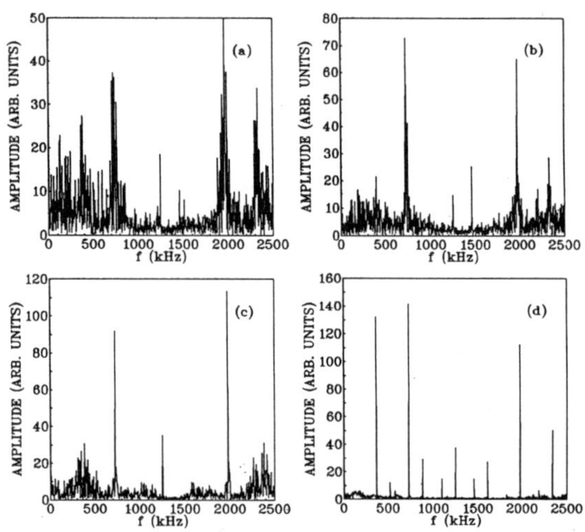

Fig.7.15: Evolution of the power spectra of the auto-oscillation observed at point $h = 1.10Oe$ and $H_0 = 1750Oe$, with increasing amplitude δH of the field modulation with frequency $f_1 = 1480kHz$. The spectra in (a), (b), (c) and (d) were obtained with $\delta H = 0, 140, 287$ and $435mOe$, respectively. The spectrum (d) corresponds to a periodic signal with fundamental frequency $f_0 = 740kHz$ and a subharmonic component at $f_0/2$ superimposed to another frequency of $1975kHz$.

above this value the spectrum becomes progressively cleaner, with sharp lines characteristic of a periodic signal. This is shown in Fig.7.15(b), (c) and (d) for $\delta H = 140$, 287 and $435mOe$ respectively. The spectrum in (d) corresponds to a periodic signal with fundamental frequency $f_0 = 740kHz$ and a subharmonic component at $f_0/2$, characteristic of a period doubled oscillation, superimposed to another signal with frequency $1975kHz$.

The control of chaos is further demonstrated by the variation of the information dimension D_1 and the metric entropy K of the attractor with increasing field modulation, measured as described in Section 7.5. Fig.7.16 shows that for $\delta H < 200mOe$, $D_1 \simeq 3.5$ and $K \simeq 0.20$, characterizing a fully chaotic regime. Strikingly, for a modulation amplitude above the critical value $\delta H^* \simeq 200mOe$, D_1 and K decrease with increasing δH towards the values $D_1 = 1$ and $K = 0$, characteristic of a periodic trajectory. This results confirms that the control of chaos is achieved by a small periodic modulation of the applied field. In regard to the data of Figs.7.15 and 7.16 we note that with the application of the field modulation the auto-oscillation enters a steady-state regime only after a time τ has ellapsed. The spectra of Fig.7.15 and the time-series used to obtain D_1 and K were calculated from the steady-state portion of signal. The delay time τ decreases rapidly with increasing amplitude modulation δH, in qualitative agreement with the results of Ott et al [56]. However we could not obtain a critical exponent relating τ and the increment amplitude $\varepsilon = \delta H - \delta H^*$

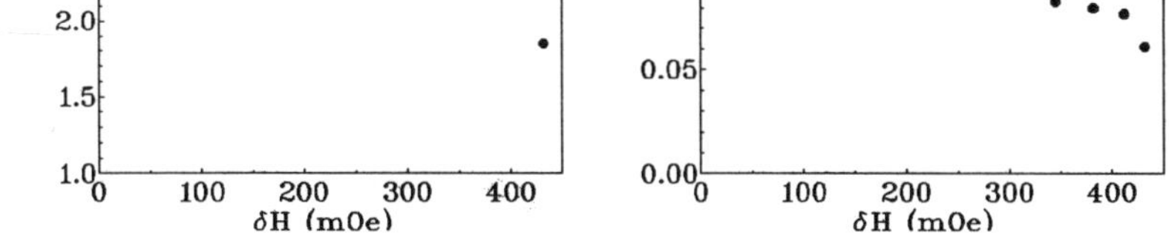

Fig.7.16: Information dimension $D(0)$ (a) and metric entropy $K(0)$ (b) vs field modulation amplitude δH. These quantities have been obtained with the embedding technique from the time delayed digitized signals. Note that for $\delta H > 200mOe$ the values of $D(0)$ and $K(0)$ decrease towards the values $D(0) = 1$ and $K(0) = 0$ appropriate for periodic signals.

because of the difficulty in measuring τ and δH^* with adequate precision. Finally in Fig.7.17 we show the variation of the critical amplitude δH^* necessary to control chaos with the modulation frequency f_1, measured at point A in Fig.7.3(b). The measurements were done by scanning f_1 at constant δH and observing the change from chaotic to periodic behavior directly at the oscilloscope. Notice that δH^* has minima at values commensurate with the fundamental frequency f_0, i.e. $f_1 = mf_0/n$, increasing rapidly as f_1 departs from these values. Hence, the boundaries between the chaotic and controlled regions have appearance of "tongues", similar to those observed in phase-locking phenomena [58]. This is not surprising since we expect the chaotic motion to be suppresssed when the unstable orbit becomes phase-locked to the external parameter modulation. Although phase-locking phenomena has been widely studied, it seems that this is the first demonstration that it can be used

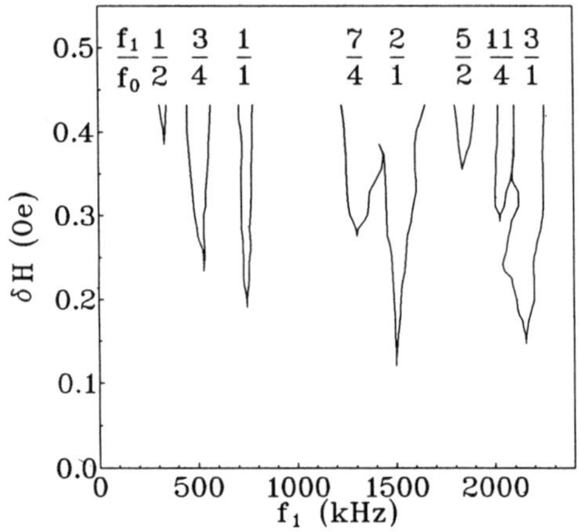

Fig.7.17: Critical modulation amplitude δH^* vs modulation frequency f_1. The boundaries between the chaotic and controlled regions have minima at values of f_1 commensurate with the fundamental frequency f_0 at ratios indicated at the top.

to control chaos. Finally we note that it has been possible to suppress the chaotic behavior of spin-wave instabilities over a wide region in the phase diagram of Fig.7.3(b) using a small modulation in the biasing magnetic field, not only with a sinusoidal waveshape, but also with a square wave or pulse modulation. In some points of the h_c *vs* H_0 diagram only a specific wave shape results in a controlled signal, whereas in others they are all equally effective. In each case, however the resulting periodic signal has a different waveform, corresponding to a different trajectory. This confirms that by using a carefully chosen small parameter perturbation, it is possible to control a chaotic trajectory and create a variety of attracting periodic orbits, as proposed by Ott, Grebogi and Yorke [56].

7.9 CONCLUDING REMARKS

We have presented a review of experimental and theoretical investigation of nonlinear dynamic phenomena in spin waves in bulk YIG samples pumped by microwave fields. All of the nonlinear effects observed in other physical systems are present here. A unique feature of this system is that the nonlinear coupling parameters can be calculated from well-known microscopic interactions, providing a theoretical framework to study the dynamics. However, since many spin wave modes are pumped above the SI threshold, the problem of identifying the modes involved in the dynamics has proved to be a difficult one. The work in recent years has led to the following understanding of the origin of the auto-oscillations.

As the microwave pumping field exceeds the threshold for parametric spin-wave excitation h_c, many modes are pumped with population (10^{17}) much higher than the thermal values ($\sim 10^3$). If the field is not much higher than h_c, these modes are degenerate with frequency $\omega_k \simeq \omega_p/2$ and propagate at an angle close to that of the SI threshold, $\theta_k = \pi/2$ in parallel pumping and $\theta_k = \pi/4$ in subsidiary resonance for fields $H_0 < H_c$. Initially all the modes are at steady state values, corresponding to fixed point solutions of the spin-wave equations. However, at such high populations, the effective microwave field that each mode exerts on the others due to the magnon-magnon interaction, is comparable in intensity with the direct microwave pumping of those modes, which tends to create a dynamic competition between them. This leads to a new instability, and a bifurcation occurs when the pumping field exceeds a second threshold h_c'. Above h_c' the modes involved in the dynamics are not individual k-magnon modes, but rather collective spin waves which grow together coupled by the nonlinear interaction [36].

Not all of the nearly degenerate modes with $\omega_k \simeq \omega_p/2$, or even those pumped above threshold, become involved in the dynamics of the self-oscillation. This is so because the spin-wave interaction parameters vary considerably in magnitude and sign in the spin-wave manifold, especially with the propagation angle θ_k. Thus, only the modes in some small regions in k-space will have the nonlinear parameters appropriate for the lowest auto-oscillation threshold. As we showed in Section 7.7, the threshold for auto-oscillation is reduced considerably when one considers the pumping of spin-wave pairs with opposite wave vectors differing by π/d, in addition to the usual pumping of pairs with equal wave vectors.

As the pumping field is increased further, new bifurcations may occur, leading to period doubling in the self-oscillation and eventually to chaos. Although there is no theory available for this regime, the measurement of the attractor dimension at the onset of chaos and the agreement of the two-mode calculations with experiments indicate that only two collective modes are involved even at high pumping levels. Of course, as the pumping field increases in the chaotic region, other modes may develop, increasing the number of degrees of freedom [29,52,59].

In closing we mention some directions of the current research. The understanding of the detailed nature of the auto-oscillations has been attracting attention both from theoreticians and experimentalists. On the theoretical side powerful mathematical tools to treat nonlinear dynamical systems are being applied to understand the mode reduction. Experimentally more systematic investigations are being conducted, both in bulk samples and thin films. Another area of interest is the characterization of the strange attractors by quantities such as the fractal dimension Kolmogorov entropy, and multifractal singularity spectra and Lyapunov exponents. Finally, one area of growing interest is high dimensional chaos and turbulence. These examples indicate that research in nonlinear spin-wave phenomena is likely to stay vigorous for years to come.

The authors acknowledge helpful discussions with Prof. Jair Koiller and Dr. A. Cascon. This work has been supported by FINEP, CNPq, CAPES and FACEPE.

REFERENCES

1. See for example: P. Berge, Y. Pomeau and C. Vidal, *Order within Chaos* (John Wiley, New York; Herman, Paris, 1984); H.G. Schuster, *Deterministic Chaos* (Physik-Verlag, Weinheim, 1984).
2. R.W. Damon, Rev. Mod. Phys. **25**, 239 (1953).
3. N. Bloembergen and S. Wang, Phys. Rev. **93**, 72 (1954).
4. H. Suhl, J. Phys. Chem. Solids **1**, 209 (1957).
5. F.R. Morgenthaler, J. Appl. Phys. **31**, 95S (1960).
6. E. Schlomann, J.J. Green, and U. Milano, J. Appl. Phys. **31**, 386 S (1960).
7. T.S. Hartwick, E.R. Peressini, and M.T. Weiss, J. Appl. Phys. **32**, 223 S (1961).
8. S. Wang, G. Thomas and Ta-Lin Hsu, J. Appl. Phys. **39**, 2719 (1968).
9. G. Thomas and G. Komoriya, J. Appl. Phys. **46**, 883 (1974).
10. V.S. L'vov, S.L. Musher, and S.S. Starobinets, Zh. Eksp. Teor. Fiz. **64**, 1074 (1973) [Sov. Phys-JETP **37**, 546 (1973)]; V.E. Zakharov, V.S. L'vov, and S.S. Starobinets, Usp. Fiz. Naul **114**, 609 (1974) [Sov. Phys. Usp. **17**, 896 (1975)]; V.I. Ozhogin and A. Yu. Yakubovskii, Z. Eksp. Teor. Fiz. **67**, 287 (1974) [Sov. Phys. JET P **40**, 144 (1975)].
11. For a review, see, e.g. J. P. Eckmann, Rev. Mod. Phys. **53**, 643 (1981).
12. See for example, *Laser Physics*, ed. J.D. Harvey and D.F. Walls (Springer-Verlag 1983).
13. See for example, *Statistical Physics and Chaos in Fusion Plamas*, ed. C.P.W. Horton Jr. and L.E. Reichl (John Wiley, New York, 1984).
14. T. Braun, J.A. Lisboa, R.E. Francke, and J.A.C. Gallas, Phys. Rev. Lett. **59**, 613 (1987).
15. C.D. Jeffries, Phys. Scr. (Sweden), **T9**, 11 (1985).

16. K. Nakamura, S. Ohta and K. Kawasaki, J. Phys. **C15**, L143 (1982); S. Ohta and K. Nakamura, J. Phys. **C16**, L605 (1983).

17. G. Gibson and C. Jeffries, Phys. Rev. A **19**, 811 (1984).

18. S.M. Rezende, F.M. de Aguiar and O.F. de Alcantara Bonfim, J. Mag. Mag. Materials **54-57**, 1127 (1986).

19. F.M. de Aguiar and S.M. Rezende, Phys. Rev. Lett. **56**, 1070 (1986).

20. M. Mino and H. Yamazaki, J. Phys. Soc. Jpn. **55**, 4168 (1986).

21. P. Bryant, C. Jeffries, and K. Nakamura, Phys. Rev. Lett. **60**, 1185 (1988); Phys. Rev. **A38**, 4223 (1988).

22. S.M. Rezende, F.M. de Aguiar, and A. Azevedo, in *Magnetic Excitations and Fluctuations II*, edited by U. Balucani, S.W. Lovesey, M.G. Rasetti and V. Tognetti (Springer-Verlag, Berlin 1987), p.79; Phys. Rev. **B39**, 9448 (1989).

23. T.L. Carrol, L.M. Pecora, and F.L. Rachford, Phys. Rev. Lett. **59**, 2891 (1987).

24. H. Benner, F. Rodelsperger, H. Seitz, and G. Wiese, J. Physique **C8**, 1603 (1988).

25. R. Boscaino and R.N. Mantegna, J. Physique **C8**, 2033 (1988).

26. H. Yamazaki, J. Phys. Soc. Japan **53**, 1155 (1984).

27. F. Waldner, D.R. Barberis, and H. Yamazaki, Phys. Rev. A **31**, 420 (1985).

28. A.I. Smirnov, Zh. Eksp. Teor. Fiz. **90**, 385 (1986) [Sov. Phys.-JETP **63**, 222 (1986)].

29. H. Yamazaki, M. Mino, H. Nagashima, and M. Warden, J. Phys. Soc. Jpn. **56**, 742 (1987).

30. P.E. Wigen, H. Doetsch, Y. Ming, L. Baselgia, and F. Waldner, J. Appl. Phys. **63**, 4157 (1988).

31. R.D. Mc Michael and P.E. Wigen, Phys. Rev. Letters **64**, 64 (1990).

32. G.M. Dudko and A.N. Slavin, Fiz Tverd, Tela **31**, 114 (1989).

33. X.Y. Zhang and H. Suhl, Phys. Rev. **A32**, 2530 (1985).

34. S.M. Rezende, O.F. de Alcantara Bonfim, and F.M. de Aguiar, Phys. Rev. B **33**, 5153 (1986).

35. S.M. Rezende and F.M. de Aguiar, Rev. Bras. Fis. **16**, 324 (1986); Proc. IEEE **78**, 893 (1990); Physica **A163**, 232 (1990).

36. H. Suhl and X.Y. Zhang, Phys. Rev. Lett. **57**, 1480 (1986); Phys. Rev. **B38**, 4893 (1988).

37. T.L. Gill and W.W. Zachary, J. Appl. Phys. **61**, 4130 (1987).

38. S.P. Lim and D.L. Huber, Phys. Rev. **B37**, 5426 (1988).

39. F. Waldner, J. Phys. **C 21**, 1243 (1988).

40. F. Keffer, *Spin Waves*, in Handbuch der Physik, Vol.18b, ed. by S. Flugge (Springer, Berlin 1966).

41. M. Sparks, *Ferromagnetic Relaxation Theory*, Mc Graw-Hill (New York 1964).

42. R.M. White, *Quantum Theory of Magnetism*, Springer-Verlag (New York 1983).

43. A.I. Akhiezer, V.G. Baryakhtar, and S.V. Peletminskii, *Spin Waves*, (North Holland, Amsterdam, 1968).

44. R.M. White and M. Sparks, Phys. Rev. **130**, 632 (1963).

45. C.E. Patton and W. Jantz, J. Appl. Phys. **50**, 7082 (1979).

46. See J. Guckenheimer and P. Holmes, *Nonlinear Oscillations, Dynamical Systems, and Bifurcations of Vector Fields* (Springer-Verlag, New York, 1983).

47. S.M. Rezende, A. Azevedo, J. Koiller, and A. Cascon, J. Appl. Phys. **69**, 5430 (1991).

48. M.J. Feigenbaum, J. Statist. Phys. **19**, 25 (1978).

49. A. Arneodo, P. Coullet, C. Tresser, A. Libchaber, J. Maurer, and D. d'Humières, Physica **6D**, 385 (1983).

50. D. Ruelle and F. Takens, Commun. Math. Phys. **20**, 167 (1971).

51. P. Grassberger and I. Proccacia, Phys. Rev. Lett. **50**, 346 (1983) and Physica **9D**, 189 (1983).

52. T.L. Carrol, L.M. Pecora, and F.J. Rachford, J. Appl. Phys. **69**, 5630 (1991).

53. M. Mino, H. Yamazaki, and K. Nakamura, Phys. Rev. **B40**, 5279 (1989).

54. R. Badii and A. Politi, Phys. Rev. Lett. **52**, 1661 (1984) and J. Stat. Phys. **40**, 725 (1985).

55. R. Badii and G. Broggi, Physik-Institut der Universitat Zurich Software Report Nº6, May 1987 (unpublished).

56. E. Ott, C. Gregori, and J.A. Yorke, Phys. Rev. Lett. **64**, 1196 (1990).

57. Antonio Azevedo and Sergio M. Rezende, Phys. Rev. Lett. **66**, 1342 (1991).

58. M. Warden and F. Waldner, J. Appl. Phys. **64**, 5386 (1988).

59. A. Cascon, J. Koiller, and S.M. Rezende, Physica D, to appear.

CHAPTER 8

PARAMETRIC INSTABILITIES AND AUTO-OSCILLATION BEHAVIOR OF MAGNETOSTATIC MODES IN MAGNETIC FILMS

P.E. WIGEN

Department of Physics, Ohio State University,
Columbus, Ohio 43210-1106, USA

Contents

8.1. INTRODUCTION

Magnetism has contributed significantly to our understanding of microscopic and macroscopic interactions in nature. Magnetic excitations, such as spin-waves and magneto-static waves [1-4], are active in a variety of devices involving microwave applications. These excitations produce high power limitations due to their nonlinear effects [5-8] and the understanding of these nonlinearities is important for both fundamental and applied reasons. Magnetic resonance has become a phenomenon which bridges the gap between linear and nonlinear dynamics, providing a nonlinear system simple enough that many contributions are easily controlled and readily understood, yet complex enough to exhibit interesting dynamics. Many recent experimental [9-30] and theoretical [31-36] studies in magnetic systems show these materials to be particularly valuable systems for non-linear investigations and modeling.

For many nonlinear systems, the equations of motion and the nature of the interactions that give rise to the instabilities are very complicated and often critical approximations are made or serious truncation procedures are invoked. It has been the goal of investigators to consider "simpler" systems and magnetic materials are particularly promising candidates.

In magnetic materials the non-linear responses occur in the magnetic resonance of a homogeneous bulk media for which the microscopic Hamiltonian, which describes the magnetic interactions, is available to generate the equations of motion. These equations include the nonlinear coupling to all the degenerate spin-wave modes in the spin-wave band from which it is possible to predict a threshold pumping power for the excitation of auto-oscillations and bifurcation routes to chaos as a function of the *rf* drive field. In the experiments, the auto-oscillations were observed to have a frequency in the range of 10^4 - 10^7 Hz and as the pumping power is increased, period doubling, quasiperiodicity and chaos have been observed by many investigators in a number of magnetic materials.

Exciting and important experimental results have been obtained in antiferromagnetic [9-13] and bulk ferromagnetic materials [14-27] and they have stimulated considerable activity in the theoretical models [31-36]. However, the models of bulk materials give only qualitative agreement with the very beautiful and precise data because the driven modes are complicated by their coupling to all of the 10^{10} degenerate spin-wave modes. In the analysis, the coupling is normally truncated to a limited number of these degenerate modes and as a result the agreement between the model and the experimental results has been only qualitative. It is desirable to investigate a system in which the number of modes are limited and can be varied allowing the study of a system where the truncation process would not be a severe limitation.

A thin magnetic film at perpendicular resonance is just such a system [28-30]. The magneto-static modes excited by the *rf* field lie at the bottom of the spin-wave band where there are no degenerate short wavelength spin-waves. In addition, by changing the geometric shape and size of the film (diameter and thickness) the positions and intensities of the various magnetostatic modes can be modified. The variation of the periodic oscillations (auto-oscillations in the output voltage at the detector of the spectrometer) and the routes to chaos can be studied systematically.

A new model based on the Hamiltonian associated with the nondegenerate magnetostatic modes has been developed [28-30]. The results predict an initial onset of auto-oscillation within a dB of the power observed in the experiment and the auto-oscillation frequencies are within 10% to 15% of those observed.

8.2. BRIEF REVIEW OF BULK MAGNETIC SYSTEMS

8.2.1. Experimental Results

8.2.1.1. Review of Magnetic Resonance

A detailed derivation of the dispersion relation for the spin-waves is given in reference [1]. If the applied magnetic field is along a crystal direction of high symmetry, the dispersion relation in bulk materials for spin-waves whose wavelength is short in comparison to sample dimensions, but long in comparison to the unit cell dimension, is given by

$$\omega_k^2/\gamma^2 = (H_o - 4\pi N_z M_z + H_A + Dk^2) X$$

$$(H_o - 4\pi N_z M_z + H_A + Dk^2 + 4\pi M_z \sin^2\theta_k), \qquad 8.1$$

where ω_k is the classical precession frequency for a spin-wave of wave vector k, N_z and M_z are the demagnetization factor and the magnetization respectively in the (z) direction of the applied field H_o, γ is the gyromagnetic ratio, H_A is the effective magnetic anisotropy field, Dk^2 is the effect of exchange interactions between the spins, and θ_k is the angle a spin-wave is propagating with respect to the magnetization direction. For pure YIG at room temperature, $4\pi M = 1760$ Oe, $H_A \simeq 60$ Oe and $D = 5.4 \times 10^{-9}$ Oecm^2. For $g = 2$, $\gamma = 2.8$ GHz/kOe.

In the earliest experiments, nonlinear spin-wave dynamics were observed under a variety of excitation conditions [3-8]. Because a uniform driving field will couple only to the very long wavelength excitations, short wavelength spin-waves must be generated parametrically, either through interactions with the uniform precession under perpendicular pumping, as in the 1st and 2nd order Suhl instabilities [1], or through the coupling of an elliptical precession to an alternating field directed along the z-axis in parallel pumping [7].

These geometrical arrangements of the pumping field are shown in Fig. 8.1. In the normal configuration, called perpendicular pumping (see Fig. 8.1a), the pumping field is perpendicular to the external field. For low powers a linear response of the system is observed but with increasing power a critical drive field is observed where two $k = 0$ spin-waves are annihilated and a Suhl pair of $\pm k \neq 0$ spin-waves are excited through a four magnon interaction.

Another perpendicular pumping resonance condition shown in Fig. 8.1b occurs when the pumping frequency is at twice the spin-wave precession frequency. The *rf* field will couple to the $k = 0$ mode far from resonance. At some critical value of the rf field strength a single $k = 0$ spin-wave will be annihilated and again a Suhl pair of $\pm k \neq 0$ spin-waves will be parametrically excited in a three magnon process known as subsidiary resonance [5].

A third condition involves a parallel pumping configuration shown in Fig. 8.1c. In this geometry, a critical value in the rf field will parametrically excite a Suhl pair of $\pm k \neq 0$ spin-waves at half the drive frequency [7].

Once excited by any of these three processes, spin-wave pairs interact through four magnon terms resulting in auto-oscillations and routes to chaos at higher pumping powers. In each of these geometries, the driven modes are degenerate with large numbers of additional spin-wave modes and the dynamical interactions are difficult to model and evaluate in detail. Nevertheless, in the last decade there has been considerable interest in the nonlinear dynamics of parametrically pumped spin-waves.

8.2.1.2. Antiferromagnetic Materials

The earliest attempts to analyze the non-linear response of magnetic materials in terms of a non-linear chaotic theory were made by Yamazaki for parallel pumped magnons in the antiferromagnet, $CuCl_2.2H_2O$ [9]. This paper contains the first report of the observation of an "auto-oscillation" in the

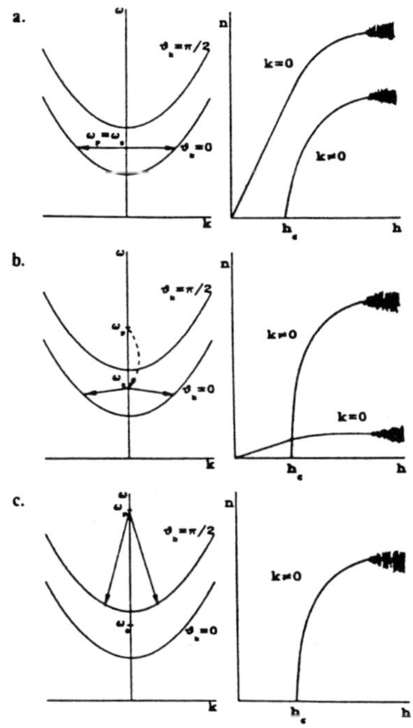

Figure 8.1. Various geometrical arrangements of the pumping field for ferromagnetic resonance. The spin-wave band on the left indicates the coupling of the driving fields at ω to the various Suhl pair $\pm$ k spin-wave modes. θ_k is the angle between the direction of the spin-wave propagation and that of the magnetization. The graph on the right shows the population densities of the various magnons as a function of the drive field, h. a) Second order Suhl instability for perpendicular pumping; b) first order Suhl instability for perpendicular pumping; c) parallel pumping.

absorption signal with increasing power as well as subharmonics attributed to a period doubling type bifurcation. This work was followed by that of Waldner and his colleagues [10, 11] on the antiferromagnetic material $(CH_2NH_3)_2CuCl_4$. They also used the parallel pumping geometry but found a transition to chaos by "irregular periods." The next study in the antiferromagnetic $CsMnF_3$ was done by Smirnov [12, 13]. As the absorbed power was increased, the density of spin-waves increased at the center of the sample. This increase resulted in a shift of the magnon dispersion relation and at sufficient power, a periodic redistribution of the density of the spin-waves was observed.

8.2.1.3. Ferrimagnetic Materials

The initial investigations into the route to chaos in ferrimagnets were obtained from spheres of gallium doped yttrium iron garnet (YIG) investigated at the main resonance by Gibson and Jeffries [14]. The second order Suhl instability was observed and at higher powers the details of the onset of auto-oscillations in the magnetization were investigated. The observation of a subharmonic route to chaos for parallel pumped spin-wave was reported by Rezende, et al. [15-18].

Details of auto-oscillation and chaos in a YIG sphere at subsidiary resonance were initially reported in a series of papers by Bryant and Jeffries [19-21]. In a quite complete study they reported regions and boundaries of coupled behavior of the magnetization and a wide variety of phenomena observed in their work includes:

1) fine structure - the excitation of single spin-wave modes;
2) dynamics in the fine structure regime involving the interaction of a small number of spin-wave modes;
3) low frequency relaxation oscillations (about 1 MHz);
4) periodic spiking;
5) high amplitude collective oscillations;
6) abrupt emergence of high frequency noise;
7) hysteresis at the Suhl threshold in which the system jumps from a quiescent to a turbulent state;
8) periodic doubling routes to chaos;
9) systematic increase in auto-oscillation frequency with pumping power.

While most experiments at the main resonance involve the second order process, experiments were performed by Carroll, et al. [22-25] investigating first order Suhl instabilities by using applied fields and pumping frequencies low enough that the spin-wave manifold includes spin-waves with frequencies equal to half the pumping frequency. This greatly reduces the pumping power required to observe the nonlinearities.

Yamazaki, et al. [26, 27], have investigated the response of the magnetization vector in the chaotic regime in terms of strange attractors using ferromagnetic platelets of $(CH_2NH_3)_2CuCl_4$ under the parallel pumping configuration. The chaotic behavior was shown to be deterministic through the use of return maps and the signal was analyzed in terms of the fractal dimension and an estimate of the Kolmogorov entropy of the strange attractor was obtained. The largest Lyapunov exponent was found by iterating a one-dimensional function fit to a return map.

From these experiments, it can be concluded that magnetic systems have a number of advantages as a model system for investigating nonlinear dynamics. The range of auto-oscillation frequencies observed is well suited for the use of electronic instruments and computers in the detection, storage and analysis of data. In addition, many material parameters can be systematically adjusted. It is easy to control and maintain the sample temperature; it is possible to vary the drive frequency over 1 to 2 orders of magnitude; with a variation of the orientation of the sample in the magnetic field or the chemical composition the influence of various anisotropy energies can be investigated. The geometry of the sample is another important variable as the demagnetization factor influences the distribution and position of the spin-wave band with respect to the resonance modes. Thus, it is possible to investigate the chaotic properties in magnetic systems under a wide variety of conditions that can be readily controlled in the experiment.

8.2.2. Theoretical Models

While the experiments in bulk materials have produced a number of beautiful results, the nonlinear dynamics of parametrically excited spin-waves are governed by several intrinsic factors and the theoretical analysis is quite complicated due to the large number of degenerate, on the order of 10^{10}, spin-waves associated with the spin-wave band. In bulk materials, there are a multitude of spin-wave modes which have the same frequency, and while some spin-waves have lower excitation thresholds than others, at powers above the excitation threshold, where auto-oscillations are observed, many modes can be involved in the dynamics of the system. A further complication is that long wavelength excitations must obey boundary conditions at the surfaces of the sample, and the plane wave form of spin-waves do not conform to the boundary conditions of a sphere, which, because of its uniform shape demagnetization fields, is the most commonly used bulk sample geometry.

The initial attempt to model the behavior of the magnetization vector in the regime beyond the Suhl threshold [1] was made by L'Vov and Zakharov [31, 32]. Their theoretical model, known as the S-theory, is based on a microscopic Hamiltonian for the system and limits the coupling to pairs of degenerate ± k Suhl spin-waves. This model forms the foundations of the present theories used to explain the auto-oscillatory response in bulk magnetic systems [31-36]. Their approach is reviewed briefly below.

Using the Holstein and Primakoff representation [37] the Hamiltonian for noninteracting spin-waves has the form

$$\boldsymbol{H}^{(2)} = \sum_k \omega_k b_k b_k^* \,, \qquad 8.2$$

where

$$\omega_k^2/\gamma^2 = [H_o - 4\pi M N_z + Dk^2 + 4\pi M \sin^2\theta_k]^2 - \frac{4\pi M^2 \sin^4\theta_k}{4}. \qquad 8.3$$

The variables b_k can be interpreted as classical variables which satisfy Hamilton's equations

$$\frac{db_k}{dt} = -i\frac{\partial \boldsymbol{H}}{\partial b_k^*} = -i\omega_k b_k \qquad 8.4$$

and in the linear (low power) regime the information essential to the establishment of the normal modes of the spin-wave system is available, i.e., all the terms in the Hamiltonian are known.

To investigate the dynamics of the magnetization vector in the nonlinear regime, the coupling of the excited modes to the nearly degenerate Suhl pairs of ± k spin-waves are included in the theory by appropriate three magnon and four magnon interaction terms in the Hamiltonian.

Using a random phase approximation the appropriate form of the three magnon interaction Hamiltonian important in the first order Suhl process (subsidiary resonance) indicated in Fig. 8.1b is

$$\mathbf{H}^{(3)} = \frac{1}{2}\sum_{k} (A_{0k} b_0^* \; b_k \; b_k + c.c.)\,, \qquad 8.5$$

where A_{0k} is the coupling terms between the k = 0 mode and the Suhl pairs of ± k spin-waves.

The four magnon interaction is responsible for limiting the growth of the parametrically driven spin-waves and for interactions between spin-waves. Using the random phase approximation to decouple the magnons, Zakarov, et al. [31, 32] constructed a four magnon interaction Hamiltonian for Suhl pair spin-waves that consists of two parts:

$$\mathbf{H}_{int}^{(4)} = \sum_{k,k'} T_{k,k'} b_k b_k^* b_{k'} b_{k'}^* + \left| \frac{1}{2} \sum_{k,k'} S_{k,k'} b_k b_{-k} b_{k'}^* b_{-k'}^* + c.c \right|. \qquad 8.6$$

The $T_{kk'}$ term has the effect of renormalizing the frequency while the $S_{k,k'}$ term involves the annihilation of one Suhl pair of ± k spin-waves and the creation of another Suhl pair of ± k' spin-waves.

In the S-theory, when a pumping field is applied above the threshold values, an unstable growth of Suhl pairs of spin-waves is predicted. As the population density increases, energy is transferred between the excited mode and all the degenerate spin-wave modes in the spin-wave band via the four magnon interaction. The evaluation of the equations of motion for all these modes is much too complicated to evaluate. To model the system on a computer, the system must be truncated to a limited number of modes, and a choice must be made about which modes to include.

Nakamura, et al. [33-35], was the first to truncate the system to investigate the behavior when the excited mode is coupled to only one or two sets of the degenerate Suhl spin-waves. By choosing a limited number of Suhl pairs, the behavior of the system can be selectively determined and the results compared with the experimental data.

Computer models of spin-wave dynamics have produced results that are qualitatively similar to experimental results, but the modeled auto-oscillations have frequencies that are orders of magnitude higher than the experimental frequencies [12, 36]. A detailed review of these results are given by L'Vov and Prozorova in reference 38.

As indicated, truncating the system to a few modes in the spin-wave manifold having about 10^{10} degenerate spin-waves is a severe limitation to the analysis and a major weakness in the analytical process for bulk systems. It has been argued that an entire manifold of modes pumped above threshold should be considered. In an effort to extend these models, Lim and Huber [39, 40] have applied the S-theory formalism to a simulation of a finite spin system with coupling to up to one hundred modes.

In another approach X.Y. Zhang and H. Suhl [41-43] have taken into account the excitation of the entire degenerate spin-wave manifold and employed center-manifold techniques to explain the empirical result that the dynamics of the system is low-dimensional. This elegant approach provides a qualitative understanding of the low-dimensional attractor.

8.2.3. Strange Attractors

When a seemingly chaotic signal is observed either from experiments or from numerical integrations of the equation of motion, one is faced with the question of how to characterize the signal. Is the signal chaotic and "how" chaotic is it? The "degree of chaos" can be described in terms of a number of quantitative measures including fractal dimensions, Lyapunov exponents and Kolmogorov entropy [26, 27, 44-46]. In using this approach an n-dimensional phase space is constructed in which the attractor can exist. Ideally, this space will have axes corresponding to the amplitude and phase of each of the modes involved in the nonlinear behavior. Usually, only the time dependent signal V(t) is available. To embed this signal in a multidimensional phase space of dimension n, the n-vector components are created by using the method of time delays where $x_1(t)=V(t)$, $x_2(t)=V(t+\tau)$, ... $x_n(t)=V(t+(n-1)\tau)$.

Strange attractors [47-48], while embedded in a phase space of integer dimension, have fractal characteristics which are characterized by a non-integer fractal dimension. The fractal dimension can be calculated by using a method outlined by P. Grassberger and I. Procaccia [49] or Y. Termonia and Z. Alexandrowicz [50]. This procedure was used by Yamazaki, et al. [26, 27] to determine the fractal dimension of chaotic signals in the parallel pumped ferromagnetic $(CH_3NH_3)_2CuCl_4$ as a function of pumping power.

Another quantitative measure of a strange attractor is the average rate of the exponential divergence of nearby trajectories within the attractor [51]. These exponents are called the Lyapunov exponents.

The Kolmogorov entropy, K, is another measure of the degree of chaos. For analytically defined models, K can be estimated from the Lyapunov exponents, but it is very difficult to determine K directly from a measured time signal. Grassberger and Procaccia have developed a technique for obtaining the Renyi entropies which are related to the Kolmogorov entropy. Of all the terms in the series, the term K_2 is chosen due to its ease of calculation from a time series [52].

Another aspect of chaos that can be investigated in the magnetic materials is "crisis" or "critical" instability that occurs when two attractors "touch" [53, 54]. In the garnet film geometry, each magnetostatic mode has its own basin of attraction. As the power is increased these basins will approach each other, touch and cross over. On such occasions an instability or intermittency would be expected [53]. Such effects have been observed in coupled lattice maps [54]. These thin film geometries are a system where the regions of the attractors can be controlled and these effects can be evaluated in detail.

8.3. RESONANCE IN THIN FILM DISCS

8.3.1. Magneto-Exchange Modes

As indicated in the previous Section, the S-theory has a critical weakness in that the uniform precession frequency lies well within the spin-wave manifold at $\omega/\gamma = H_o$ and is degenerate with a large number of Suhl pairs of spin-waves and the truncation process to a few modes is a severe limitation to the analytical process. In the normally magnetized slab the uniform precession frequency lies at the bottom of the spin-wave manifold given by $\omega/\gamma = H_o - 4\pi M_z$. In this geometry, the normal modes are the long wavelength magneto-exchange modes. Some of these modes can be coupled directly to a uniform pumping field and a typical spectrum at low powers is shown in Fig. 8.2a. The FMR spectrum consists of a series of resonances down field from the main resonance, the spacing of which can be controlled by selecting appropriate sample dimensions. At higher pumping powers, these modes exhibit nonlinear behavior including foldover effects (Fig. 8.2b) and auto-oscillations [28, 30].

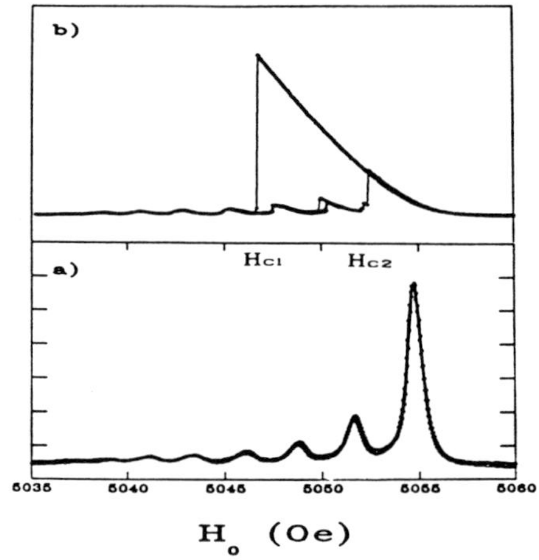

Figure 8.2. a) The absorption spectrum of sample 1 at -30 dB pumping power.
b) The absorption spectrum of the same sample showing foldover at -12.5 dB pumping power.

In a thin film, the magnitudes of both the exchange interaction and the *rf* magnetostatic interactions are greatly reduced. The exchange interaction is reduced by having small wavevectors, and the magnetostatic interaction is reduced because spins which contribute dipole fields in the case of a bulk sample are missing from the regions above and below the film.

The long wavelength magneto-exchange modes are determined by the finite size of the sample through boundary conditions. The condition of the surfaces of the sample determines whether the spins at the surface are pinned, free, or whether they experience a condition between these two extremes. Spin-waves directed normal to the film, the z-direction, form standing waves which, because of the surface pinning conditions, can have only discrete wavevectors [2]. The magneto-exchange spectrum is organized into exchange branches according to these wavevectors. The lowest wavevector branch is referred to as the 'main' branch [55].

The in-plane modes must also obey boundary conditions at the edges of the sample. Because of nonuniform demagnetization fields in non-ellipsoidal disks, the demagnetization fields near the edge of a film fall off rapidly, producing an effectively pinned condition. The in-plane components of the excitation form standing waves with discrete wavevectors, k_t, which depend on the shape of the film sample. In circular samples, the standing waves take the form of Bessel functions. Diagrams of a few of the lowest wavevector modes are shown in Fig. 8.3. Only those modes with no azimuthal dependence have net dipole moments that couple to a uniform driving field. The other modes will not appear in a low-power FMR spectrum, and are therefore referred to as 'hidden' modes.

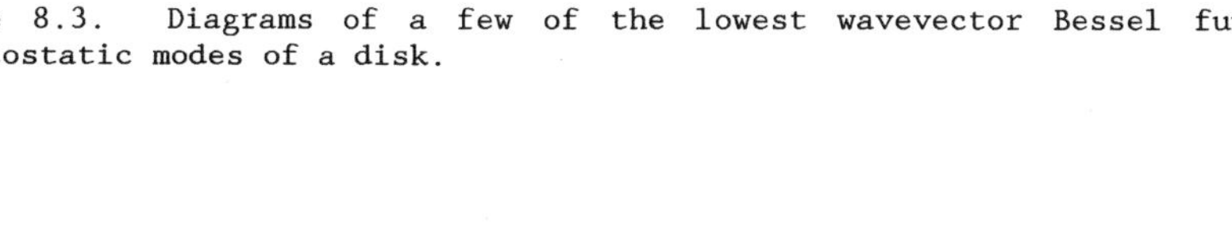

Figure 8.3. Diagrams of a few of the lowest wavevector Bessel function magnetostatic modes of a disk.

The dispersion relation for the main branch of a normally magnetized thin film having unpinned spins at the film surfaces is given by

$$\omega/\gamma = H_o - 4\pi M_z + \pi M_z k_t d, \qquad 8.7$$

where k_t is the component of the wavevector in the plane of the film and d is the thickness of the film. The last term is due to *rf* demagnetization fields similar to that in the case of bulk spin-waves except that planes of magnetic 'charge' cannot be formed because of the limited thickness of the film. For the magnetostatic mode, gradients in the *rf* magnetization produce strips of magnetic 'charge' in the film lying perpendicular to the in-plane wavevector is as shown in Fig. 8.4. Crudely speaking, since the 'charge' density per unit length of a strip is proportional to the thickness of the film, and because the fields due to a line charge fall off as 1/r, the effects of the *rf* demagnetization fields are reduced by a factor on the order of d/λ from the bulk spin-wave case.

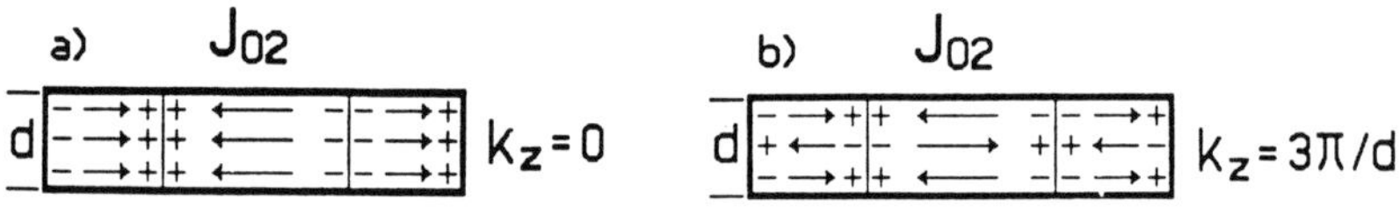

Figure 8.4. Schematic illustration of the tansverse magnetization for the J_{02} Bessel function and for (a) $k_z = 0$, and (b) $3\pi/d$. The pluses and minuses signify magnetic sources of the demagnetization field h^d.

The longest wavelength magnetostatic modes have wavelengths which are on the order of the size of the sample, so d/λ is typically on the order of 10^{-3}. For YIG, $4\pi M = 1750$ G, and typical magnitudes of the *rf* demagnetization fields are on the order of 2 G. Since these fields are small compared to the applied field and the shape demagnetization field, the precession of the spins can be assumed to be circular.

For the higher order spin-wave branches, the *rf* magnetostatic contribution to the dispersion relation is decreased from the main branch value by a factor on the order of the ratios of the respective k_z values as shown in Fig. 8.4.

Because of this, the magnetostatic modes of a higher order branch are often unresolved, and appear in the spectra as a single resonance.

For thick films of YIG, d > 1μm, the exchange energy of the z-directed standing waves, Dk_z^2, is small compared to the magnetostatic energy of the in-plane standing waves, and the spectrum appears to consist of only magnetostatic modes of the main branch, although it consists of many nearly degenerate exchange branches. In the case of a thin film where d < 1μm, the magnetostatic modes may be unresolved and the spectrum appears to consist of exchange modes, although each exchange mode is really a superposition of nearly degenerate magnetostatic modes. This is sometimes referred to as spin-wave resonance [2]. In the intermediate range, both the exchange and the magnetostatic energies are important. By using a film that is thin enough to separate the exchange branches and small enough to separate the magnetostatic modes, the spectrum will consist of well resolved magneto-exchange modes for which the degeneracy between exchange branches and magnetostatic modes is lifted as observed in Fig. 8.2.

The essential physics behind the nonlinear behavior of the magneto-exchange modes is that the effective static field at the samples, H_o-$4\pi M_z$, determines the resonant frequencies of precession which are modified when amplitudes of precession angle become large and the z-component decreases. Since the wavelength of the excitations is much greater than the thickness of the film, each point in the film behaves as though it were in an infinite film, and the local amplitude dependent shape demagnetization field is given by $-4\pi M_z(r)$.

There are several advantages to the thin film geometry for investigations into nonlinear dynamics. Primary among these is the fact that the low wavevector modes lie at the minimum of the bulk spin-wave manifold, and are therefore not degenerate with a large number of short-wavelength spin-waves. In the main magneto-exchange branch, there are only a few modes with a given frequency. Despite the fact that the spatial forms of the magneto-exchange modes are not as analytically convenient as plane waves, the simplicity of the nonlinear interactions, and the low number of involved modes make the system feasible to study.

8.3.2. Hamiltonian Theory

The equations of motion for the interacting magneto-exchange modes can be derived from the classical torque equations or by using a Hamiltonian formalism.

The method consists of writing the energy density of the system in terms of the magnetization and then making a transformation to canonical variables which is the classical equivalent of the Holstein-Primakoff transformation [37]. Damping terms are added to the equation and finally, estimates are made of the relative importance of various terms.

8.3.2.1. Basic Hamiltonian

The Hamiltonian $\boldsymbol{H}$, is made up of the following terms:

$$\boldsymbol{H} = \boldsymbol{H}_{Zeeman} + \boldsymbol{H}_{demag} + \boldsymbol{H}_{dd} + \boldsymbol{H}_{e} + \boldsymbol{H}_{pump}. \qquad 8.8$$

The first term is the Zeeman interaction of the magnetization with the static applied magnetic field, H_o, which is applied perpendicular to the surface of the film defined to be the z-direction. The second term is the interactions of the magnetization with the demagnetization field in the film. The third term represents the transverse dipole-dipole interactions associated with the spatial dependence of the transverse component of the magnetization which produces the magnetostatic splitting of the magneto-exchange modes, the fourth term includes the spin-wave exchange interactions arising from the standing spin-wave modes across the thickness of the film and the final term includes the interaction of the magnetization with the microwave frequency pump field. While both $\boldsymbol{H}_{demag}$ and $\boldsymbol{H}_{dd}$ arise from dipole-dipole interactions between the magnetic moments in the sample, they are separated because they differ in magnitude and in character. $\boldsymbol{H}_{demag}$ arises from the fields which are essentially static while $\boldsymbol{H}_{dd}$ arises from the demagnetization fields which oscillate with the precession frequency. The contribution of these terms to the equations of motion will be considered separately below.

The expression for the first two terms in the Hamiltonian, Eq. 8.8, will have the form

$$\boldsymbol{H}_{Zeeman} + \boldsymbol{H}_{demag} = - H_o M_z + 2\pi M_z^2. \qquad 8.9$$

In order to obtain the equations of motion of the magnetic vector, M, it is necessary to transform the energy terms in Eq. 8.8 into a set of basis vectors a and a^* which are canonical for arbitrary values of M_z. In terms of the magnetization vector, the transformations have the form

$$M_z = M_s - \gamma a a^* \qquad 8.10$$

and

$$M^+ = M_x + iM_y = a(2\gamma M_s - \gamma^2 a a^*)^{1/2}. \qquad 8.11$$

This is the classical equivalent of the Holstein-Primakoff transformation [36, 37].

The equation of motion for the base variables a is then determined by the relation

$$\frac{da}{dt} = -i\frac{\partial \boldsymbol{H}}{\partial a*}. \qquad 8.12$$

Expressed in terms of a and a^*, the first two terms of the Hamiltonian become

$$\boldsymbol{H}_{static} + \boldsymbol{H}_{demag} = Const + \gamma[(H_o - 4\pi M_s)aa^* + 2\pi\gamma a^* a^* aa]. \qquad 8.13$$

Of course, a will have some spatial dependence and therefore it is expanded in terms of orthogonal functions $m_i(\vec{r})$ so that

$$a = \sum_i a_i m_i(\vec{r}) \qquad 8.14$$

and

$$\int_v d\vec{r}\; m_i^*(\vec{r})\, m_j(\vec{r}) = v\delta_{ij}, \qquad 8.15$$

where v is the volume of the sample. When the Hamiltonian is written in terms of the expansion of a, and an integration is performed over the sample volume, the first two terms of the Hamiltonian become

$$\boldsymbol{H}_{static}+\boldsymbol{H}_{demag}=Const+\gamma\left[(H_o-4\pi M_s)\sum_i a_i^* a_i+2\pi\gamma\sum_{ijkl} A_{ijkl}a_i^* a_j^* a_k a_l\right] \quad 8.16$$

with the mode interaction parameters given by

$$A_{ijkl}=\frac{1}{V}\int_V d\vec{r}\; m_i^*(\vec{r})\, m_j^*(\vec{r})\, m_k(\vec{r})\, m_l(\vec{r})\,. \quad 8.17$$

8.3.2.2. Dipole-dipole interactions

The degeneracy of the normal modes of the system is lifted by the *rf* demagnetization fields, h^d, in the sample, yielding the $\boldsymbol{H}_{dd}$ term of the Hamiltonian. The *rf* demagnetization fields can be calculated from the requirement that $\nabla\cdot\vec{B}=0$ giving

$$\nabla\cdot\vec{h}^d = -4\pi\nabla\cdot\vec{M}_t. \quad 8.18$$

where M_t is the transverse *rf* component of the magnetization.

It can be seen from this that h^d and M_t are linearly related, and this will be sufficient for the present discussion. More detailed calculations have been carried out by Sparks [55].

Taking $M^+ = M_x + iM_y$ and $h^{d+} = h_x^d + ih_y^d$, h^d is expanded in terms of the orthogonal functions $m_i(\vec{r})$

$$h^d=\sum_i h_i^d m_i(\vec{r})\,. \quad 8.19$$

Using the linear relation between h^d and M^+ and choosing the orthogonal spatial functions such that the *rf* demagnetizing tensor, D, is diagonal and real, $\boldsymbol{H}_{dd}$ will become

$$\boldsymbol{H}_{dd} = 2\gamma M_s \sum_i D_{ii} a_i a_i^* - \frac{\gamma^2}{2} \sum_{ijkl} \left(D_{ii} A_{ijkl} a_i^* a_j^* a_k a_l + c.c. \right). \qquad 8.20$$

The quadratic term shifts the resonant field of the modes by $2M_sD_{ii}$, which is on the order of 1 to 10 Oe and determined by the aspect ratio of the film [55]. A comparison of this shift with the sample magnetization shows that $D_{ii}/2\pi = 10^{-2}$ or 10^{-3}. Furthermore, $D_{ii}/2\pi$ is the ratio of the magnitudes of the quartic terms arising from the dipole-dipole interactions and the shape demagnetization energy. This allows the quartic terms of $\mathbf{H}_{dd}$ to be dropped.

The magnitude of the term D_{ii} is determined by the spatial dependence of the transverse component of M in the direction of the film normal [55]. This in turn will depend on the nature of the magnetic surface interactions at each of the film surfaces [2]. For the purpose of simplicity in this review, only the extreme cases of a perfectly natural boundary (free boundary condition) and perfectly pinned (infinite surface energy) boundary condition will be considered. For the natural condition, the *rf* components of the magnetization will have no variation along the z-direction and the effective volume magnetic charge will have a maximum value as shown in Fig. 8.4. For the perfectly pinned condition the normal modes will have integral half wavelengths across the film thickness. In the fundamental mode the amplitude of the magnetic precession will be decreased at the surface thus changing the interaction form and the dispersion relation for the magnetostatic modes arising from D_{ii}. For the higher order modes in the pinned condition, the sign of the effective surface charge alternates across the film thickness, further decreasing the magnitude of the D_{ii} term. As a result, for higher order modes the magnetostatic modes are often not resolved.

8.3.2.3 Exchange Energy

The thickness of the film is often of such a dimension that the standing modes across the film thickness may have significant exchange energies. To evaluate this contribution to the energy of the system, the Hamiltonian has the form

$$\boldsymbol{H}_{hex} = (2A/M^2)\nabla_z^2 M. \qquad 8.21$$

For this purpose, only the z component of the Laplacian operator is required because the radial dimensions of the sample are such that the wavelength of the transverse variations of the modes is so long that contributions to the exchange energies are negligible and the *rf* dipole-dipole energies dominate. Evaluating the terms in Eq. 8.21 introduce a quadratic term to the Hamiltonian which will be proportional to γDk_z^2. The effect of this term is to shift all of the magnetostatic modes of a given exchange branch by an effective magnetic field equal to Dk_z^2 but will not contribute to the higher order terms in the Hamiltonian.

Including the first four terms in Eq. 8.8, the Hamiltonian in terms of the basis vectors becomes

$$\boldsymbol{H} = Const + \gamma\left[(H_o - 4\pi M_s + Dk_z^2 + 2M_s D_{ii})\, a_i a_i^* + 2\pi\gamma \sum_{ijkl} A_{ijkl} a_i^* a_j^* a_k a_l\right] + \boldsymbol{H}_{pump} \qquad 8.22$$

8.3.3. Pumping Term

The sample is also subjected to a microwave frequency field, h^p, which couples to the transverse magnetization of the sample. If the *rf* field is polarized in the x-direction and oscillates at a frequency ω,

$$\boldsymbol{H}_{pump} = M_x h^p \cos(\omega t). \qquad 8.23$$

Writing M_x in terms of M^+ and M^-, and making the transformation to the canonical variables a and a^*,

$$\boldsymbol{H}_{pump} = \frac{h^p}{2}(a + a^*)(2\gamma M_s - \gamma^2 aa^*)^{\frac{1}{2}} \cos(\omega t). \qquad 8.24$$

As with the other terms in the Hamiltonian, the next step is to expand in terms of the orthogonal spatial functions and integrate over the volume of the sample. The result gives

$$\boldsymbol{H}_{pump}=h^{p}\sqrt{\frac{\gamma M_s}{2}}\left[\sum_i (a_i I_i+a_i^* I_i^*)-\frac{\gamma}{4M_s}\sum_{ijk}(B_{ijk}a_i a_j a_k^*+c.c)\right]\cos(\omega t) \quad 8.25$$

where

$$I_i = \frac{1}{V}\int_V d\vec{r}\; m_i(\vec{r}) \quad 8.26$$

is the normalized dipole moment of mode i, and

$$B_{ijk} = \frac{1}{V}\int_V d\vec{r}\; m_i(\vec{r})\, m_j(\vec{r})\, m_k^*(\vec{r}) \quad 8.27$$

is another interaction parameter. It will be shown below that due to differences in the size of the coefficients of these cubic pumping terms and the quartic shape anisotropy terms, the cubic terms are generally much smaller for regions of interest and can be neglected.

8.3.4. Damping

In a real sample, the energy stored in a single magneto-exchange mode decays with time. This process can occur by excitation of spin-waves and lattice phonons through crystalline anisotropy, lattice defects, and scratches and etch pits in the surface [56]. Modelling such an enormous number of degrees of freedom seems impractical. Instead, a phenomenological damping term is usually added to the equations of motion. The form of the damping used is the Landau-Lifshitz damping term given by

$$\left(\frac{d\vec{M}}{dt}\right)_{damp} = -\alpha\vec{M}\times(\vec{M}\times\vec{H}) \; . \quad 8.28$$

Other workers in the field have used linear damping, and this will prove to be sufficient for the thin film case. It will prove desirable to express this damping in terms of the canonical variables a and a^*. If H is taken to be $H_o - 4\pi M_z$, the internal field along the z-direction,

$$\left(\frac{dM_z}{dt}\right)_{damp} = \alpha\,(H_o - 4\pi M_z)\,(M_x^2 + M_y^2)\,. \qquad 8.29$$

Using Eqs. 8.10 and 8.11 to express this damping rate in terms of a and a^*, expanding a by using Eq. 8.14 and integrating over the sample volume, the damping of the complex amplitude of the ith mode is

$$\left(\frac{da_i}{dt}\right)_{damp} = -\gamma\Gamma a_i - \frac{\gamma^2\Gamma}{2M_s}\frac{H_o - 12\pi M_s}{H_o - 4\pi M_s}\sum_{jkl} A_{ijkl}a_j^* a_k a_l\,. \qquad 8.30$$

where $\Gamma = \alpha\,(H_o - 4\pi M_s)\,M_s/\gamma$ is the experimentally measured linewidth of the magneto-exchange modes (full width at half maximum is 2Γ).

This expression is added to the equations of motion to model the damping of the modes.

8.3.5. Linear Behavior

The linear behavior of the magneto-exchange modes can be obtained by keeping only terms of the Hamiltonian of order a_i^2 or less.

$$\boldsymbol{H}_{quad} = \gamma\sum_i\left[(H_o - 4\pi M_s + 2M_s D_{ii})\,a_i a_i^* + h^p\sqrt{\frac{M_s}{2\gamma}}\,(a_i I_i + a_i^* I_i^*)\cos(\omega t)\right]. \qquad 8.31$$

From this Hamiltonian the equations of motion are given by

$$\frac{da_i}{dt} = -i\gamma\left[(H_o - 4\pi M_s + 2M_s D_{ii} - i\Gamma)\,a_i + h^p I_i^*\sqrt{\frac{M_s}{2\gamma}}\,\frac{1}{2}\,(e^{i\omega t} + e^{-i\omega t})\right]. \qquad 8.32$$

From this equation of motion, the resonant frequency of a_i is seen to be $\gamma\,[H_o - 4\pi M_s + Dk_z^2 + 2M_s D_{ii}]$. However, in most experiments, the frequency of the driving field is held constant and H_o is varied. It is more convenient, therefore, to define a resonant field, H_i^{res}

$$H_i^{res} = \omega/\gamma + 4\pi M_s - Dk_z^2 - 2M_s D_{ii}. \tag{8.33}$$

Using this notation, the fixed points of Eq. 8.32 are found by looking for steady state solutions of the form $a_i = a_i^o e^{-i\omega t}$ and dropping the counter-rotating $e^{-i\omega t}$ pumping term. The fixed points of Eq. 8.32 are given by

$$a_i^o = \sqrt{\frac{M_s}{2\gamma}} \frac{h^p I_i^*/2}{(H_o - H_i^{res} - i\Gamma)}. \tag{8.34}$$

This expression will be useful for estimating the relative size of some of the non-linear terms in the equations of motion.

8.3.6. Nonlinear Equations of Motion

The equations of motion for the interacting magneto-exchange modes in a thin film at perpendicular resonance are obtained from the full Hamiltonian with the addition of damping terms. The Hamiltonian, including dipole-dipole interactions and pumping is

$$\boldsymbol{H} = \gamma\left[\sum_i (H_o - 4\pi M_s + 2M_s D_{ii}) a_i a_i^* + 2\pi\gamma \sum_{ijkl} A_{ijkl} a_i^* a_j^* a_k a_l\right] \tag{8.35}$$

$$+ h^p \sqrt{\frac{\gamma M_s}{2}} \left[\sum_i (a_i I_i + a_i^* I_i^*) - \frac{\gamma}{4M_s} \sum_{ijk} (B_{ijk} a_i a_j a_k^* + c.c)\right] \cos(\omega t).$$

This expression can be simplified by comparing the magnitudes of the nonlinear terms and discarding those terms which are not important.

Assuming that there are N significantly excited modes, the magnitude of *a* for which the quartic and cubic terms are of the same magnitude can be estimated.

Setting the cubic and quartic terms equal to each other and assuming that B_{ijk} and A_{ijkl} are both on the order of one,

$$\sqrt{\frac{\gamma M_s}{2}}\frac{\gamma}{2M_s}N^3a^3 \approx 2\pi\gamma^2N^4a^4, \tag{8.36}$$

which gives

$$a \approx \sqrt{\frac{M_s}{2\gamma}}\frac{h^p/2}{N2\pi M_s}. \tag{8.37}$$

Comparing this estimate of the amplitude of a mode with the amplitude of a mode assuming linear equations of motion given by Eq. 8.34, the cubic and quartic terms of the Hamiltonian are seen to have comparable amplitudes when the modes are driven at a field which is 1000G away from their resonant fields. Closer to resonance, where the experiments are carried out, the amplitudes of the modes will be larger and the quartic terms will dominate the cubic terms. For this reason, the cubic pumping terms can be dropped.

Dropping the cubic pumping terms, the equations of motion obtained from the Hamiltonian are

$$\frac{da_n}{dt} = -i\gamma\left[(H_o - 4\pi M_s + D_{nn})a_n + 4\pi\gamma\sum_{jkl}A_{njkl}a_j^*a_ka_l + h^p\sqrt{\frac{M_s}{2\gamma}}I_n^*\cos(\omega t)\right]$$

$$- \gamma\Gamma a_n - \frac{\Gamma\gamma*^2}{2M_s}\frac{H_o - 12\pi M_s}{H_o - 4\pi M_s}\sum_{jkl}A_{njkl}a_j^*a_ka_l, \tag{8.38}$$

where terms describing the effects of damping have been added. Again comparing the magnitudes of the nonlinear terms, the ratio of the nonlinear terms associated with damping to the nonlinear terms associated with the shape anisotropy is

$$\frac{\Gamma}{8\pi M_s}\frac{H_o - 12\pi M_s}{H_o - 4\pi M_s}. \qquad 8.39$$

For typical experimental values of H_o, Γ, and M, this ratio is much smaller than one. Again, the shape anisotropy term is dominant and the nonlinear damping terms are dropped.

Finally, it is convenient to make a transformation to a slowly varying set of unitless variables, c_i , such that $a_i = \sqrt{M_s/2\gamma}\, c_i e^{-i\omega t}$. Making this change of variables and ignoring terms which oscillate at angular frequencies of 2ω, the equations of motion reduce to the form.

$$\frac{dc_i}{dt} = -i\gamma\left[(H_o - H_i^{res} - i\Gamma)\,c_i + \frac{h^p}{2}I_i^* + 2\pi M_s\sum_{jkl} A_{ijkl}c_j^* c_k c_l\right]. \qquad 8.40$$

These are the equations of motion which are used to model the behavior of the interacting magneto-exchange modes.

8.3.7. Spatial Forms

The equations of motion in the preceding section depend on the spatial form of the magneto-exchange modes through I_i and A_{ijkl}. The only restriction placed on the spatial functions $m_i(r)$ so far is that they are othogonal and diagonalize D_{ii}. In practice it is difficult to find functions for a finite film that will diagnonalize D_{ii} exactly. To diagonalize D_{ii}, a complete, orthogonal set of functions that obey the boundary conditions at the edges of the samples would be used as a basis to calculate D_{ii}. This matrix would then be diagonalized and the eigenvectors found. The resulting linear combinations would be analytically awkward and could hinder a physical understanding of the system. For analytical and conceptual convenience, the functions $m_i(r)$ will be approximated by orthogonal functions suitable for a given sample shape while relaxing the restrictions that D_{ii} be diagonal.

For a rectangular film, the spatial functions are of the form

$$m_{pq}(r) = 2\sin(p\pi x/L)\sin(q\pi y/W) \qquad 8.41$$

where p and q are integers, and the sample extends from x = 0 to x = L and y = 0 to y = W. In this expression, it has been assumed that the spins are unpinned at the top and bottom surfaces of the film, and the modes of interest are all in the main branch. Otherwise, m_{pq} would be multiplied by a z-dependant term.

For a circular film, the spatial functions are of the form

$$m_{\nu,s}(r) = \frac{1}{J_{\nu+1}(x_{\nu,s})} J_\nu(x_{\nu,s}\; r/R)\begin{pmatrix}\sin(\nu\phi)\\ \cos(\nu\phi)\end{pmatrix}. \qquad 8.42$$

where $x_{\nu,s}$ is a zero of the Bessel function J_ν and R is the sample radius. Diagrams of a few of the lowest wavevector modes with ν,s indices are given in Fig. 8.2.

In rectangular films, only those modes where both p and q are odd will appear in the low-power spectra since only these modes will have a nonzero coupling, I_1, to the pumping field. If either p or q or both are even in rectangular films, or if $\nu \neq 0$ in a circular film, the mode will not couple to the driving field and will not appear in the low-power spectra. These modes are referred to as 'hidden' modes. Although the hidden modes are not excited by the driving field, they can be parametrically pumped through interactions with modes that are driven by the pumping field.

The excitation of 'hidden' magneto-exchange modes is significant for two reasons. First, since the hidden modes attain amplitudes comparable to the amplitudes of the directly driven modes, they play an active role in the dynamics observed in thin films. Second, because the excitation of hidden modes in a circular film is analogous to the excitation of spin-waves in bulk samples, investigations in the relatively simple case of the magneto-exchange modes can lend understanding to spin-wave dynamics.

In almost every published theory of the nonlinear dynamics of parametrically excited spin-waves, the full set of nonlinear spin-wave interactions has not been used. Instead, certain interactions have been eliminated through the random phase approximation. In an infinite medium, parametrically excited $\pm k$ spin-waves are phase correlated through the pumping process, though their overall phase is not fixed. Physically, this corresponds to excitation of a standing spin-wave with nodes and antinodes that are free to slide in the sample. From this point, the argument proceeds that the spin-waves, although phase locked to their pair partners, are randomized by residual interactions such as inhomogeneities and shape defects and therefore, those terms in the equations of motion that do not have fixed phases average to zero and are unimportant [21]. This simplifies the equations of motion for spin-wave dynamics and yields the widely used S-theory.

The hidden magneto-exchange modes of a disk can be thought of as standing waves around the circumference of the sample. In order to study the excitation of hidden modes in a circular film, an $\exp(\pm i\nu\varphi)$ form for the azimuthal dependence of the modes could be used in close analogy to the form of a spin-wave, but it is more convenient to use real functions $m_i(r)$ with sine and cosine azimuthal dependence for two reasons. Although the azimuthal part of the integral in Eq. 8.17 will be slightly more difficult to calculate, A_{ijkl} will become symmetric with respect to interchange of indices. Second, the linear parts of the equations governing excitation of the hidden modes become decoupled. If the $\exp(i\nu\phi)$ basis were used, the equation for the clockwise propagating wave would include a term which is linear in the amplitude of the counter-propagating wave, as is the case for the excitation of spin-waves. The choice of sine and cosine standing waves as a basis eliminates this complication.

In the circular film, the physical analog to spin-wave pairs with random phases in bulk media is a situation with random amplitudes of the $\sin(\nu\phi)$ and $\cos(\nu\phi)$ modes. A different linear combination of these modes produces nodes and antinodes in different locations around the disk.

The parameters A_{ijkl} can be separated into two categories, the pairwise interactions, which have the form A_{iijj}, and the non-pairwise interactions which do not have this form. The pairwise interactions are the analogs of the $S_{k,k'}$ and $T_{k,k'}$ matrices in the S-theory. The non-pairwise interactions are the terms that would be thrown out in the random phase approximation.

Consider a system consisting of a pair of hidden modes labeled, i and j, being excited by a single directly pumped mode, $\nu=0$. In this case, only the pairwise interactions are nonzero as can be seen by considering the azimuthal part of the A_{ijkl} integral Eq. 8.17. In order to investigate parametric excitation in this system, the equations of motion Eq. 8.40 for the complex amplitudes of the modes are more easily interpreted if they are written separately in terms of real magnitudes and phases such that $c_i = C_i \exp(i\psi_i)$. In terms of these variables, the equations of motion for a hidden mode ($I_i = 0$) are

$$\frac{dC_i^2}{dt} = -2\gamma C_i^2\left[\Gamma + \sum_{k=0,j} A_{iikk} C_k^2 \sin 2(\psi_i - \psi_k)\right] \qquad 8.43$$

and

$$\frac{d\psi_i}{dt} = -\gamma\left[H_o - H_i^{res} + A_{iiii}C_i^2 + 2\sum_{k=0,j} A_{iikk}C_k^2\left(1+\frac{1}{2}\cos 2(\psi_i - \psi_k)\right)\right]. \qquad 8.44$$

From Eq. 8.43, the threshold for excitation of a hidden mode when $C_i, C_j \approx 0$ occurs when

$$A_{00ii}C_0^2 \sin 2(\psi_i - \psi_0) + \Gamma = 0, \qquad 8.45$$

and the minimum threshold value of C_o^2 is obtained when $\sin 2(\psi_i - \psi_o) = -1$. To meet this requirement, the hidden mode must be effectively on resonance, with an applied field given by

$$H_o = H_i^{res} - 2A_{00ii}C_0^2, \qquad 8.46$$

where the resonant field of the hidden mode has been shifted downfield by its interaction with the driven mode.

Above this threshold, the mode will grow exponentially until its growth is limited by a number of factors. As C_i grows, it will shift its own eff^ctive resonance field through a term $A_{iiii}C_i^2$, thereby changing its equilibrium phase. The large amplitude of the hidden mode will also shift the effective resonant field of the directly driven $\nu = 0$ mode, changing its phase, and will also contribute an effective damping term to the amplitude equation of the driven mode.

8.3.8. The Equilibrium State

Next, consider the growth of a degenerate pair of modes labeled i and j having $\sin(\nu\Phi)$ and $\cos(\nu\Phi)$ azimuthal dependence driven by a pumped mode, labeled 0 [57]. Because the driven mode is rotationally invariant, and because $\sin(\nu\Phi)$ transforms to $\cos(\nu\Phi)$ under a rotation of π/ν, the hidden modes will behave identically because the interactions between the driven mode and each of the hidden modes in a pair are the same. Because of this, the phases of the hidden modes will either be the same ($\psi_i = \psi_j$), or differ by π.

For small values of $\psi_i - \psi_j$,

$$\frac{d}{dt}(\psi_i - \psi_j) = \gamma\left[(H_i^{res} - H_j^{res}) + 2A_{ii00}C_0^2\sin(\psi_i + \psi_j - 2\psi_0)(\psi_i - \psi_j)\right]. \qquad 8.47$$

If the modes are degenerate, the phase difference will decay exponentially since $\psi_i = \psi_j$, and $\sin 2(\psi_i - \psi_0) < 0$ so that the modes can grow. In deriving this result, the relation $A_{iiii} = 3A_{iijj}$ has been used. This relation can be verified by computing the azimuthal part of the integral in Eq. 8.17 needed to find A_{ijkl}. Having shown that the phase difference of the excited pair decays to zero, it can be shown that the phase difference can also decay to π by noting that the equations of motion, Eqs. 8.43 and 8.44 are invariant under adding π to the phase of a hidden mode.

When $\psi_i = \psi_j$, the amplitudes of the two modes will grow independently because the growth of a hidden mode only depends on the amplitude of the other hidden mode through a term proportional to $\sin 2(\psi_i - \psi_j)$. Since these two modes have random thermal amplitudes before excitation, they will have random amplitudes before the growth limiting terms become important. The final amplitudes of the modes will be determined by the phase differences $\psi_i - \psi_0$ and $\psi_j - \psi_0$. Looking for a fixed point of the phase in Eq. 8.44, the relation

$$3A_{iijj}C_j^2 + A_{iiii}C_i^2 = const \qquad 8.48$$

is found. Since $A_{iiii} = 3A_{iijj}$, a fixed point can be obtained for arbitrary values of C_i and C_j subject to the constraint that

$$C_i^2 + C_j^2 = const. \tag{8.49}$$

Any real normalized linear combination of the sin(vϕ) and cos(vϕ) modes will behave in the same way as a pure sine or cosine mode, as such a linear combination is physically equivalent to a rotation of the disk through an arbitrary angle about its axis. A priori, there is no reason for one mode of a pair to have a larger amplitude than the other, and it is expected that upon excitation the modes will reach a neutrally stable equilibrium state described by Eq. 8.49. This state of locked phases and random, but correlated amplitudes is equivalent to the random phase state of parametrically excited $\pm k$ spin-waves.

8.4. EXPERIMENT

8.4.1. Low Power Spectra

A block diagram of a typical experimental setup used for nonlinear ferromagnetic resonance is shown in Fig. 8.5. The YIG samples are grown by liquid phase epitaxy on a GGG substrate and circular films were prepared by masking and etching.

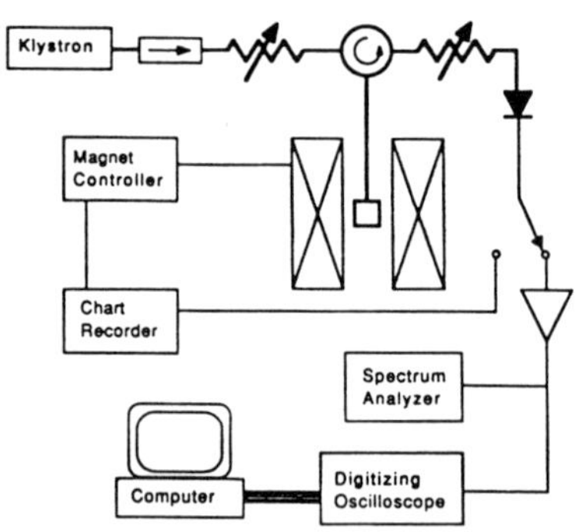

Figure 8.5. Block diagram of the ferromagnetic resonance spectrometer.

Two methods can be used to record the signal from the crystal detector. For d.c. signals such as absorption spectra, the signal is fed directly into a chart recorder. In order to record auto-oscillations, the signal is fed through an amplifier, and sent to a spectrum analyzer and a digitizing oscilloscope.

The magneto-exchange spectrum depends on the pinning conditions at the surface of the film in two ways. Although the pinning conditions at the surfaces of the film can range from pinned to free, the present discussion will be limited to the two extremes.

In the case of films where the spins are unpinned at both surfaces of the film at z=0 and z=d, the z-dependence of the mode will vary as $\cos(n\pi z/d)$. Modes for which n is greater than zero will not couple to a uniform pumping field and, therefore, the higher order exchange branches will not appear in the spectra. The main branch, n = 0, will have uniform magnetization across the thickness of the film, and will have the dispersion relation

$$H^{res} = \omega/\gamma + 4\pi M_z - \pi M_z k_t d. \tag{8.50}$$

If the spins are pinned at both surfaces of the film, the exchange branch will have a spatial dependence proportional to $\sin(n\pi z/d)$, and only the branches with n odd will couple to the driving field. If n is even, the branch will be hidden. In this case, the magnetostatic interactions between the spins will be reduced because spins at the surface cannot contribute to *rf* demagnetization fields, and the dispersion relation is given by

$$H^{res} = \omega/\gamma + 4\pi M_z - Dk_z^2 - 8M_z\frac{k_t d}{\pi}. \tag{8.51}$$

For spins that are pinned at both surfaces, the magnetostatic interactions are reduced by a factor of approximately 4/5 [55] from the unpinned case. In addition to reduced magnetostatic interactions, pinned films also exhibit a phenomenon known as branch repulsion between odd exchange branches [58]. The branch repulsion effect distorts the spectrum near the exchange branches.

If the pinning conditions are not identical on the two surfaces of a film, all of the exchange branches may couple to the pumping field and may appear in the spectra.

Data obtained from two films are reviewed here. The low power spectra for sample 1 is shown in Fig. 8.6. The linewidth of this spectrum is about 0.4 Oe. The higher order exchange branches were located at 30 Oe and 87 Oe downfield from the main resonance and they were very weak indicating that the spins are nearly free at the surfaces of the film. By contrast, the low power spectra of Sample 2 is shown in Fig. 8.7. It has a series of strong higher order branches out of the range of this figure. Fitting these spectra to Eqs. 8.50 and 8.51 respectively, it is possible to obtain the film thickness as indicated in the captions of Figs. 8.6 and 8.7.

8.4.2. Non-Linear Effects

In Fig. 8.2b, the spectrum is shown for sample 1 with the input power increased from that in 2a by 17.5 dB. This spectrum shows a hysteretic absorption curve termed 'foldover' [59] in both the main (0,1) mode and the (0,2) mode. The higher order modes display foldover at higher pumping powers. The hysteresis in the (0,2) mode can only be observed when the (0,1) mode is in its low amplitude state.

While the d.c. component of the power reflected from the cavity exhibits foldover in certain ranges of the applied field and pumping power, the system also auto-oscillates resulting in an a.c. component in the detected signal. In Figs. 8.6 and 8.7, the domains of auto-oscillation and chaotic behavior are outlined for samples 1 and 2 respectively. The domains of oscillation are divided into regions called 'fingers' which are associated with each mode in the low power spectrum. The fingers are numbered from right to left, with the first finger being associated with the (0,1) mode and the second finger associated with the (0,2) mode, and so on.

The domains of auto-oscillation also follow a hysteresis indicated in Figs. 8.6 and 8.7 by the arrows and the thinner lines within the fingers. The lower field boundary of the second finger corresponds to H_{c1} of the (0,2) mode. The thin line within the finger corresponds to H_{c2}.

The first finger in both samples has a behavior quite different from the

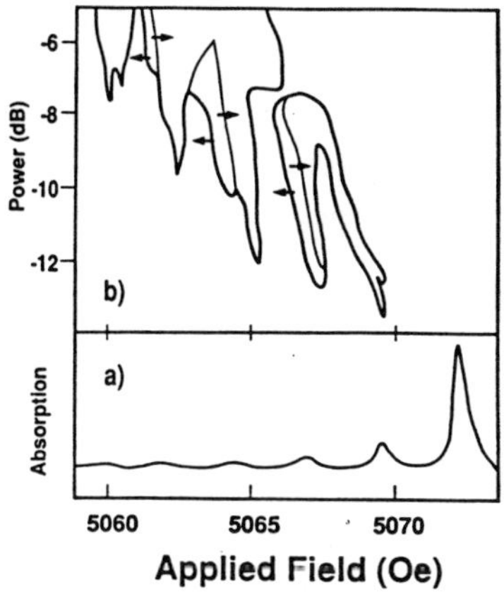

Figure 8.6. Fingers of auto-oscillation in Sample 1. d=0.56μm and R=0.30mm. Hysteresis is indicated by the arrows and thinner lines within the fingers.

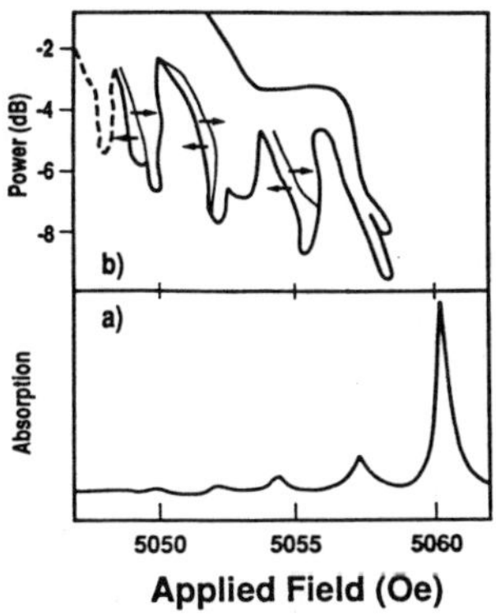

Figure 8.7. Fingers of auto-oscillation in Sample 2, d=1.6μm and R=0.55mm. Hysteresis is indicated by the arrows and thinner lines within the fingers.

rest of the fingers in that it is only observed when the main resonance is in the low amplitude state and the bias field has been increased from lower fields. At the high field boundary of the first finger, corresponding to H_{c2} of the main mode, the first mode makes a transition to a high amplitude state and the oscillations cease. If the field is then decreased, oscillations are not observed until H_{c1} is reached, and the first mode drops back to the low amplitude state.

TABLE 8.1

Auto-oscillation Frequencies
and Onset Powers Calculated for Sample 1

	Frequencies (MHz)		Onset Power (dB)	
Finger #	model	expt.	model	expt.
1	3.97	4.61	-14.5	-13.5
	1.5	1.5	-13.7	-12.5
2	--	5.22	--	12.5
	7.8	9.2	-14.0	-12.5

It should also be noted that many of the fingers are split to give two onset powers and fields, with each having a different onset frequency of auto-oscillation. The frequencies of auto-oscillation at onset for the fingers shown in Fig. 8.6 are listed in Table 8.1. Near onset, the frequencies of auto-oscillation vary only slightly as the field is swept through a finger, generally varying less than 10%, usually decreasing as the field is increased. An exception is the high field fingertip of the first finger in both samples. The auto-oscillation frequency varies rapidly, dropping from 2 MHz to 1 MHz in about 0.3 Oe. At higher power levels, often more than one type of behavior is observed within a finger indicating that the system may be hopping from one attractor to another as the applied field and pumping power are varied over ranges of a few tenths of an Oersted or a few dB. This multiple attractor effect is more prevalent in sample 2 having the higher density of normal modes than is observed in sample 1, but it appears in both samples at power levels where the fingers merge.

Finger 2 in sample 2 has quite complicated behavior of this type. For example, at 6 dB attenuation, as the applied field is increased from a value between the second and third fingers, auto-oscillations arise at 5054.4 G with a frequency of 3.18 MHz. At 5055.0 G, there is a sharp transition to a 4.35 MHz attractor which does not change significantly up to the border of the finger at 5055.4 G. This transition corresponds to the upward foldover transition of mode 2. If the static field is then decreased from the border of the finger, the frequency of auto-oscillation will increase slightly then the auto-oscillations become quasiperiodic at 5054.5 G and a period doubled 4.57 MHz oscillation is seen at 5054.3 G just before another transition to a 7.90 MHz attractor which persists until the low field boundary of the finger is reached at 5054.0 G.

Of particular interest in Fig. 8.6 is the large additional domain of auto-oscillation between the second and third fingers. The location of this finger suggests that it arises from the parametric excitation of the n=2 exchange branch which is predicted to have a resonant field between those of the (0,2) and (0,3) modes of the main branch. The frequency of auto-oscillation of the third fingers also differ between sample 1 and sample 2. While the first two fingers have similar frequencies, the third fingers have very different frequencies, suggesting that different parametrically driven modes are responsible for the auto-oscillations.

To investigate the temperature dependence of the non-linear effects, a third film was prepared and investigated in some detail [30]. By fitting the low power spectrum at room temperature, all of the material parameters were again determined. The linewidth, ΔH, was observed to be 0.8 G and the separation, δH_{ij}, between the magnetoexchange modes was 2.8 G. As the temperature is decreased to 130K, the magnetization increases producing an increase in the separation of the magnetoexchange modes as shown in Fig. 8.8. A significant increase in the linewidth was also observed. Below 130K the linewidth increases to the point where the spectrometer no longer had sufficient power to drive the modes into the auto-oscillation mode. The auto-oscillation map for this film was similar to those observed in Figs. 8.6 and 8.7 with the exception that the splitting of the first finger was much more pronounced. The onset frequency and power for each of the segments of the first finger are shown in Figs. 8.9 and 8.10 respectively. In these graphs the 1A signals are the low power and low frequency oscillation while the 1B signals are the high power and high frequency fingertip as observed in Figs. 8.6 and 8.7.

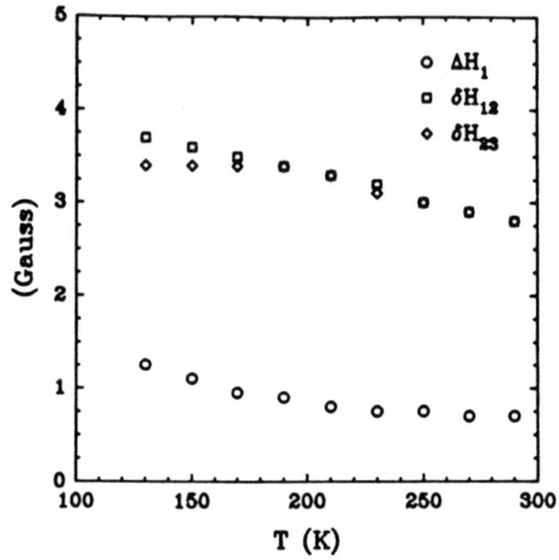

Figure 8.8. Experimental measurements of the temperature dependence of the linewidth, ΔH_1; the spacing between the (0,1) and (0,2) modes, δH_{12}; and the spacing between the (0,2) and (0,3) modes δH_{23}.

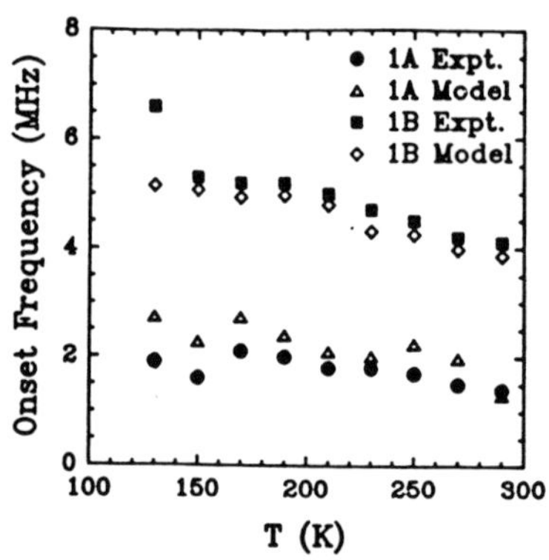

Figure 8.9. Experimental and simulation results for the temperature dependence on the onset frequency of sub-fingers 1A and 1B.

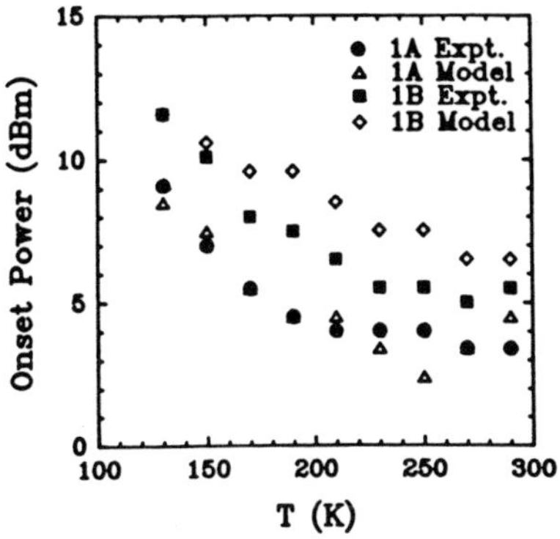

Figure 8.10. Experimental and simulation results for the temperature dependence of the onset power of sub-fingers 1A and 1B. Experimental uncertainty is ± 0.5 dB; model data is generated in 1.0 dB increments.

8.5. MODELING

Using the equations of motion in Eq. 8.40 for the magneto-exchange modes that are derived in Section 8.3, models involving various combinations of modes were constructed and integrated numerically. The results are compared to the experimental behavior of the magneto-exchange modes. The parameters used in the modeling were either calculated from the spatial dependence of the modes, or taken from the low power spectra. This provided enough information that no adjustable parameters were used in the models.

The interaction parameters were calculated from the Bessel function spatial dependence of the magneto-exchange mode given in Eq. 8.42. These functions are used as an approximation to the forms of the modes, and although they do not diagonalize the quadratic part of the Hamiltonian exactly, they do have the proper form and analytically feasible. The mode intensities, $I_{v,s}$, were calculated analytically from Eq. 8.26 from which

$$I_{o,s} = 2/x_{o,s}. \tag{8.52}$$

also

$$I_{\upsilon\neq 0,s} = .001. \tag{8.53}$$

This small intensity ascribed to the hidden modes was included to model the effects of thermal excitation and inhomogeneities in the sample and in the pumping field. Without this small coupling to the pumping field, the amplitudes of the hidden modes have fixed points at the origin, where, although the fixed point may become unstable, a mode with an amplitude numerically equal to zero will remain. Ascribing the hidden modes a small coupling to the pumping field eliminates this problem.

The other parameters in the equations of motion were taken from the low power spectra. The linewidths of the modes were taken by measuring the width of an absorption peak at half-maximum and dividing by two. The resonant fields of the driven modes were taken directly from the spectra.

The resonant fields of the hidden modes were obtained by a different method. Because the linewidth of the modes is on the order of a few tenths of an Oersted, the resonant fields of the hidden modes must be determined to within a 0.1 Oe. Rather than leave these numbers as the only adjustable parameters in the models, the resonant fields of the hidden modes were calculated in a two-stage process.

First, the resonant fields were interpolated from plots using the zeros of the Bessel functions $x_{\nu,s}$. Beyond this, additional corrections were made to take into account the fact that the demagnetization field in a disk, since it is not an ellipsoid, is not uniform. The demagnetization field falls off from $4\pi M$ at the center of a very large thin film to $2\pi M$ at the edge of the film. The driven modes, having the form of Bessel functions of order zero, have large amplitudes near the center of the film, but the hidden modes have zero amplitude at the center and are concentrated nearer the edges of the film where a nonuniform demagnetization field is expected to depress the resonant fields of the hidden modes more than it does the resonant fields of the driven modes.

The number of modes that can be included in the computer model is limited in some sense by the number of interaction parameters needed [28]. To include all of the interactions needed to model N modes, N^4 interaction parameters would be reqiured. For three or even four modes, this is not prohibitive if use is made of the fact that when real functions are used to describe the spatial dependence of a mode, A_{ijkl} is invariant under interchange of indices. Of the 81 parameters needed to model the interactions of three modes, only 15 are independent, and of the 256 interaction parameters needed for four modes, only 32 are independent.

The auto-oscillations and foldover jumps were detected by two methods. In the first method, the power absorption at a point in the field-power plane was compared with the value at the previously calculated point. If the difference was greater than a cutoff value, this was taken as an indication of either a foldover transition or the presence of auto-oscillations. As an alternative, the auto-oscillations were detected without including foldover transitions from stable fixed point to stable fixed point by calculating the sum of the squares of the time derivatives of the modes. At a fixed point, this number has a value which is very small. If the system is in an auto-oscillatory state, the sum of the squares of the time derivatives of the mode amplitudes will have a nonzero value.

The model concentrates on the first and second fingers. It was discovered through simulation that the (0,1) and (0,3) neighbors of the (0,2) mode have significant amplitudes at the applied fields and pumping powers where the second finger was expected to appear. In contrast to the linear behavior, where the modes appear to be well separated, the nonlinear interactions allow other channels for energy sharing between the driven modes, resulting in significant amplitudes for modes which are off resonance. A five mode model including the (0,1), (0,2), and (0,3) driven modes and the (1,1), and (2,1) hidden modes exhibits fingers of auto-oscillation as shown in Fig. 8.11, where the resonant fields of the modes are indicated by the vertical arrows. For simplicity, the first finger is shown for increasing values of H_o, and the second finger is shown for decreasing values of H_o.

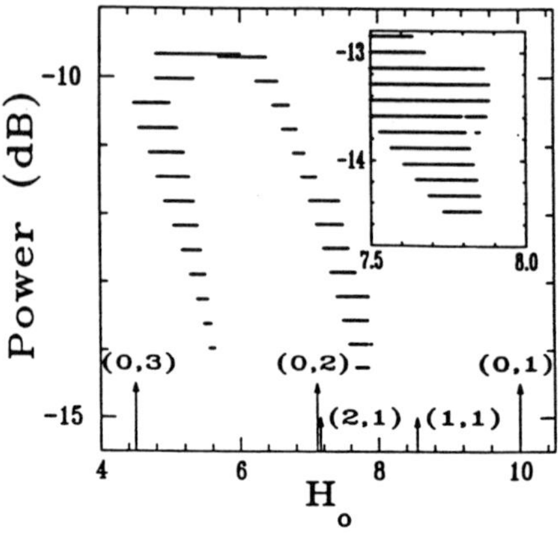

Figure 8.11. Domains of auto-oscillation for the (2,1) and (1,1) hidden modes being driven by the (0,1) (0,2) and (0.3) modes. Doubling of the first finger is shown in the inset.

The first finger was well modeled using only the four-driven-mode model (0,1; 0,2; 0,3; 0,4) except for the absence of a small second 'fingertip' on the high field side of the experimental main finger with a frequency which varies from 2.0 to 1.0 MHz as the static field is swept. Including the (1,1) mode leaves the main finger nearly unaffected, but a 1.5 MHz fingertip appears on the high field side of the main finger. This fingertip is shown in detail in the inset to Fig. 8.11. The rest of the finger is not affected, as the (1,1) mode is only excited in a narrow range just below H_{c2} of the (0,1) mode.

For the second finger of auto-oscillation the amplitude of the (1,1) mode remained small, and did not enter into the dynamics of the system. The (3,1), (4,1) or (1,2) modes were substituted for the (1,1) mode to see if they might be involved in the second finger, but they were not excited either. According to the model, then, the second finger is produced by interactions between the driven modes. Auto-oscillation frequencies and onset powers calculated from the five mode model including hidden modes are given with the experimental values from sample 1 in Table 8.1.

The second finger has an onset power which is within 2 dB of the experimental value, and the frequency of auto-oscillation in this finger is 7.8 MHz, which is about 40% higher than the experimental value of 5.22 MHz observed in the main portion of the finger. The model frequency of 7.8 MHz is much closer to the frequency of the 9.2 MHz oscillations observed in narrow regions on the low-field edge of the second finger of sample 1. The auto-oscillations do appear in the high-amplitude foldover state of the (0,2) mode, as in the experiment, so that the finger is observed with decreasing static field. The static field at onset and the finger shape depend sensitively on the value of the resonant field of the (2,1) mode, the static field of the finger 'tip' varying by 0.5 Oe as the (2,1) resonant field is changed by 0.3 Oe.

To compare with the experimental results of the temperature dependent data, the values of the film parameters were input to the model for each temperature and the onset frequency and power of the auto-oscillations were evaluated. The results are included in Figs. 8.9 and 8.10 as the solid symbols. The agreement that is observed between the experiment and the data is quite excellent.

8.6 CONCLUSIONS

The dynamical properties of the magnetic vector in ordered magnetic materials are observed to yield nonlinear effects that are directly observed, easily analyzed and readily modeled. The precession frequency of the magnetization is in the GHz range and frequencies of the auto-oscillations in these materials are observed in the 0.5 to 5 MHz range. The nonlinear equations of motion that couple the normal modes of the magnetic precession can be obtained directly from the energy Hamiltonian of the magnetic system.

In the bulk magnetic materials a major weakness exists in the comparison of the experimental data with the predictions of the model. In the experiment, the driven mode is coupled to a large number, approximately 10^{10}, of degenerate short wavelength spin-waves with each one having a different coupling coefficient. In the model calculations it is necessary to truncate the system to a limited number of these degenerate modes. This truncation procedure is a serious limitation to the full understanding of the dynamical behavior of the magnetic system in the non-linear regime.

It is possible to significantly minimize this effect by using the thin magnetic film geometry with the magnetic field applied normal to the film. In this configuration the normal modes of the system are long wavelength modes which lie at the bottom of the spin-wave band and therefore not degenerate with any of the short wavelength spin-waves. These normal modes are the magneto-exchange modes whose mode functions and mode frequencies are determined by the geometry and the dimensions of the sample. With the thin film geometry it is possible to modify the density of the normal modes and while the models still require that the system of equations be truncated to a few modes, the modes that are excluded from the analysis have frequencies that are substantially different from the driven modes and are not expected, or found, to have significant contributions to the non-linear interactions.

The equations of motion for the interactions between the magneto-exchange modes are obtained by using a Hamiltonian formalism. The method involves the writing down of the energy density in terms of the magnetization and making a transformation to canonical variables which are the classical equivalent of the Holstein-Primakoff transformation. By determining the magnetic material properties of the sample, it is possible to evaluate all of the coefficients of the energy terms in the Hamiltonian. From the low power, linear ferromagnetic resonance spectrum, it is possible to determine the positions and the loss parameters associated with each of the modes and as a result, all the remaining terms in the equations of motion are known.

In the experimental results two non-linear phenomena are observed. The first is the foldover effect and the resulting hysteresis that occur in the spectra. At somewhat higher drive fields each of the magneto-exchange modes are observed to produce an auto-oscillation in the output signal at the detector. With increasing drive fields the auto-oscillations are observed to bifurcate and enter the chaotic regime. At still higher fields the basins of attraction for each of the modes is observed to overlap with the resulting hopping of the system from one basin to another.

A comparison of the foldover effects as well as the onset drive field and the initial auto-oscillation frequencies evaluated for each of the normal modes in the numerical modeling procedure are observed to be within fifteen percent of the values observed in the experiment, including its temperature dependence over a wide range of temperatures. Thus, the magnetic film geometry has proved to be one of the best physical systems for which it is possible to have such close agreement between the proposed model and the experimental results.

8.7 ACKNOWLEDGMENTS

The author graciously acknowledges the generous assistance of a number of his students and colleagues over the past four years. In particular to Prof. Horst Doetsch and Prof. Franz Waldner who introduced the author to the topic while on a sabbatical leave to their laboratories and to R. McMichael, K. Ball, M. Ye and P. Shields for their contributions to the data, the model and the critical reading of this manuscript. In addition a special thanks to L. Zhou and S. Saxton for their assistance with the preparation of the manuscript. Finally the author recognizes the National Science Foundation for its contribution of partial support for this research program.

REFERENCES

1. H. Suhl, *J. Phys. Chem. Solids* **1** (1957) 209.
2. P.E. Wigen, *Thin Solid Films* **114** (1984) 135.
3. R.W. Damon and J.R. Eshbach, *J. Phys. Chem. Solids* **19** (1961) 308.
4. C.E. Patton, *Phys. Stat. Solidi (b)* **92** (1979) 211 and **93** (1979) 63.
5. T.S. Hartwick, E.R. Peressini, and M.T. Weiss, *J. Appl. Phys.* **32** (1961) 223S.
6. F.R. Morganthaler, *J. Appl. Phys.* **31** (1960) 95S.
7. E. Schlömann, J.J. Green, and U. Milano, *J. Appl. Phys.* **31** (1960) 386S.
8. W. Wettling, W.D. Wilber, P. Kabos, and C.E. Patton, *Phys. Rev. Lett.* **51** (1983) 1680.
9. H. Yamazaki, *J. Phys. Soc. Japan* **53** (1984) 1155.
10. F. Waldner, D.R. Barberis, and H. Yamazaki, *Phys. Rev. A* **31** (1985) 420.
11. F. Waldner, R. Badii, D.R. Barberis, G. Broggi, W. Floeder, P.F. Meier, R. Stoop, M. Warden, and H. Yamazaki, *J. Mag. Magn. Mat.* **54-57** (1986) 1135.
12. A.I. Smirnov, *Sov. Phys. - JETP* **61** (1985) 815 [*Zh. Eksp. Teor. Fiz.* **88** (1985) 1369].
13. A.I. Smirnov, *Sov. Phys. - JETP* **63** (1986) 222 [*Zh. Eksp. Teor. Fiz.* **90** (1986) 385].
14. G. Gibson and C. Jeffries, *Phys. Rev. A* **29** (1984) 811.
15. S.M. Rezende, F.M. de Aguiar, and O.F. de Alcantara Bonfim, *J. Mag. Magn. Mat.* **54-57** (1986) 1127.
16. F.M. de Aguiar and S.M. Rezende, *Phys. Rev. Lett.* **56** (1986) 1070.
17. S.M. Rezende, O.F. de Alcantara Bonfim, and F.M. de Aguiar, *Phys. Rev. B* **33** (1986) 5153.
18. F.M. de Aguiar, A. Azevedo, and S.M. Rezende, *Phys. Rev. B* **39** (1989) 9448.
19. P. Bryant, C. Jeffries, and K. Nakamura, in *The Physics of Chaos and Systems Far from Equilibrium*, ed. M. Duong-van and B. Nichols (North-Holland, Amsterdam, 1987), p. 25.
20. P. Bryant, C. Jeffries, and K. Nakamura, *Phys. Rev. Lett.* **60** (1988)

1185.

21. P. Bryant, C. Jeffries, and K. Nakamura, *Phys. Rev. A* **38** (1988) 4223.
22. T.L. Carroll, L.M. Pecora, and F.J. Rachford, *Phys. Rev. Lett.* **59** (1987) 2891.
23. T.L. Carroll, F.J. Rachford, and L.M. Pecora, *Phys. Rev. B* **38** (1988) 2938.
24. T.L. Carroll, L.M. Pecora, and F.J. Rachford, *J. Appl. Phys.* **64** (1988) 5396.
25. T.L. Carroll, L.M. Pecora, and F.J. Rachford, *Phys. Rev. A* **40** (1989) 377.
26. H. Yamazaki and M. Warden, *J. Phys. Soc. Japan* **55** (1986) 4477.
27. H. Yamazaki, M. Mino, H. Nagashima, and M. Warden, *J. Phys. Soc. Japan* **56** (1987) 742.
28. R.D. McMichael and P.E. Wigen, *Phys. Rev. Lett.* **64** (1990) 64.
29. P.E. Wigen, R.D. McMichael, and C. Jayaprakash, in *Proc. Commemorative Symposium for Prof. R. Pauthenet*, *J. Mag. Magn. Mat.* (in press).
30. P.E. Wigen, K.D. Ball, and P.J. Shields, in *Proc. First Experimental Chaos Conference* (World Scientific, London, 1992).
31. V.S. L'vov, S.L. Musher, and S.S. Starobinets, *Sov. Phys. - JETP* **37** (1973) 546 [*Zh. Eksp. Teor. Fiz.* **64** (1973) 1074].
32. V.E. Zakarov, V.S. L'vov, and S.S. Starobinets, *Sov. Phys. - Usp.* **17** (1975) 896 [*Usp. Fiz. Nauk.* **114** (1974) 609].
33. K. Nakamura, S. Ohta, and K. Kawasaki, *J. Phys. C* **15** (1982) L143.
34. S. Ohta and K. Nakamura, *J. Phys. C* **16** (1983) L605.
35. K. Nakamura, Y. Okazaki, and A.R. Bishop, *Phys. Rev. Lett.* **57** (1986) 5.
36. F. Waldner, *J. Phys. C* **21** (1988) 1243.
37. T. Holstein and H. Primakoff, *Phys. Rev.* **58** (1940) 1098.
38. V.S. L'vov and L.A. Prozorova, Chapter 4 in *Spin Waves and Magnetic Excitations I*, eds. A.S. Borovik-Romanov and S.K. Sinha (Elsevier, New York, 1988).
39. S.P. Lim and D.L. Huber, *J. Appl. Phys.* **63** (1988) 4151.

40. S.P. Lim and D.L. Huber, *Phys. Rev. B* **37** (1988) 5426.
41. H. Suhl and X.Y. Zhang, *Phys. Rev. Lett.* **57** (1986) 1480.
42. X.Y. Zhang and H. Suhl, *Phys. Rev. B* **38** (1988) 4893.
43. H. Suhl and X.Y. Zhang, *J. Appl. Phys.* **63** (1988) 4147.
44. N.H. Packard, J.P. Crutchfield, J.D. Farmer, and R.S. Shaw, *Phys. Rev. Lett.* **45** (1980) 712.
45. A.M. Frazer and H.L. Swinney, *Phys. Rev. A* **33** (1986) 1134.
46. J.P. Crutchfield and B.S. McNamara, *J. Complex Syst.* **1** (1987) 417.
47. J.-P. Eckmann and D. Ruelle, *Rev. Mod. Phys.* **57** (1985) 617.
48. F. Takens, in *Vol. 898, Lecture Notes in Mathematics*, eds. D.A. Rand and L.S. Young (Springer, Berlin, 1981), p. 366.
49. P. Grassberger and I. Procaccia, *Physica* **9D** (1983) 189.
50. Y. Termonia and Z. Alexandrowixa, *Phys. Rev. Lett.* **51** (1983) 1265.
51. A. Wolf, J.B. Swift, H.L. Swinney, and J.A. Yastano, *Physica* **16D** (1985) 285.
52. P. Grassberger and I. Procaccia, *Phys. Rev. A* **28** (1983) 2591.
53. C. Grebogi, E. Ott, and J.A. Yorke, *Physica* **7D** (1983) 181.
54. H. Chate and P. Manneville, *Physica* **32D** (1988) 409.
55. M. Sparks, *Phys. Rev B* **1** (1970) 3831, 3856, and 3869.
56. M. Sparks, *Ferromagnetic-Relaxation Theory* (McGraw Hill, New York, 1964).
57. R.D. McMichael and P.E. Wigen, *Phys. Rev. B* **42** (1990) 6723.
58. R. Henry, S.D. Brown, P.E. Wigen, and P.J. Besser, *Phys. Rev. Lett.* **28** (1972) 1272.
59. D.J. Seagle, S.H. Charap, and J.O. Artman, *J. Appl. Phys.* **57** (1985) 3076; K.D. McKinstry, C.E. Patton, and Kogekar, *J. Appl. Phys.* **58** (1985) 925; K. Gnatzig, H. Dötsch, M. Ye, and A. Brockmeyer, *J. Appl. Phys.* **62** (1987) 4839.

CHAPTER 9

NONLINEAR DYNAMICS OF PROPAGATING SPIN WAVES IN MAGNETIC FILMS

A.N. SLAVIN*, B.A. KALINIKOS†, and N.G. KOVSHIKOV†

* Department of Physics, Oakland University,
Rochester, Michigan 48309-4401, USA

† St. Petersburg Institute of Electrical Engineering,
197376 St. Petersburg, Russia

Contents

9.1. INTRODUCTION

Ideas from the modern *nonlinear mechanics* and modern *theory of nonlinear waves* have been recently adopted in the physics of magnetic phenomena. In many papers published in the last decade considerable progress has been made in theoretical studies of nonlinear magnetization waves (or spin waves) in media which can be described by the Landau-Lifshitz equation (see, for example, [1-7] and the literature cited therein). The appearance of new concepts such as those of a *soliton* or a *strange attractor*, has stimulated a number of new experiments designed to study nonlinear wave processes in magnetic crystals [8-12] (see also literature cited in Chapters 7 and 8 of this book).

As happened earlier with the pioneering experimental investigations of spin wave parametric instabilities (see review article [13]), rare-earth garnets and especially yttrium-iron garnet (YIG) turned out to provide the perfect testing ground for experimental verification of these new physical ideas and concepts.

The progress in thin film technology made possible industrial production of high quality magnetic films which are, in our view, among the most promising classes of media for the experimental investigation of linear and nonlinear properties of spin waves and for the modeling of nonlinear wave processes in dispersive media in general. The relative simplicity of the techniques for the excitation and detection of spin waves, the richness of their dispersion characteristics, the accessibility of waves from the surface of the film (e.g., by means of Brillouin light scattering [14]), and the weak attenuation per unit propagation distance all make magnetic films convenient objects for investigating spin wave processes.

Single-crystal epitaxial yttrium iron garnet (YIG) films stand out from other materials with the necessary homogeneity of properties and low magnetic losses. The room-temperature magnetic dissipation parameter of these films has an extremely low value (ΔH = 0.1-0.3 Oe at microwave frequencies) so that it is easy to achieve directional propagation of microwave spin waves over distances of few or even few tens of millimeters (which amounts to hundreds and thousands of the wavelengths of a traveling wave). Moreover, the low magnetic losses and the readily attained high specific microwave powers in films when the absolute value of the power is still low make it easy to carry out nonlinear wave experiments. It is worth noting, that the dispersion characteristics of the excited waves and, therefore, also the magnitude and sign of the nonlinearity of YIG films depend strongly on the magnitude and direction of the external static magnetic field, which makes it possible to vary these quantities within a wide range.

Another advantage of magnetic films as a medium for nonlinear wave experiments is the possibility of direct excitation of *propagating* spin waves with finite value of in-plane wave number Q by a localized pumping magnetic field of a microstrip antenna. By increasing the input pumping signal we can achieve the nonlinear regime of spin wave propagation and can study the evolution of the spin wave profile at different distances from the point of excitation.

In this chapter we summarize theoretical, experimental, and numerical results on the nonlinear dynamics of dipole-exchange spin waves propagating in magnetic films with *strongly pinned* surface spins. We restrict our attention to the case when the three-wave parametric decay processes (or first order Suhl processes) [see Eq. (1.131) in Chapter 1] are prohibited by conservation laws and the nonlinear dynamics of spin waves in the film is governed by the *four-wave parametric processes* (or second-order Suhl processes) [see Eqs. (1.134), (1.155) in Chapter 1].

We show that in this case, using a YIG film sample at a room temperature under the input microwave power level below 0.1 W, it is possible to observe a very rich and interesting nonlinear dynamics of spin waves which includes *formation of spin wave envelope solitons, multi-soliton regime of spin wave propagation, modulational instability of spin waves, and formation of a spin wave strange attractor through the destruction of two-frequency quasi-periodic motion.* We also show that all this complicated nonlinear dynamics can be described (at least qualitatively) in the framework of a relatively simple model of a nonlinear Schrödinger equation (NSE) with dissipation.

9.2. SPECTRUM AND PROPAGATION CHARACTERISTICS OF DIPOLE-EXCHANGE SPIN WAVES IN MAGNETIC FILMS WITH PINNED SURFACE SPINS

The experimental investigations reported below were based on specific ("film") properties of the dispersion of spin waves. We used in our experiments both volume dipole-exchange spin waves propagating in perpendicularly magnetized films ($\theta = 0$ in Fig. 1.5) and quasi-surface dipole-exchange spin waves (or transverse spin waves) propagating perpendicularly to the direction of bias magnetic field in tangentially magnetized films ($\theta = \pi/2$, $\phi = \pi/2$ in Fig. 1.5). The dipole-exchange spectrum of spin waves in these particular cases is discussed in detail in Chapter 2 of this book (see Sections 2.8 and 2.9). We shall briefly review below the main features of this spectrum.

Dipole-exchange spin wave modes propagating in a magnetic film of thickness L are the nonuniform plane waves (see Eqs. (1.57),(1.143)) having the in-plane component Q of wave number which assumes continuous values and the transverse component $q_{\xi n}$ which assumes discrete values. In the case of totally pinned surface spins, which we are considering below, $q_{\xi n} = n\pi/L$, where $n = 1,2,3, \ldots$, is the mode number. Each mode has a different distribution of the variable magnetization along the film thickness which depends on the mode number n and satisfies the electrodynamic and exchange boundary conditions (Eqs. (1.60), (1.61)). These distributions are given by the functions $\Phi_n(\xi)$ in Eq. (1.87) (see also Section 2.4 in Chapter 2 for more details).

The spectrum of spin waves in magnetic films is of a discrete multi-mode type and in the cases of perpendicular and tangential magnetization is given by the approximate "diagonal" dispersion equations (2.29) and (2.104) respectively. The dispersion branches of propagating spin wave modes start at the frequencies of spin wave resonance in the film (see, e.g., Eq. (2.97)). The lowest spin wave mode ($n = 1$) in a magnetic film has the biggest group velocity and its dispersion branch in sufficiently thick films ($L > 0.2$ μm) crosses branches corresponding to higher spin wave modes. For the films of the thickness of several μm the crossing points lie in the longwave region of the spectrum ($QL < 1$). Near the crossing points the dipole-dipole hybridization of dispersion branches takes place and the dipole gaps are formed in the spectrum of the film. This effect is described by the secular dispersion equations (2.95) and (2.105) obtained in the first order of the classical perturbation theory.

The width $\Delta\omega_{nn'}$ of dipole gaps in the spectrum can be calculated using Eqs. (2.98), (2.99) and Eqs. (2.106)-(2.109) for the cases of perpendicular and tangential magnetizations correspondingly. The numerical estimation for YIG films of the thickness of several μm having *strongly pinned* surface spins shows that $\Delta\omega_{nn'}/2\pi$ = 20-30 MHz which is much larger than the typical spin wave relaxation frequency in these films ($\Gamma/2\pi = \gamma\, \Delta H/2\pi \cong$ 1-2 MHz for high quality epitaxial YIG films), so the dipole gaps in the spectrum could be easily observed experimentally. Contrary to that, in the case of unpinned surface spins the width of the dipole gaps is much smaller than the relaxation frequency and the gaps are not observed in experiments [15].

Note that, according to Eqs. (2.95) and (2.105), in the perpendicularly magnetized films with pinned surface spins the lowest ($n' = 1$) spin wave mode interacts only with symmetrical (odd numbered) higher volume spin wave modes ($n = 3,5,7, \ldots$), while in tangentially magnetized films the interaction takes place with all higher quasi-surface spin wave modes ($n = 2,3,4, \ldots$), so there must be two times more dipole gaps in the spectrum of a tangentially magnetized film than in the case of perpendicular magnetization.

It is physically clear that the fastest of the spin wave modes excited at the operating frequency determines the transfer of the energy of a microwave signal through the magnetic film. For example, the attenuation of a spin wave $\exp[i(Q\zeta - \omega t)]$ which has traveled a distance $\zeta = l$ in the film is (in decibels)

$$A = 20 \log[\exp(Q''l)], \tag{9.1}$$

where Q'' is the imaginary part of the in-plane wave number $Q = Q' + iQ''$. In the approximation of weak dissipation, when $\omega'' \ll \omega'$ holds (where ω' and ω'' are the real and imaginary components of the spin wave eigenfrequency), we can easily show that $Q'' \cong \omega''/(\partial\omega'/\partial Q') \cong \omega''/v_g$.

Therefore, the relationship (9.1) can be rewritten in the form

$$A = 20 \log\left[\exp\left(\frac{\omega''}{v_g} l\right)\right], \tag{9.2}$$

which shows that a reduction in the group velocity increases the attenuation of the signal transferred by a spin wave. These conclusions are confirmed by the experimental data.

The experimental linear transmission-loss characteristics (i.e., the frequency dependences of the attenuation of a microwave signal transmitted through the film by spin waves in the linear regime) of perpendicularly (a) and tangentially (b) magnetized YIG film samples of the thicknesses $L = 5.6$ μm and $L = 5.3$ μm having strongly pinned surface spins are presented in Fig. 9.1 a,b. One can see a series of transmission minima (attenuation maxima) on each of the characteristics Fig. 9.1 a,b. Note that the frequency separation between the attenuation maxima in a tangentially magnetized film (Fig. 9.1b), where the lowest spin wave mode is hybridized with all higher modes, is roughly two times smaller than in perpendicularly magnetized film (Fig. 9.1 a), where hybridization takes place only near the points of intersection of dispersion branches corresponding to the odd-numbered spin wave modes.

Figure 9.2 a shows a dipole-exchange spectrum of odd-numbered (symmetric) spin wave modes in a perpendicularly magnetized YIG film of thickness $L = 5.8$ μm ($\alpha = 3.1 \times 10^{-12}$ cm^2, $\omega_M/2\pi = 4.9$ GHz, $\omega_H/2\pi = 4.312$ GHz) calculated from the dispersion equations (2.94) and (2.95) in the case of pinned surface spins. Near the dipole gaps in the spectrum the spin wave group velocity changes rapidly with frequency variation from a relatively large value corresponding to the lowest mode to a value which is approximately n^2 times smaller and which corresponds to a higher mode with a number n, i.e., the group velocity exhibits strong dispersion.

Figure 9.2 b shows the frequency dependence of the group velocity $v_g(\omega)$ calculated from Eqs. (2.94) and (2.95) of the fastest waves within the selected frequency intervals. It follows from Eq. (9.2) that this curve describes also the frequency dependence of the spin wave attenuation as it propagates in a film. The inset in Fig. 9.2 c shows the $\omega(Q)$

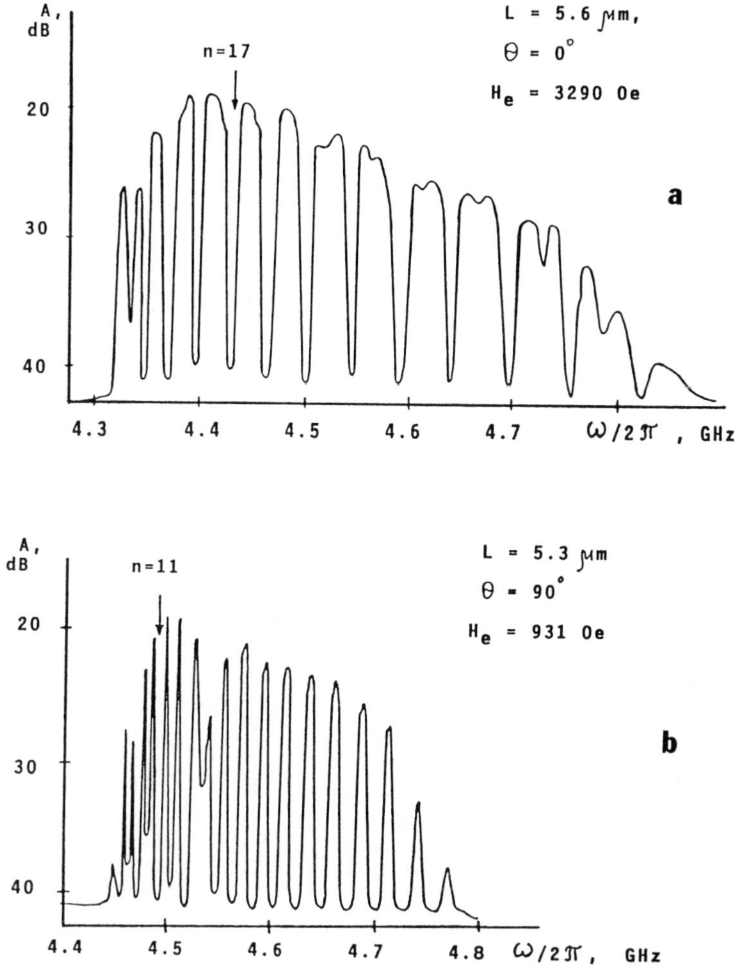

Fig. 9.1. Experimental linear transmission-loss characteristics of: (a) the volume dipole-exchange spin waves in a perpendicularly magnetized YIG film (θ = 0, L = 5.6 μm, ΔH = 0.2 Oe, H_0 = 3290 Oe) and (b) the quasi-surface dipole-exchange spin waves in a tangentially magnetized YIG film ($\theta = \pi/2$, $\phi = \pi/2$, L = 5.3 μm, ΔH =0.2 Oe, H_0= 931 Oe) with strongly pinned surface spins. Arrows show the dipole gaps (n = 17 in (a) and n = 11 in (b)) in the spectra near which our nonlinear experiments were performed.

and $v_g(\omega)$ curves on an enlarged scale in the vicinity of a gap characterized by $n' = 1$ and $n = 17$ where the main part of our nonlinear experiments was performed. The experimental transmission-loss characteristic of the same film is presented in Fig. 9.2 d. The very good agreement between the frequency positions of the dipole gaps found by calculations (Fig. 9.2 b) and in experiment (Fig. 9.2 d), and also the excitation of only the symmetric (odd-numbered) spin wave modes, allowed us to assume that surface spins in the sample chosen for further nonlinear experiments were pinned quite strongly.

We conclude this discussion of the dispersion properties of spin waves by stressing once again that, in the vicinity of the dipole gaps, the spectrum of spin waves of films with pinned surface spins has pronounced frequency intervals of strong dispersion. It is in these intervals that the nonlinear phenomena described below are observed.

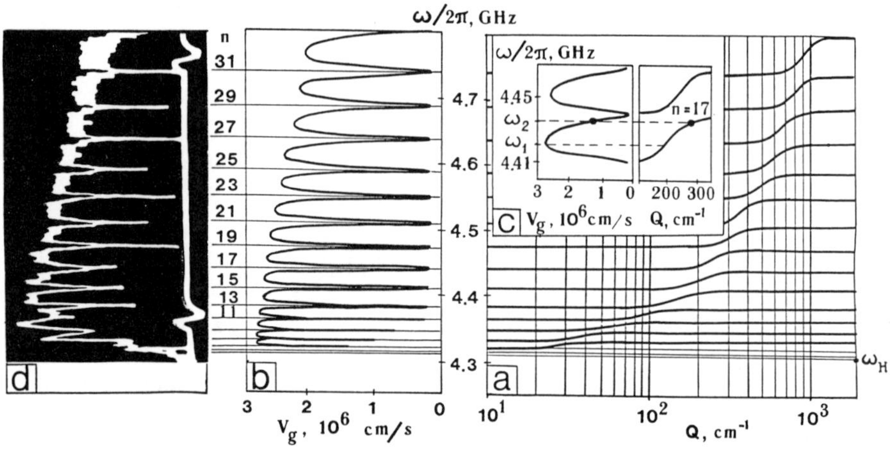

Fig. 9.2. Frequency spectrum (a,c) and group velocity (b,c) of odd-numbered volume spin waves and (d) experimental linear transmission-loss characteristic for these waves in a perpendicularly magnetized ($\theta = 0$) YIG film of thickness $L = 5.8$ μm, with pinned surface spins when $\omega_H/2\pi = 4312$ MHz, $\omega_M/2\pi = 4900$ MHz.

9.3. THEORETICAL DESCRIPTION OF NONLINEAR SPIN WAVE DYNAMICS IN MAGNETIC FILMS IN THE NON-DECAY SPECTRAL REGION

It was shown in Sections 1.5 and 1.6 of Chapter 1 that the nonlinear dynamics of spin waves in magnetics can be described in the framework of the classical Hamiltonian formalism developed in [6,16] for infinite ferromagnetic medium and extended for the case of a ferromagnetic film of finite thickness in our paper [17].

However, the actual calculations of the Hamiltonian coefficients V, S, and T [see, e.g., Eqs. (1.132), (1.133), (1.154)] in magnetic films for the given orientation of the bias magnetic field and the given direction of spin wave propagation usually involve lengthy numerical computations and were done only for several particular cases [18-20].

Fortunately, in the non-decay part of the spin wave spectrum [where the the subharmonic $\omega/2$ of the spin wave eigenfrequency ω lies below the lower boundary ω_H of the spin wave spectrum ($\omega/2 < \omega_H$), and the three-wave parametric decay processes (1.131) are prohibited by conservation laws] there is a simple approximate method which allows us to introduce the dependence of the spin wave eigenfrequency ω on the dimensionless spin wave amplitude ψ [Eq. (1.118)] into the dispersion equation for spin waves, to estimate the value of the coefficient of nonlinear interaction N ($N \cong T$) without lengthy numerical computations, and to derive the evolution equation for the spin wave envelope directly from the nonlinear spin wave dispersion equation [21-24].

Substituting Eq. (1.117) for M_0 in Eqs. (2.95) and (2.105) we obtain a nonlinear dispersion equations of the form

$$\omega = f(Q, q_{\xi n}, q_{\xi n'}, |\psi|^2) \tag{9.3}$$

for the hybridized dipole-exchange spin waves in perpendicularly and tangentially magnetized magnetic films correspondingly. Note that, in accordance with the electrodynamic boundary conditions [Eqs. (1.60)], in a perpendicularly magnetized film $\omega_H = \gamma H_i = \gamma(H_0 - 4\pi M_0)$, while in a tangentially magnetized film $\omega_H = \gamma H_i = \gamma H_0$. Thus ω_H is dependent on the squared modulus of the spin wave amplitude $|\psi|^2$ in a perpendicularly magnetized film and is independent of $|\psi|^2$ in a tangentially magnetized film. This difference leads to the opposite signs of the nonlinear frequency shift $N = \partial\omega/\partial|\psi|^2$ in these two cases.

We can use the "method of envelopes", developed in [21,22] and based on the ideas of geometrical optics, for the analysis of the nonlinear spin wave dispersion equation (9.3). We assume that dimensionless variable magnetization $\psi(\xi,\zeta,t)$ of the hybridized spin wave mode near the dipole gap (n',n) can be represented in the form of a narrow wave packet propagating in the film plane along the ζ axis with carrier wave number Q_0, carrier frequency $\omega(Q_0)$, and a slowly varying envelope $\varphi(\zeta,t)$:

$$\psi(\xi,\zeta,t) = \varphi(\zeta,t)[c_n\Phi_n(\xi) + c_{n'}\Phi_{n'}(\xi)]\, e^{i[Q_0\zeta - \omega(Q_0)t]} , \tag{9.4}$$

where the coefficients c_n, $c_{n'}$ are determined in the zero approximation of the classical perturbation theory when the degeneracy of eigenfrequencies ω_{nQ}, $\omega_{n'Q}$ is taken into account (see, e.g., Eq.(2) on page 139 in [25]). In the case of perpendicular magnetization the explicit expressions for the coefficients c_n, $c_{n'}$ can be written in the form

$$c_n = \left\{ \frac{1}{2} \left[1 \pm \frac{\omega_{nQ}^2 - \omega_{n'Q}^2}{\left[\left(\omega_{nQ}^2 - \omega_{n'Q}^2\right)^2 + 4\omega_M^2\, \Omega_{nQ}\, \Omega_{n'Q}\, P_{nn'}^2 \right]^{1/2}} \right] \right\}^{1/2} ,$$

$$c_{n'} = \pm \left\{ \frac{1}{2} \left[1 \mp \frac{\omega_{nQ}^2 - \omega_{n'Q}^2}{\left[\left(\omega_{nQ}^2 - \omega_{n'Q}^2\right)^2 + 4\omega_M^2\, \Omega_{nQ}\, \Omega_{n'Q}\, P_{nn'}^2 \right]^{1/2}} \right] \right\}^{1/2} , \tag{9.5}$$

where the unperturbed spin wave eigenfrequency ω_{nQ} is determined by Eq. (2.94) [see also Eqs. (2.95)-(2.96)].

Using the assumption of the spectral narrowness of the spin wave packet Eq. (9.4) in the Q-space ($\kappa \ll Q_0$, where κ is the spectral width of the packet) we can expand the nonlinear spin wave dispersion equation (9.3) in a Taylor series near the point $Q = Q_0$, $\varphi = 0$. Taking into account small magnetic dissipation ω_r we get

$$\omega(Q,|\varphi|^2) = \omega(Q_0) + v_g\kappa + \frac{D}{2}\kappa^2 + \ldots + N\,|\varphi|^2 + \ldots + i\omega_r , \tag{9.6}$$

where

$$v_g = \left(\frac{\partial \omega}{\partial Q}\right)_{Q = Q_0,\ |\varphi|^2 = 0} , \quad D = \left(\frac{\partial^2 \omega}{\partial Q^2}\right)_{Q = Q_0,\ |\varphi|^2 = 0} \tag{9.7a}$$

$$N = \left(\frac{\partial \omega}{\partial |\varphi|^2}\right)_{Q=Q_0,\ |\varphi|^2 = 0} , \quad \omega_r = \omega'' \cong \frac{\partial \omega}{\partial \omega_H}\Gamma \tag{9.7b}$$

Now in accordance with [21,22] we substitute in Eq. (9.6) the frequency deviation $\Delta\omega$ and the wave number deviation κ by the operators

$$\begin{aligned} \Delta\omega &= [\ \omega(Q, |\varphi|^2) - \omega(Q_0)] \quad \longrightarrow \quad i\ \partial/\partial t, \\ \kappa &= [Q - Q_0] \quad \longrightarrow \quad -i\ \partial/\partial\zeta\ , \end{aligned} \tag{9.8}$$

As a result we get a nonlinear evolution equation for the envelope $\varphi(\zeta, t)$ of the spin wave packet in the form (1.172) where the nonlinear four-wave coefficient T is substituted by the nonlinear frequency shift N, (9.7b):

$$\left[i\left(\frac{\partial}{\partial t} + v_g \frac{\partial}{\partial \zeta} \right) + \frac{D}{2}\frac{\partial^2}{\partial \zeta^2} - N\ |\varphi|^2 + i\omega_r \right] \varphi(\zeta, t) = 0 \tag{9.9}$$

Numerical calculations performed in [18] for the case of a perpendicularly magnetized YIG film show that the nonlinear frequency shift N calculated in the longwave part ($QL < 1$) of the spin wave spectrum using Eqs. (9.3), (9.7b) gives a good approximation (of about 5% accuracy) for the nonlinear four-wave Hamiltonian coefficient T, and the calculated value of N can be used instead of T in evaluations of the parameters of spin wave envelope solitons in magnetic films.

It is well known [21,22,26] that in the lossless case ($\omega_r = 0$) the envelope of a finite amplitude wave described by Eq. (9.9) is unstable to longitudinal perturbations (and can therefore split into envelope solitons in the direction of propagation) if the Lighthill criterion [27] is observed:

$$D\ N < 0\ . \tag{9.10}$$

Criterion (9.10) requires opposite signs of dispersion and nonlinearity in the medium and guarantees competition between the spreading effects of dispersion and self-steepening effects of nonlinearity on the envelope of the initial spin wave. Envelope instability resulting from this competition is usually called *modulational* instability.

In the dissipative medium ($\omega_r \neq 0$) "pure" envelope solitons do not exist, but if the dissipation is sufficiently small long-lifetime soliton-like nonlinear excitations of the type

$$\varphi(\zeta, t) = \varphi_0 e^{-2\omega_r t} \operatorname{sech}\left[|N/D|^{1/2} \varphi_0 e^{-2\omega_r t} (\zeta - V_0 t)\right], \tag{9.11}$$

where φ_0 and V_0 are independent arbitrary constants denoting the amplitude and velocity of a soliton, can be obtained using the perturbation theory (see, e.g., [26]). The criteria for sufficient smallness of the dissipation can be written in the form

$$\omega_r \ll |N| \; |\varphi|^2, \tag{9.12}$$

$$\omega_r \ll |D/2| \; \kappa^2, \tag{9.13}$$

For example, when the input spin wave radio pulse of duration τ having carrier frequency $\omega(Q_0)$ propagates in the magnetic film the spectral width $\kappa = \kappa_{in}$ of this pulse in the Q-space can be estimated as

$$\kappa_{in} = \frac{2\pi}{\tau \, v_g} \, . \tag{9.14}$$

The criteria (9.12), (9.13) guarantee that the influence of dissipation on the envelope of the input wave is small compared to the influences of competing dispersion and nonlinearity.

Another criterion should be observed to ensure that the distance l of propagation of the input pulse in the dispersive dissipative medium is big enough for the four-wave process of parametric instability Eq. (1.155) (responsible for the spin wave modulational instability and formation of envelope solitons) to develop [28]:

$$l \gg \frac{v_g}{(N \, |\varphi|^2 - \omega_r)} \, . \tag{9.15}$$

We must point out that the condition Eq. (9.15) is independent of the duration τ of the input pulse. Therefore, in the pulsed regime, modulational instability and formation of envelope solitons may be observed also when the input pulse "does not fit" between the transmitting and the receiving antennae, i.e., when $l < \tau v_g$.

Formation of n_s envelope solitons from a rectangular pulse of a finite duration τ occurs also provided the amplitude $|\varphi|$ of the input pulse exceeds a critical amplitude governed by the condition [29]

$$|\varphi|^2 > \frac{(2n_s - 1)^2 \kappa_{in}^2 |D|}{16 |N|} . \tag{9.16}$$

We note that in soliton theory, when a lossless dispersive media is considered, the necessary criterion of modulational instability and envelope soliton formation is the Lighthill criterion (9.10). In any real media dissipation should be taken into account and the fulfillment of the criteria (9.12), (9.13) is equally significant for the existence of envelope solitons given by Eq. (9.11). It is clear that in media with a given dissipation, criterion (9.12) can be observed simply by increasing the input pulse amplitude. It is also clear that, in order to satisfy the criterion Eq. (9.13), we must find the regions of high dispersion $|D|$ in the spin wave spectrum of our experimental sample. In our experiments we found such spectral regions of high dispersion (where both criteria Eqs. (9.10), (9.13) are fulfilled) near the dipole gaps in the dipole-exchange spin wave spectrum of YIG films with pinned surface spins [8,30].

We use below the nonlinear Schrödinger equation with dissipative term (9.9) as our main mathematical model for the description of the experimentally observed nonlinear dynamics of propagating spin waves in the non-decay part of the dipole-exchange spectrum.

9.4. MEASUREMENT TECHNIQUES

The propagation of nonlinear spin waves in YIG films was investigated using a prototype delay line (Fig. 9.3). This prototype consisted of two alumina plates on which transmitting and receiving spin wave antennae of width W = 30 μm and length up to 0.5 cm, together with microstrip supply lines, were formed by photolithography. These plates were attached to a metal base consisting of two movable halves which made it possible to vary continuously the distance between the transmitting and receiving antennae from 0.1 to 2.5 cm. A YIG film sample was placed on top of the antennae and kept in place by a pressure device. The delay line was placed inside a gap of a permanent magnet which generated a magnetic field H_0 of the required intensity and direction.

The value of the bias magnetic field (H_0 = 3290 Oe in the case of perpendicular magnetization and H_0 = 931 Oe in the case of tangential magnetization) was chosen in such a way that three-wave decay processes [Eq. (1.131)] were prohibited by conservation laws. Thus, the nonlinearity of the propagating spin waves was determined by the four-wave processes [Eq. (1.134)].

Measurements were made at room temperature under pulsed and continuous excitation.

A microwave signal, reaching the transmitting antenna of the delay line from a klystron oscillator via a microwave amplifier, was used to excite spin waves. The maximum signal power beyond the amplifier could reach 1 W. The profile of the input microwave signal was monitored by deflecting part of it, identifying it, and applying it to one of the inputs of a dual-trace oscilloscope having operational bandwidth of Δf = 350 MHz. The output microwave signal, transmitted by the YIG film and recorded using the receiving antenna of the delay line, was also rectified and applied to the other input of the oscilloscope.

The microwave spectrum of the signals at the input and output of the delay line was recorded using a spectrum analyzer. The power of the input and output microwave signals was measured with a thermistor power meter to the accuracy within 20%. The frequency measurements were made employing an electronic digital voltmeter.

In the pulsed regime of spin wave excitation, a microwave signal supplied by the klystron oscillator reached the microwave amplifier after passing first through a modulator. Modulation was imposed by pulses of 3 ns to 10 μs duration and a repetition frequency 10^2 - 10^5 Hz, supplied by a generator of rectangular pulses. The delay of a

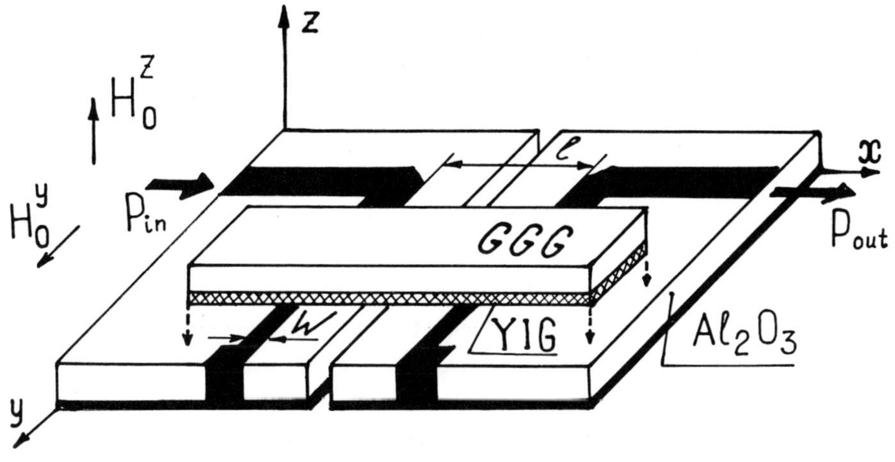

Fig. 9.3. Prototype delay line

microwave rf pulse and the reduction of its amplitude during the propagation in the YIG film between the transmitting and receiving antennae of the delay line were displayed on the oscilloscope screen.

The investigated samples were single-crystal YIG films of thickness L = 3-16 μm characterized by low magnetic losses (dissipation parameter ΔH = 0.2 - 0.3 Oe). These films were grown on gadolinium-gallium garnet (GGG) substrates of the orientation <111> by the method of liquid phase epitaxy. For our experiments on the nonlinear dynamics of highly dispersive dipole-exchange spin waves we selected YIG film samples having strongly pinned surface spins. The criterion for judging the state of surface spins was the transmission-loss characteristic of a film, i.e., the frequency dependence of the attenuation of the microwave signal transmitted by the film (see Fig. 9.1 and Fig. 9.2 d). This characteristic was investigated using standard microwave meters with sweep generators at the incident microwave power levels less than 5 μW, which ensured that the measurements were carried out under linear conditions.

9.5. EXPERIMENTAL RESULTS

9.5.1. Pulsed regime. Spin wave envelope solitons

The experiments under pulsed excitation were performed on both perpendicularly and tangentially magnetized YIG films of thickness L = 5.8 μm and L = 5.3 μm, respectively. Rectangular input radio pulses of duration τ (τ = 180 ns in the perpendicularly magnetized film and τ = 1000 ns in the tangentially magnetized film) having carrier frequency in the vicinity of one of the dipole gaps in the spectrum (n = 17 and n = 11, correspondingly) were introduced to the input antenna, and propagated in the film for the distance l (l = 0.4 cm and l = 1.4 cm, correspondingly) and were registered at the output antenna (Fig. 9.3). The power of the input radio pulses was varied within the range P_{in} = 0.05-150 mW.

The results of the pulse experiments for the cases of perpendicular and tangential magnetization are presented in Figs. 9.4 (A)-(I) and 9.5 (A)-(I), respectively. The upper traces in the oscillograms in Figs. 9.4 and 9.5 show the positions and the forms of the input pulses, and the lower traces demonstrate the shapes of the envelope and the positions of the output spin wave pulses. Figures 9.4 (J) and 9.5 (J) show, on an enlarged scale, parts of the transmission-loss characteristics of the prototype delay line near the dipole gaps in the spectrum having numbers (n' = 1, n = 17) and (n' = 1, n = 11), respectively.

Oscillograms (A), (B), and (C) were taken at a constant level of input power P_{in} when the carrier frequency varied from the spectral point A of low dispersion situated between the gaps to the spectral point C of high dispersion situated on one of the slopes of spectral gap [see Figs. 9.4 (J) and 9.5 (J)] . It can be seen that at point A the usual delayed output pulse is obtained. When the carrier frequency is shifted into the region of high dispersion the spin wave pulse spreads and its amplitude decreases [Figs. 9.4 (B)-(C) and 9.5 (B)-(C)].

Oscillograms (C)-(F) in Figs. 9.4 and 9.5 were obtained at a fixed carrier frequency situated on the slope of the dipole gap in the spin wave spectrum of the film where the Lighthill criterion, Eq. (9.10), is fulfilled. We note, that in perpendicularly (Fig. 9.4) and tangentially

(Fig. 9.5) magnetized films the Lighthill criterion is fulfilled on the opposite slopes of the spectral gaps because of the opposite signs of the nonlinear frequency shift *N* (see the commentary to Eq. (9.3)). It can be seen that, with the increase of the input power, the spin-system nonlinearity compensates dispersion, and at a certain input power level the high and narrow peak (τ_s = 120 ns and τ_s = 360 ns for perpendicular and tangential magnetization, correspondingly) appears in the place of the delayed pulse, i.e., the *spin-wave envelope soliton* is formed [see

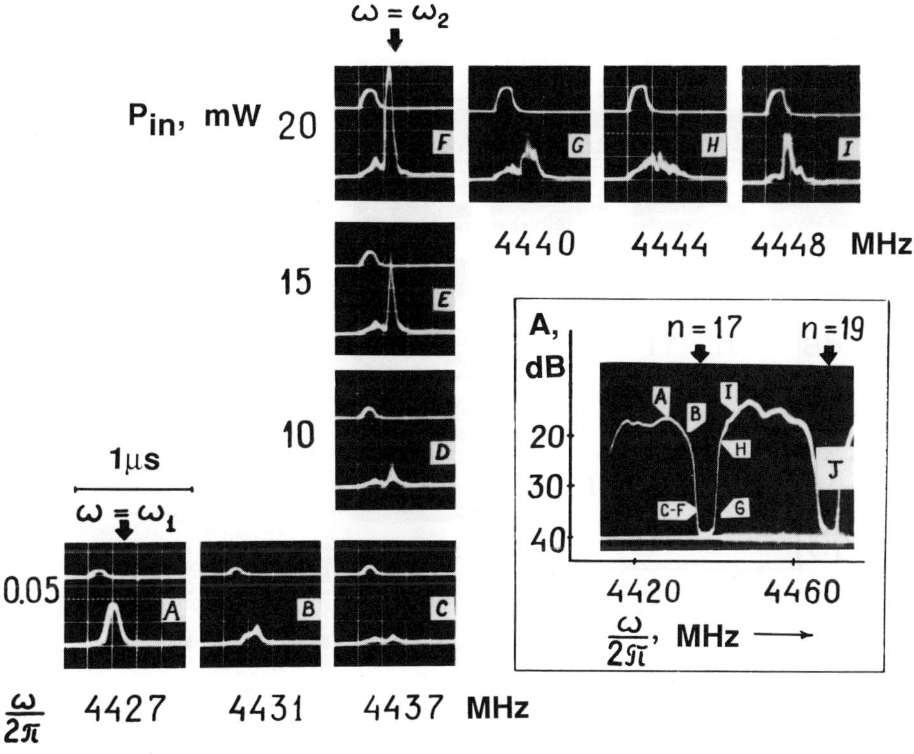

Fig. 9.4. (A)-(I) Formation of the envelope soliton from the input pulse signal (τ = 180 ns) in perpendicularly magnetized YIG film sample (L = 5.8 μm); (J) dipole gap in the dipole-exchange spin wave spectrum (n' = 1, n = 17) where formation of solitons was observed.

Figs. 9.4 (E)-(F) and 9.5 (E)-(F)]. We call this peak the spin wave envelope soliton in accordance with the physical definition of a soliton as a stable nonlinear excitation emerging as a result of a balance between a spreading effect of dispersion and a self-steepening effect of nonlinearity.

It should be pointed out that we observed the formation of solitons only in the spectral regions of high dispersion where the Lighthill criterion is fulfilled [points (C)-(F) in Figs. 9.4 and 9.5]. In the regions of high dispersion of the opposite sign situated on the opposite slopes of the gaps [points (G) and (H) in Figs. 9.4 (J) and 9.5 (J)] the

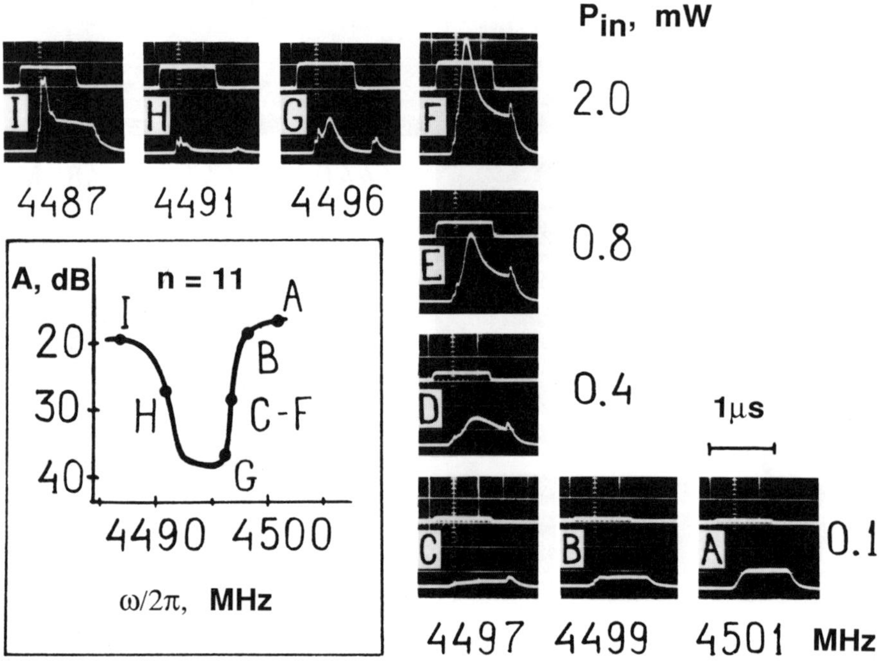

Fig. 9.5. (A)-(I) Formation of the envelope soliton from the input pulse signal (τ = 1000 ns) in the tangentially magnetized YIG film sample (L = 5.3 μm); (J) dipole gap in the dipole-exchange quasi-surface spin wave spectrum (n' = 1, n = 11) where formation of solitons was observed.

formation of solitons did not take place [see oscillograms (G) and (H) in Figs. 9.4 and 9.5].

With an increase of amplitude or/and duration of the input pulse at the points (C)-(F), the spin wave soliton profile fragmented, and the multisoliton regime of spin wave propagation was observed [28,31].

This is illustrated by Fig. 9.6 where the oscillograms of the envelopes of the output pulse are presented in the case when the power of the input radio-pulse of duration τ = 180 ns propagating in a perpendicularly magnetized film was increased from 20 to 120 mW. The distance between the input and output antennae was l = 0.4 cm and the carrier frequency of the input pulse corresponded to the point (C)-(F) in Fig. 9.4 J. It is clear that at P_{in} = 120 mW three soliton-like maxima are observed on the profile of the output pulse.

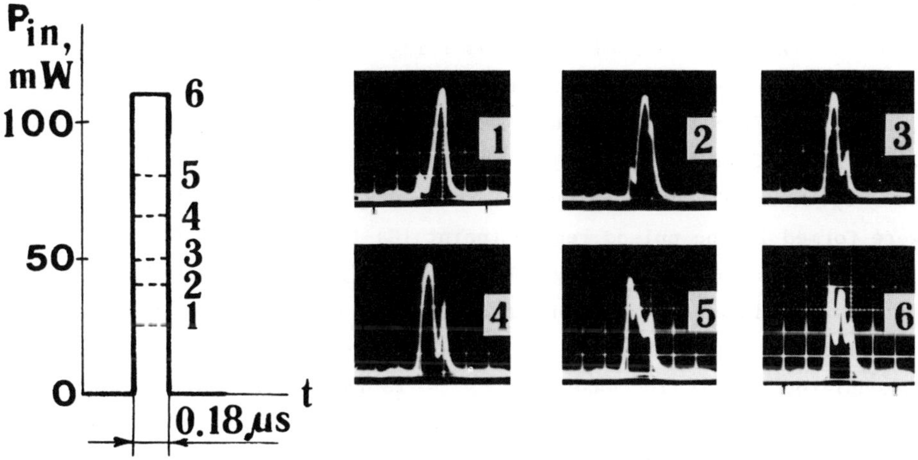

Fig. 9.6. The oscillograms of the envelope of a pulse signal at the output of the prototype delay line (l = 0.4 cm) obtained for the input pulse of duration τ = 0.18 μs when increasing values of theinput power P_{in} (the values in parentheses give the calculated values of square modulus of the dimensionless input amplitude $|\varphi|^2 \times 10^3 \sim P_{in}$):
1. 25 mW (2.0); 2. 40 mW (3.2); 3. 50 mW (4.0); 4. 65 mW (5.2); 5. 80 mW (6.4); 6. 110 mW (8.8).

9.5.2. Continuous regime. Modulational instability of spin waves and transition to chaos

In the case of continuous excitation we studied the propagation of dipole-exchange spin waves of finite amplitude with the frequency lying in the vicinity of one of the dipole gaps of a perpendicularly magnetized YIG film [9]. The aim of the measurements carried out in the continuous regime was to determine whether an unmodulated traveling spin wave with frequency lying in the parts of the spectrum where envelope soliton formation was observed in the pulse regime was modulationally unstable when the amplitude was increased.

The results of the measurements are presented in Fig. 9.7 (1)-(8). This figure gives the frequency spectra and the corresponding oscillograms of the envelope of a microwave signal transmitted through the perpendicularly magnetized YIG film, which were obtained as the input power P_{in} of an unmodulated continuous input signal gradually increased. The lower traces in the oscillograms (1)-(8) correspond to the zero level of the video signal. The oscillograms and spectrograms in Fig. 9.7 were obtained for a YIG film of thickness $L = 5.8$ μm when the distance between the antennae was $l = 0.4$ cm. The input signal frequency $\omega_c/2\pi = 3.453$ GHz was selected to have the same position in the spin wave spectrum of the film as the frequency at which envelope solitons were formed in the pulsed regime (point (C)-(F) in Fig. 9.4).

The first pair of photographs [Fig. 9.7 (1)] was obtained under conditions of linear propagation of spin waves when the receiving antenna detected a wave of constant amplitude. When the power of the input unmodulated signal increased to $P_{in} = 13$ mW a spin wave of constant amplitude became unstable during its propagation in the YIG film. The receiving antenna then recorded a modulated spin wave with an asymmetric frequency spectrum in which the amplitude of the low-frequency satellite $f_c - f_m$ was considerably greater than the amplitude of the high-frequency satellite $f_c + f_m$, and the modulation frequency was $f_m = 5.2$ MHz [Fig. 9.7 (2)].

At the input power level of $P_{in} = 16$ mW a second satellite of frequency $f_c - f'_m$, ($f'_m = 6.2$ MHz) appeared in the spectrum of the output

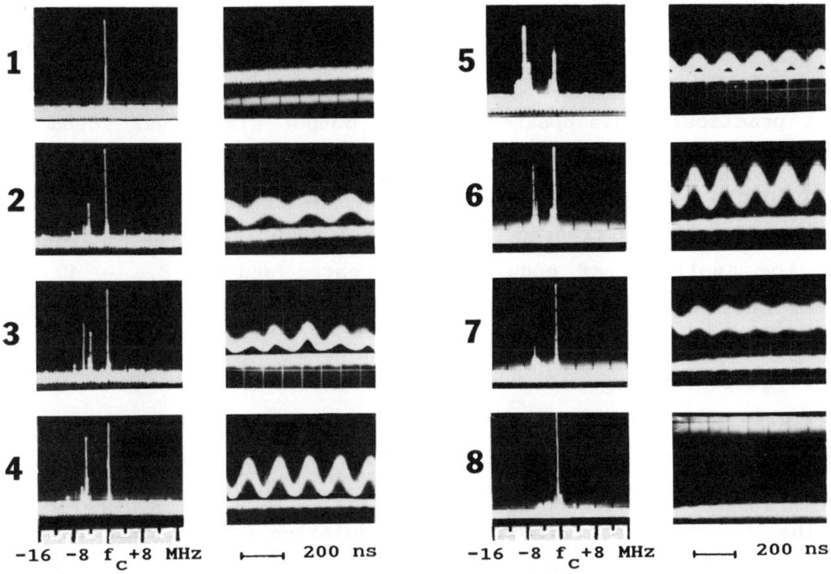

Fig. 9.7. The spectrograms (left) and the oscillograms (right) of a continuous signal at the output of the prototype delay line obtained for the input signal frequency corresponding to the point (C)-(F) in Fig. 9.4 (J) and different input powers P_{in} : 1. 10 mW; 2. 12 mW ; 3. 16 mW ; 4. 18 mW; 5. 20 mW; 6. 22 mW; 7. 30 mW; 8. 40 mW.

signal [Figs. 9.7 (3)]. The beat between the two satellites f_m and f'_m created a second (lower) frequency component $F_m = f'_m - f_m = 1$ MHz in the spectrum of the spin wave modulational instability.

A further increase in P_{in} increased the modulation coefficient M [see Fig. 9.7 (4)], and at $P_{in} = 20$ mW there was an abrupt change from a regime characterized by the modulation coefficient $M < 1$ [Fig. 9.7 (4)] to an "over modulation" regime with $M > 1$ [Fig. 9.7 (5)]. In the latter regime the minima of the signal envelope at the output of the YIG film reached zero level and the energy concentrated in the low-frequency side band of the signal spectrum exceeded the energy at the carrier frequency f_c [Fig. 9.7 (5)].

A further increase in P_{in} again resulted in an abrupt change to the regime with $M < 1$ [Fig. 9.7 (6)]. Then at $P_{in} = 30$ mW the modulation coefficient M decreased, the frequency spectrum of the modulation became wider and continuous, and the profile of the envelope of the output signal exhibited strong noise distortions [Fig. 9.7 (7)]. With the increase in P_{in} the coherent modulation of the spin wave envelope at the output practically disappeared and the output signal was found to be chaotically modulated. The spectrum components of modulation then merged into a single noise-like peak and the average frequency of this chaotic modulation decreased. With the further increase of the input power the amplitude of modulation decreased and at P_{in} = 40 mW we observe only weak chaotic modulation in the spectrum of the output signal [Fig. 9.7 (8)].

In comparing Figs. 9.7 (1)-(8) we should bear in mind that the sensitivity of the instruments used in our apparatus decreased in inverse proportion to P_{in} so that in Fig. 9.7 (8) the amplitude scale was four times greater than in Fig. 9.7 (I).

The observed chaotization of the modulation spectra of the modulationally unstable finite-amplitude spin waves led us to the assumption that a spin wave *strange attractor* may be formed in the phase space of the system when the power P_{in} of the input signal is increased.

To study the transition from coherent to chaotic modulation in greater detail and to elucidate how this regime is connected to the regime of spin wave envelope soliton formation described in the previous section we performed a series of experiments with relatively long input pulse signals when the pulse duration τ was much larger than the characteristic period of modulational instability $T_m = 1/f_m \cong 200$ ns. Again we used a perpendicularly magnetized YIG film (L = 5.8 μm) and the carrier frequency $f_c = \omega_c/2\pi$ of the input pulse of duration τ = 8 μs was situated at the region of high negative dispersion where the Lighthill criterion (9.10) is observed [point (C)-(F) in Fig. 9.4 (J)]. The distance l between the antennae was chosen equal to l = 0.4 cm.

The oscillograms and the spectrograms of the output pulsed signal corresponding to different values of input power 5 mW < P_{in} < 25 mW are presented in Fig. 9.8. The oscillograms of the envelopes of the output pulses corresponding to P_{in} = 12 mW and P_{in} = 15 mW are also shown on an enlarged scale in Fig. 9.9.

When the input power is below the threshold of modulational instability we get the unmodulated envelope of the output pulse and only one (carrier) frequency f_c in the output signal spectrum [Fig. 9.8 (1)]. This regime corresponds to the stable equilibrium point in the phase space of the system.

At the threshold of modulational instability [$P_{in} \cong 12$ mW, see Fig. 9.9 (1)] several soliton-like pulses appear on the envelope of the output pulse. At this level of input power the increment of the four-wave parametric process (1.155) responsible for the modulational instability of the propagating spin wave is comparable with the magnetic dissipation in the film, so the modulation is heavily damped. This regime corresponds to the stable focus in the phase space of the system.

At $P_{in} = 14 - 15$ mW [Figs. 9.8 (2) and 9.9 (2)] we have a well-developed single-frequency modulation of the output pulse envelope and a frequency satellite $f_c - f_m$ (where $f_m = 5.3$ MHz) appears in the spectrum of the output signal. A symmetric frequency component $f_c + f_m$ did not appear in the spectrum of the output signal, because it was located in a region with a much lower group velocity and, consequently, a stronger spatial attenuation (see Eq. (9.2) in this chapter and Fig.8 in [28]). Two different parts can be distinguished in the form of envelope of the output pulse: the transition part ($\Delta t \cong 900$ ns) consisting of several soliton-like pulses, and the regular part where a stationary regime of modulation is established. The appearance of the satellite $f_c - f_m$ in the output signal spectrum corresponds to the formation of a stable limit cycle in the phase space of the system.

With the increase of the input power up to $P_{in} = 16$ mW, as in the case of continuous excitation (Fig. 9.7), a second (lower) frequency component $F_m \cong 1.1$ MHz appeared in the spectrum of the modulational instability [Fig. 9.8 (3)]. This low frequency quasi-periodic "modulation of modulation" can be interpreted as a formation of a two-dimensional torus in the phase space of the system.

At $P_{in} = 20$ mW the modulation spectrum becomes more complicated [Fig. 9.8 (4)] and at $P_{in} = 25$ mW a broad peak of chaotic modulation appears in the spectrum [Fig. 9.8 (5)]. We assume that the appearance

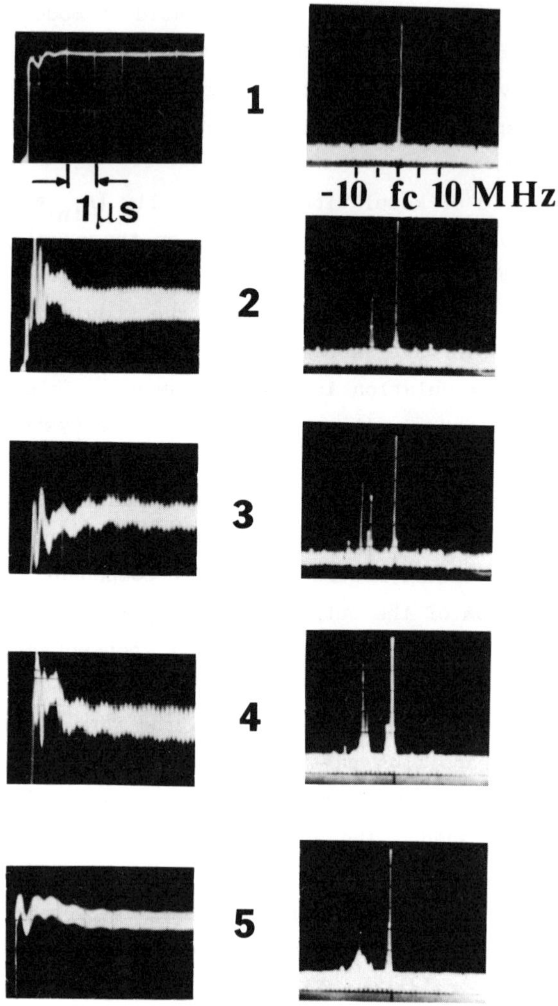

Fig. 9.8. The oscillograms of the envelope (left) and the spectrograms (right) of the output pulse signal (τ = 8 μs) transmitted through the perpendicularly magnetized YIG film (L = 5.8 μm, l = 0.4 cm) for different values of the input power P_{in} (the values in parentheses give the calculated values of square modulus of the dimensionless input amplitude $|\varphi|^2 \times 10^3 \sim P_{in}$): 1. 5 mW (0.4); 2. 14 mW (1.12); 3. 16 mW (1.28); 4. 20 mW (1.60); 5. 25 mW (2.00).

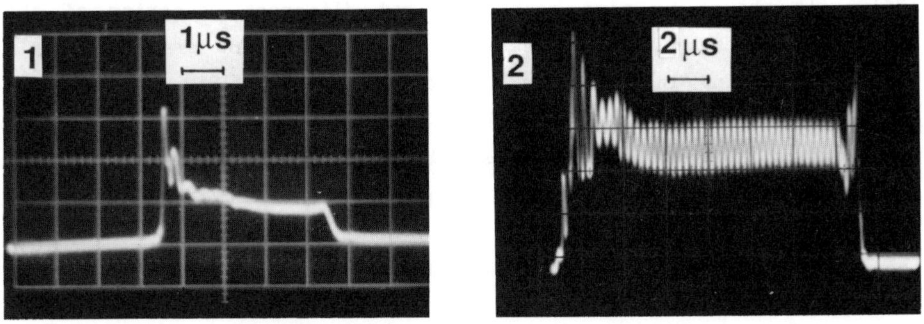

Fig. 9.9. The oscillograms of the envelope of the output pulse signal (τ = 8 μs) transmitted through the perpendicularly magnetized YIG film (L = 5.8 μm, l = 0.4 cm) for two different values of the input power P_{in} (the values in parentheses give the calculated values of the square modulus of the dimensionless input amplitude $|\varphi|^2 \times 10^3 \sim P_{in}$): 1. 12 mW (0.96); 2. 15 mW (1.20).

of chaotic modulation in the spectrum corresponds to the formation of a *spin wave strange attractor* in the phase space. With the further increase of the input power the amplitude of chaotic modulation decreases and at $P_{in} > 40$ mW the output signal becomes practically unmodulated as in the case of continuous excitation [Fig. 9.7 (8)].

A similar development of spin wave modulational instability with modulation frequencies of the order of several MHz was observed in our experiments on different YIG film samples with pinned surface spins having thickness L = 3.5 - 8.0 μm.

We note that *modulational instability* is not the only possible type of instability of finite amplitude spin waves propagating in magnetic films. Other types of four-wave instabilities, like *four-wave decay instability* [32] and *kinetic instability* [33], were observed for spin waves propagating in thin (L < 1 μm) YIG films in the spectral regions of relatively low dispersion between the dipole gaps in the spectrum

[34]. It should be also mentioned that different scenarios of the transition to chaos in a system of purely dipolar low dispersive spin waves propagating in obliquely magnetized YIG films had been investigated experimentally in [35].

However, the *modulational instability* observed in our above described experiments [9,28,36] has several characteristic features that distinguish it from the instabilities of other types. It is observed in the spectral regions of high dispersion where, in the pulsed regime of spin wave excitation, the formation of envelope solitons takes place. It has an asymmetrical frequency spectrum and exhibits a large modulation depth of the output signal. Modulational instability has the lowest power threshold and the packet of spin waves taking part in this process [see Eq. (1.155)] is spectrally narrow ($\kappa \ll Q_0$). This instability also demonstrates a distinct scenario of transition to chaos through the destruction of two-frequency quasi-periodic motion similar to the scenario of Ruelle-Takens [37].

We show in the next section that the above described development of the spin wave modulational instability which occurs with the increase of the input signal amplitude can be effectively described in the framework of a model of a nonlinear Schrödinger equation with dissipation, Eq. (9.9).

9.6. NUMERICAL MODELING OF SPIN WAVE NONLINEAR DYNAMICS IN FILMS

9.6.1. Formation of spin wave envelope solitons

To model the process of formation and propagation of dipole-exchange spin wave envelope solitons in YIG films described in Section 9.5.1, we solved the nonlinear Schrödinger equation with dissipation [Eq. (9.9)] numerically under the following boundary and initial conditions for the spin wave envelope $\varphi(\zeta, t)$ [18,38,39] :

$$\varphi(\zeta=0, t) = \begin{cases} \varphi_0 \, , & \text{when} \quad 0 < t < \tau \, , \\ 0 \, , & \text{when} \quad t > \tau \quad ; \end{cases} \tag{9.17}$$

$$\varphi(\zeta=l_0, t) = 0 \; ; \tag{9.18}$$

$$\varphi(\zeta, t=0) = \begin{cases} \varphi_0 , & \text{when} \quad \zeta = 0 , \\ 0 , & \text{when} \quad \zeta \neq 0 . \end{cases} \tag{9.19}$$

The conditions Eqs. (9.17)-(9.19) mean that the rectangular input pulse of amplitude φ_0 and duration τ is applied at the point $\zeta = 0$ at the moment $t = 0$ and then propagates in the media for the distance l_0. The full distance of propagation was chosen to be sufficiently long ($l_0 = 4$ cm) that the propagating pulse vanishes at the final point $\zeta = l_0$ due to the dissipation.

A purely implicit difference scheme was used in the numerical solution of Eq. (9.9). The system of linear difference equations was solved by the sweep method and the nonlinear term was refined by the method of simple iterations [40].

We calculated the time profiles of the spin wave pulse envelope $|\varphi(l,t)|^2$ at a fixed point $\zeta = l$ ($0 < l < l_0$) in the media (where the receiving antenna was situated in our experiments) for different values of the input pulse amplitude φ_0.

The calculations were made for the case of a perpendicularly magnetized YIG film of thickness $L = 5.8$ μm with pinned surface spins in three spectral points: A, C, and G. The point A corresponds to the case when the carrier frequency f_c of the input pulse is situated in the region of low dispersion between the dipole gaps in the dipole-exchange spin wave spectrum of the film [see point A in Fig. 9.4 (J)]. The points C and G correspond to the positions of the carrier frequency in the region of high dispersion on the opposite slopes of the spectral gap, so that at the point C the Lighthill criterion is satisfied and at the point G it is not satisfied [see Fig. 9.4 (J)].

The values of the spectral derivatives v_g and D in Eq. (9.9) for all three points were calculated using the secular dispersion equation, Eq. (2.95), (see Fig. 8 in [28]) and were also independently verified from the experimental transmission-loss characteristic Figs. 9.1 (a) and 9.2 (d) using Eq. (9.2) and the theory of dipole-exchange spin wave excitation in magnetic films [41]. The nonlinear coefficient N was calculated from the same dispersion equation [see Eq. (9.3)] where the dependence of the spin wave frequency ω on the square modulus of the spin wave amplitude $|\varphi|^2$ was introduced using the method described in Section 1.5.1 of Chapter 1.

The duration of the input pulse τ = 180 ns and the position of the point l = 0.4 cm where the output pulse was registered were chosen in accordance with the experimental values in [8,28,39] (see Figs. 9.4 and 9.6). All the values for the parameters of Eq. (9.9) which were used in the numerical calculations are presented in Table 9.1.

The threshold value of $|\varphi|^2$ for the four-wave process Eq. (1.155) responsible for envelope soliton formation and modulational instability of spin waves in the case under consideration can be estimated using Eq. (1.163) as follows:

$$|\varphi|^2_{th} = \frac{\omega_r}{T_{12}} \cong \frac{\omega_r}{N} = 2 \times 10^{-4} \ . \tag{9.20}$$

Hence, at the point A (see Table 1) we have chosen $|\varphi|^2_0 = 8 \times 10^{-5}$ to have a linear regime of spin wave propagation, and at the points C and G with the increase of $|\varphi|^2_0$ the process of spin wave propagation becomes nonlinear.

The results of our numerical calculations in comparison with the experimental results from Figs. 9.4 and 9.6 are presented in Figs. 9.10 and 9.11. The curve in the upper right-hand corner of Fig. 9.10 is the part of the experimental transmission-loss characteristic obtained under continuous excitation on the YIG film of thickness L = 5.8 μm near the dipole gap (n′ = 1, n = 17) in the dipole-exchange spin wave spectrum of the film (see also Figs. 9.1 a and 9.2 d). This curve qualitatively corresponds to the frequency dependence of the spin wave group velocity $v_g(Q)$ near the gap in accordance with Eq. (9.2). The spectral points A, C, and G where experiments and numerical calculations were performed are shown on this curve.

The experimental oscillograms of the output spin wave pulse envelopes (left) corresponding to different values of the input pulse power P_{in} and the numerically calculated time profiles of the output spin wave pulse envelopes $|\varphi(l,t)|^2/|\varphi(l,t)|^2_{max}$ (right) are shown in Fig. 9.10.

In the point A, where dispersion is low, the usual delayed linear spin wave pulse is obtained (see Fig. 9.10 A). In the point C the input linear pulse spreads due to high dipole-exchange dispersion [see Fig. 9.10 C(1)]. With the increase of the input pulse amplitude nonlinearity compensates the spreading effect of dispersion and a spin wave envelope soliton is formed [Fig. 9.10 C(2), C(3)]. With further increase of the input pulse amplitude multisoliton regime of spin wave propagation takes place [Fig. 9.10 C(4), B(5)]. Note that at the point G where the disp-

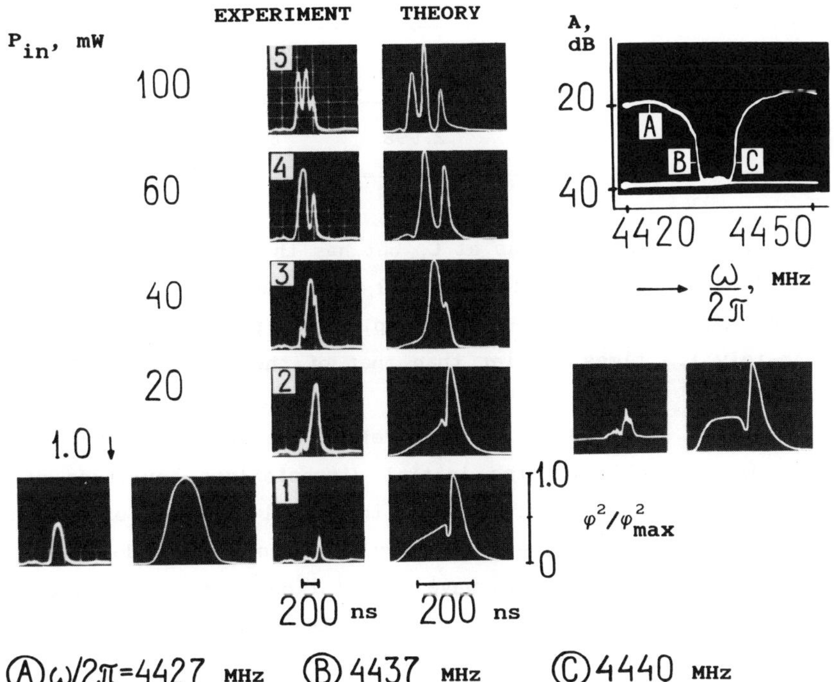

Fig. 9.10. Comparison between the experimental oscillograms (left) of the output spin wave pulses obtained on a YIG film sample of thickness $L = 5.8$ μm for $l = 0.4$ cm near the dipole gap ($n' = 1$, $n = 17$) and the numerically calculated profiles (right) of the output spin wave pulses obtained from Eq. (9.9). The values of the square modulus of the dimensionless input amplitude $|\varphi|_0^2 \times 10^3$ used in the numerical calculations are: A - 0.08; C(1) - 0.08; C(2) - 0.9; C(3) - 1.2; C(4) - 1.6; C(5) - 2.5 ; G - 1.2. The fragment in the upper right-hand corner shows a part of the linear transmission-loss characteristic of the film near the spectral gap and the positions of the spectral points A, C, and G in which measurements and calculations were performed.

Table 9.1

Parameters of Eq. (9.9) in the spectral points A, C, and G near the dipole gap (n' = 1, n = 17) in the dipole-exchange spin wave spectrum of the perpendicularly magnetized 5.8 μm-thick YIG film.

	$V_g \cdot 10^{-6}$, cm/s	$D \cdot 10^{-5}$, cm^2/s	$N \cdot 10^{-10}$, s^{-1}	$\omega_r \cdot 10^{-6}$, s^{-1}	$\lvert\varphi\rvert_0^2 \cdot 10^3$
A	3.6	- 0.025	3.014	6.0	0.08
C	1.8	- 3.0	3.000	6.0	(0.08-3.0)
G	1.8	+ 3.0	2.998	6.0	1.2

ersion is of the same value as at C, but has the opposite sign, no spin wave solitons are formed (Fig. 9.10 G), and the maximum observed (and calculated) amplitude of the output spin wave pulse at this point is approximately ten times smaller than that of the soliton profile (Fig. 9.10 C(3)).

To compare experimental and theoretical results quantitatively it is necessary to calculate the initial dimensionless amplitude of the spin wave envelope $|\varphi|_0^2$ corresponding to the given value of the input power. In the linear regime this calculation could be made using the theory of dipole-exchange spin wave excitation in ferromagnetic films [41]. For the conditions of our experiments [8,28,39] the calculation shows that that $|\varphi|_0^2$ increases proportionally with the increase of P_{in} and for P_{in} = 1 mW is equal to $|\varphi|_0^2 = 8 \times 10^{-5}$.

In the nonlinear regime (when P_{in} > 2.5 mW and $|\varphi|_0^2 > 2 \times 10^{-4}$) the linear theory of spin wave excitation [41] could be used only for rough estimations. Therefore for P_{in} > 2.5 mW there is no quantitative agreement between the value of $|\varphi|_0^2$ obtained by estimation from the experimentally measured value of P_{in} for a certain spin wave output pulse profile and the value of $|\varphi|_0^2$ in numerical calculations corresponding to the similar form of the spin wave pulse profile. For example, in the case of a single-soliton profile when the value of P_{in} lies between 20 mW and 40 mW [see Fig. 9.6 (1)-(2) and Fig. 9.10 C(2), C(3)], the theory [41] gives $1.6 < |\varphi|_0^2 \times 10^3 < 3.2$, while in numerical modeling similar single-soliton profiles are obtained in the much narrower amplitude interval $0.8 < |\varphi|_0^2 \times 10^3 < 1.2$.

Nevertheless, as is clear from Fig. 9.10, the forms and the durations of the calculated output pulse profiles are quite similar to that observed experimentally.

We find that in the numerical modeling in accordance with the experimental observations:

(i) spin wave envelope solitons are formed only in the spectral regions of high negative dispersion near the gaps in the dipole-exchange spin wave spectrum where the Lighthill criterion, Eq. (9.10), is satisfied;

(ii) with the increase of the amplitude of the input pulsed signal (duration τ = 180 ns) the multisoliton regime of spin wave propagation takes place;

(iii) no spin wave envelope solitons are formed in the spectral regions of low dispersion between the dipole gaps where the dissipation criterion, Eq. (9.13), is not fulfilled even though the Lighthill criterion, Eq. (9.10), is satisfied in these regions.

So, the mathematical model of a nonlinear Schrödinger equation with dissipation, Eq. (9.9), gives a good theoretical description of all the qualitative features of the process of dipole-exchange spin wave soliton formation and propagation in YIG films which was observed under pulsed excitation.

9.6.2 Modulational instability of spin waves and formation of a strange attractor

To model the process of spin wave propagation in YIG films under continuous (or long pulse $\tau \gg T_m$) excitation we used the above described numerical model, Eq. (9.9), with the initial conditions of Eq. (9.19) and slightly different boundary conditions [38]:

$$\varphi(\zeta=0, t) = \varphi_0 , \qquad (9.21)$$

$$\varphi(\zeta=l_0, t) = 0 . \qquad (9.22)$$

These conditions mean that an input signal in the form of a rectangular step of amplitude φ_0 is applied at the point $\zeta = 0$ at the moment $t = 0$ and then propagates in the media for the distance l_0. The full distance of propagation, as in the previous case, was chosen to be sufficiently long ($l_0 = 4$ cm) that the propagating spin wave signal vanishes at the final point $\zeta = l_0$ due to the dissipation. The numerical calculations were done for the case of a perpendicularly magnetized YIG film of the thickness $L = 5.8$ μm when the carrier frequency ω_c of the input signal was situated in the spectral point C (see Table 1). The calculations were done in the interval of the input signal amplitudes $0.08 < |\varphi|_0^2 \times 10^3 < 40$.

The time profiles of the output signal envelopes $\varphi(\zeta=l, t)$ along with the phase portraits of the system [i.e., the dependences of the time derivative $d\varphi(\zeta=l, t)/dt$ of the output envelope on the value of the envelope $\varphi(\zeta=l, t)$ itself] were calculated numerically at a distance l = 0.4 cm from the input antenna. Thus, all the parameters of our numerical model corresponded to the conditions of our experiments performed under continuous excitation [see Figs. 9.7 - 9.9].

Figure 9.11 shows the results of our numerical analysis. A comparison of the calculated (Fig. 9.11) and experimental (see, e.g., Fig. 9.8) results shows that, when the amplitude of the input signal is sufficiently small so that $|\varphi|_0^2 \times 10^3 < 0.9$, both the numerical modeling and the laboratory experiment show that there is no modulation of the output envelope [see Figs. 9.11 (1) and 9.8 (1)].

If $|\varphi|_0^2 \times 10^3 = 0.96$, it follows from the calculations that the spin wave envelope at the output begins to oscillate at a frequency $f_{m\ num} = 5.9$ MHz and the amplitude of these oscillations decays with time. In the phase space such motion corresponds to a stable focus [see Fig. 9.11 (2)]. A similar regime of damped modulation is observed experimentally [see Fig. 9.9 (1)].

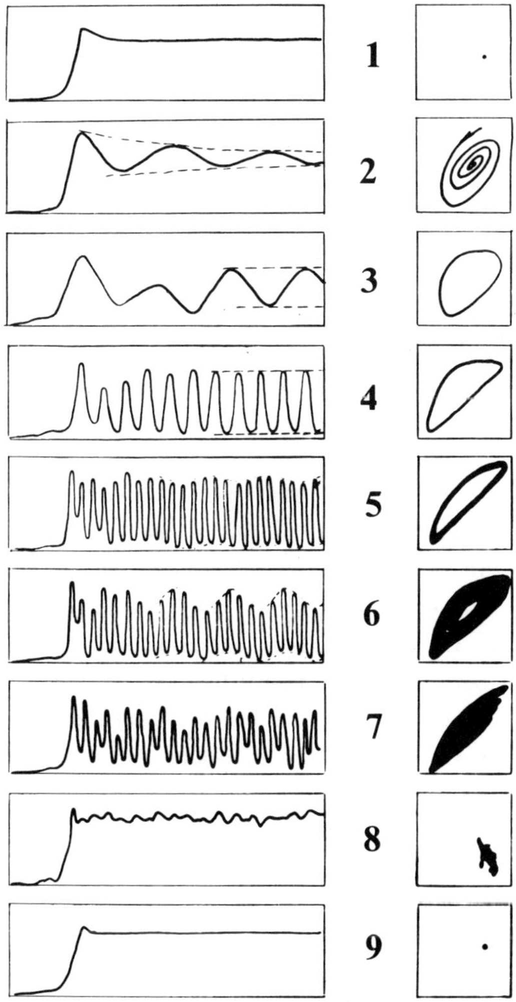

Fig. 9.11. Results of a numerical analysis. The time realizations of an output signal envelope (on the left) and the phase portraits (on the right) were obtained using Eq. (9.9) at a distance of l= 0.4 cm from the input antenna for the input signal with the envelope in a form of a rectangular step of the amplitude φ_0 for increasing values of the square modulus $|\varphi|^2_0 \times 10^3$: 1. 0.8; 2. 0.9; 3. 1.1; 4. 2.1; 5. 5.6; 6. 5.8; 7. 6.0; 8. 15.0; 9. 40.0.

When the input signal amplitude reaches the value $|\varphi|_0^2 \times 10^3 = 1$, undamped modulation of a frequency $f_{m\ num} = 6.5$ MHz is attained and it corresponds to a stable limit cycle in the phase space [Fig. 9.11 (3)]. A similar regime with a modulation frequency $f_{m\ exp} = 5.3$ MHz is observed in the laboratory experiment [Figs. 9.8 (2) and 9.9 (2)]. A further increase of the input signal amplitude in the range $1 < |\varphi|_0^2 \times 10^3 < 5.5$ in the numerical calculations results in a rapid increase of the modulation frequency $f_{m\ num}$. For example, at $|\varphi|_0^2 \times 10^3 = 2.1$, the calculated value of the frequency $f_{m\ num}$ is 14 MHz, whereas at $|\varphi|_0^2 \times 10^3 = 5.5$ the frequency $f_{m\ num}$ becomes 40 MHz.

At the point where $|\varphi|_0^2 \times 10^3 = 5.6$, the single-frequency periodic modulation ($f_{m\ num} = 41$ MHz) loses its stability and the modulation spectrum in our numerical calculations exhibits a second (lower) frequency component $F_{m\ num} = 8$ MHz. This regime corresponds to the formation of a two-dimensional torus in the phase space [see Fig. 9.11 (5) and (6)]. It is interesting to note that although the absolute values of the modulation frequencies f_m and F_m at the threshold of formation of a two-dimensional torus in the numerical calculations [41 and 8 MHz; Fig. 9.11 (5)] and the laboratory experiment [5.3 and 1.1 MHz; Fig. 9.8 (3)] differ considerably, the frequency ratios f_m/F_m are practically the same and approximately equal to 5. An increase in $|\varphi|_0^2$ in the range of existence of the torus hardly affects the ratio in the numerical experiment, although the absolute values of the modulation frequencies increase. The range of stability of the torus in the numerical experiment is narrow. Already for $|\varphi|_0^2 \times 10^3 = 6$ the envelope demonstrates chaotic behavior with a depth of modulation close to 100% and a *strange attractor* appears in the phase space of the system [Fig. 9.11 (7)]. In the laboratory experiment the modulation also becomes chaotic and a spin wave strange attractor is formed [see Fig. 9.8 (4), (5) and Fig. 9.7 (7)], but the depth of chaotic modulation does not exceed 30%.

A further increase in $|\varphi|_0^2$ in both numerical [Fig. 9.11 (8), (9)] and laboratory experiments [Fig. 9.7 (8)] reduces the depth of chaotic modulation and eventually suppresses such modulation.

The qualitative pattern of development of the modulational instability obtained in our numerical calculations on increase in $|\varphi|_0^2$ consists of the appearance of modulation (a focus transforming into a cycle), followed by a modulation at a single frequency (cycle), quasi-periodic two-frequency modulation (two-dimensional torus), chaotic modulation (strange attractor), and suppression of modulation. This scenario of modulational instability development is similar to the scenario of Ruelle-Takens [37] and is not affected by variation of the parameters of Eq. (9.9) within certain limits. For example, if v_g, D, and ω_r are varied by 10-20 %, no qualitative changes appear in the behavior of output spin wave envelope, and the ratio of the modulation frequencies f_m/F_m remains practically constant and equal to approximately 5. The subsequent change of the regimes in the numerical experiment is a smooth process resulting from soft bifurcations.

The above presented numerical results show that the nonlinear Schrödinger equation with dissipation Eq. (9.9) not only describes the formation of spin wave envelope solitons in the pulse regime of spin wave excitation, but also predicts correctly a qualitative pattern of the development of modulational instability of dipole-exchange spin waves in magnetic films on increase in the input signal amplitude in the continuous excitation regime. At the instability threshold the model of Eq. (9.9) also gives quantitative estimates of the threshold amplitude of the input signal $|\varphi|_{0\ \mathrm{num}}^2 = 10^{-3}$ and of the initial frequency $f_{\mathrm{m\ num}} = 6.5$ MHz of self-modulation, which are in good agreement with the results of the laboratory experiment ($|\varphi|_0^2 = 1.1 \times 10^{-3}$ and $f_{\mathrm{m\ exp}} = 5.3$ MHz).

The differences between the behavior of the theoretical model of Eq. (9.9) and the experimental results (such as an increase in the modulation frequency on increase in $|\varphi|_0^2$, the greater depth of chaotic modulation and a much greater - in experiments - amplitude interval of the existence of modulation) are in our opinion due to the fact that a real dipole-exchange spin wave spectrum in a magnetic film depends in a fairly complex manner on the power of the input signal. Therefore, the coefficients of Eq. (9.9) may change greatly as a result of a change in the input signal power. We also would like to stress again that the theoretical values of the dimensionless spin wave amplitudes φ_0 obtained for given values of the input power P_{in} using the linear theory of spin wave excitation in magnetic films [41] above the threshold of four-wave parametric process, Eq. (9.20), can be used only as rough estimations, and a development of a nonlinear theory of spin wave excitation in films is necessary for the quantitative comparison of numerical and experimental results.

9.7. CONCLUSION

The above presented experimental results have demonstrated that epitaxial YIG films are superb model media for experiments on nonlinear wave dynamics. Many interesting nonlinear wave phenomena, like formation of envelope solitons, four-wave parametric modulational instability and formation of a strange attractor, can be observed in one experimental sample of an YIG film with pinned surface spins at room temperature and at relatively low levels of the input microwave power ($P_{in} < 0.5$ W).

We have chosen a simple model of a nonlinear Schrödinger equation with dissipation, Eq. (9.9), to describe theoretically all this rich and complicated nonlinear spin wave dynamics. In our theoretical approach we did not make any assumptions about the number of modes taking part in the nonlinear interaction and about their interaction constants.

Our mathematical model for numerical calculations is a partial differential equation [Eq. (9.9)] - a system with an infinite number of degrees of freedom characterized by four parameters: group velocity v_g, dispersion D, nonlinearity N, and dissipation ω_r. The only serious assumption which we used in derivation of this equation was the assumption of spectral narrowness of the packet of interacting spin waves. We did not use any adjustable parameters. All the parameters of our numerical model were calculated or measured for the conditions of our particular experimental situation and were not changed in the course of numerical calculations. The only input characteristic which was varied in the process of numerical calculations at a given spectral point is the dimensionless input spin wave amplitude φ_0. However, as is clear from Figs. 9.10, 9.8, and 9.11, the numerical modeling of the finite amplitude dipole-exchange spin wave propagation in YIG films performed using the model of Eq. (9.9) gives a full qualitative description of the observed nonlinear spin wave dynamics.

We think that magnetic films themselves, as well as layered [43] and periodic [44] structures based on these films, have a great potential for future experimental investigations of nonlinear spin wave dynamics. We believe that further experiments performed in magnetic films, magnetic multi-layered and periodic structures by means of microwave spectroscopy and Brillouin light scattering (or by combination of both these experimental techniques) will reveal many new nonlinear wave phenomena which will not only be interesting from the fundamental point of view, but could also be of considerable practical importance for microwave signal processing and communications technology [45].

REFERENCES

1. A.I. Akhiezer and A.E. Borovik, *Sov. Phys.-JETP* **25** (1967) 332 [*Zh. Eksp. Teor. Fiz.* **52** (1967) 508].
2. E.V. Volzhan, N.P. Giorgadze, and A.D. Pataraya, *Sov.Phys.- Solid State* **18** (1976) 1487 [*Fiz. Tverd. Tela* **18** (1976) 2546].
3. A.M. Kosevich, B.A. Ivanov, and A.S. Kovalev, *Magnetic Solitons: A New Type of Collective Excitation in Magnetically Ordered Crystals*, in Soviet Scientific Reviews, Sec. A, Physics Reviews, **6** (1985) 161, ed. I.M. Khalatnikov.
4. A.M. Kosevich, B.A. Ivanov, and A.S. Kovalev, *Nonlinear Magnetization Waves: Dynamic and Topological Solitons* (Naukova Dumka, Kiev, 1988).
5. V.S. L'vov, Ch. 5 in *Solitons*, ed. S.E. Trullinger, V.E. Zakharov, and V.L. Pokrovsky (Elsevier, Amsterdam, 1986).
6. V.S. L'vov, *Nonlinear Spin Waves*, (Nauka, Moscow, 1987) (in Russian).
7. K. Nakamura, S. Ohta, and K. Kawasaki, *J. Phys. C* **15** (1982) L143.
8. B.A. Kalinikos, N.G. Kovshikov, and A.N. Slavin, *Sov. Phys.- JETP Lett.* **38** (1983) 202 [*Pis'ma Zh. Eksp. Teor. Fiz.* **38** (1983) 413].
9. B.A. Kalinikos, N.G. Kovshikov, and A.N. Slavin, *Sov. Tech. Phys.*

Lett. **10** (1984) 392 [*Pis'ma Zh. Tekh. Fiz.* **10** (1984) 936].

10. V.S. Gornakov, L.M. Dedukh, and V.I. Nikitenko, *Sov. Phys.- JETP Lett.* **39** (1984) 236 [*Pis'ma Zh. Eksp. Teor. Fiz.* **39** (1984) 199].
11. G. Gibson and C. Jeffries, *Phys. Rev. A* **19** (1984) 811.
12. H. Yamazaki, *J. Phys. Soc. Jpn.* **53** (1984) 1155.
13. R.W. Damon, in *Magnetism: A Treatise on Modern Theory and Materials*, eds. G.T. Rado and H. Suhl (Academic Press, New York, 1963) **1**, 551.
14. G. Srinivasan, C.E. Patton, and P.R. Emtage, *J. Appl. Phys.* **61** (1987) 2318.
15. B.A. Kalinikos, N.G. Kovshikov, and A.N. Slavin, *Phys. Rev. B* **42** (1990) 8658.
16. V.E. Zakharov, V.S. L'vov, and S.S. Starobinets, *Sov. Phys. - Usp.* **17** (1975) 896 [*Usp. Fiz. Nauk* **114** (1974) 609].
17. A.N. Slavin and B.A. Kalinikos, *Sov. Phys. Tech. Phys.* **32** (1987) 1446.
18. A.N. Slavin and G.M. Dudko, *J. Magn. Magn. Materials* **86** (1990) 115.
19. V.P. Lukomskii and Yu.V. Sedletskii, *Ukrainskii Fiz. Zh.* **31** (1986) 1379.
20. B.A. Kalinikos and A.N. Slavin, *IEEE Trans. on Magnetics* **27** 5444 (1991).
21. G.B. Whitham, *Linear and Nonlinear Waves* (Wiley, New York, 1974).
22. V.I. Karpman, *Nonlinear Waves in Dispersive Media* (Pergamon, New York, 1973).
23. V.P. Lukomskii, *Ukrainskii Fiz. Zh.* **23** (1978) 134.
24. A.K. Zvezdin and A.F. Popkov, *Sov. Phys.- JETP* **57** (1983) 350 [*Zh. Eksp. Teor. Fiz.* **84** (1983) 606].
25. L.D. Landau and E.M. Lifshitz, *Quantum Mechanics Non-Relativistic Theory*, 2nd edn. (London, Pergamon, 1965).
26. G.L. Lamb, Jr., *Elements of Solitons* (John Wiley, New York, 1980).
27. M.J. Lighthill, *J. Inst. Math. Appl.* **1** (1965) 269.
28. B.A. Kalinikos, N.G. Kovshikov, and A.N. Slavin, *Sov. Phys.- JETP* **67** (1988) 303 [*Zh. Eksp. Teor. Fiz.* **94** (1988) 159].
29. V.E. Zakharov, S.V. Manakov, S.P. Novikov, and L.P. Pitaevskii,

Theory of Solitons: Inverse Scattering Problem Method (Nauka, Moscow, 1980) (in Russian).

30. B.A. Kalinikos, N.G. Kovshikov, P.A. Kolodin, and A.N. Slavin, *Solid State Comm.* **74** (1990) 989.
31. B.A. Kalinikos, N.G. Kovshikov, and A.N. Slavin, *Sov. Phys.- Solid State* **27** (1985) 135 [*Fiz. Tverd. Tela* **27** (1985) 226].
32. V.E. Zakharov, V.S. L'vov, and S.S. Starobinets, *Sov. Phys. - Solid State* **11** (1970) 1658 and 2368 [*Fiz. Tverd. Tela* **11** (1969) 2047 and 2922].
33. A.V. Lavrinenko, V.S. L'vov, G.A. Melkov, and V.B. Cherepanov, *Sov. Phys.-JETP* **54** (1981) 542 [*Zh. Eksp. Teor. Fiz.* **54** (1981) 1022].
34. P.E. Zil'berman, S.A. Nikitov, and A.G. Temiryazev, *Sov. Phys.-JETP Lett.* **42** (1985) 110 [*Pis'ma Zh. Eksp. Teor. Fiz.* **42** (1985) 92].
35. G.M. Dudko, G.T. Kazakov, A.V. Kozhevnikov, and Yu.A. Filimonov, *Sov. Tech. Phys. Lett.* **13** (1987) 306 [*Pis'ma Zh. Tekh. Fiz.* **13** (1987)].
36. B.A. Kalinikos, N.G. Kovshikov, and A.N. Slavin, *J. de Physique* **49**-C8 (1988) 1583.
37. D. Ruelle and F. Takens, *Commun. Math. Phys.* **20** (1971) 167.
38. G.M. Dudko and A.N. Slavin, *Sov. Phys.- Solid State* **31** (1989) 979 [*Fiz. Tverd. Tela* **31** (1989) 114].
39. B.A. Kalinikos, N.G. Kovshikov, and A.N. Slavin, *J. Appl. Phys.* **67** (1990) 5633.
40. B.P. Demidovich, I.A. Maron, and E.Z. Shuvalova, *Numerical Methods for Analysis,* (GIFML, Moscow, 1962) (in Russian).
41. V.F. Dmitriev and B.A. Kalinikos, *Sov. Phys. J.* **31** (1988) 432 [*Izvestiya VUZov- Fizika* **31** (1988) 24].
42. B.A. Kalinikos, N.G. Kovshikov, and A.N. Slavin, *J. Appl. Phys.* **69** (1991) 5712.
43. A.N. Slavin and P.A. Kolodin, *J. Mag. Magn. Materials* (1992), to be published.
44. N.N. Chen, A.N. Slavin, and M.G. Cottam, *IEEE Trans. on Magnetics* (1992), to be published.
45. J.D. Adam and S.N. Stitzer, *Appl. Phys. Lett.* **36** (1980) 485.

SUBJECT INDEX